Theory and Principles of Digital Filters and Signal Processing

Theory and Principles of Digital Filters and Signal Processing

Edited by **George Pilato**

NY RESEARCH
PRESS

New York

Published by NY Research Press,
23 West, 55th Street, Suite 816,
New York, NY 10019, USA
www.nyresearchpress.com

Theory and Principles of Digital Filters and Signal Processing
Edited by George Pilato

International Standard Book Number: 978-1-63238-449-2 (Hardback)

Printed in the United States of America.

Contents

Preface

The world is advancing at a fast pace like never before. Therefore, the need is to keep up with the latest developments. This book was an idea that came to fruition when the specialists in the area realized the need to coordinate together and document essential themes in the subject. That's when I was requested to be the editor. Editing this book has been an honour as it brings together diverse authors researching on different streams of the field. The book collates essential materials contributed by veterans in the area which can be utilized by students and researchers alike.

The theories and principles of digital filters and signal processing are explained in this extensive book. Digital filters along with signal processing, are being engaged in new technologies and information systems, and are implemented in diverse areas and applications. They are used with no costs and they can be adjusted for various cases fittingly with great flexibility and dependability. This book presents latest advancements in digital filters and signal process techniques covering different cases studies. They present the key fundamental nature of the subject, with the basic approaches to the most recent mathematical models that are being engaged worldwide.

Each chapter is a sole-standing publication that reflects each author´s interpretation. Thus, the book displays a multi-facetted picture of our current understanding of application, resources and aspects of the field. I would like to thank the contributors of this book and my family for their endless support.

Editor

Spectral Analysis of Exons in DNA Signals

Noor Zaman, Ahmed Muneer and
Fausto Pedro García Márquez

Additional information is available at the end of the chapter

1. Introduction

DNA is found in blood cells carrying nucleus. The DNA is isolated from blood through a series of different procedures including heat shock, thermal change and applications of different chemicals etc. DNA sequence contains chromosomes which further contains genes over them. The genes have regions which could translate to protein and the regions which don't perform any contribution in protein production. Both kinds of regions are made-up of nucleotides characterized as Adenine, Thymine, Cytosine and Guanine. The order of these nucleotides determines the traits, habits and livings of all species. Since with the exponential growth of biological data, there is an enormous amount of such data that needs to be translated to protein. A successful translation would result in knowing important information about species.

Comparative analysis of computational techniques employed over genetic datasets has given very interesting results. We are able to identify species from each other on the behalf of DNA properties. A true correct conversion takes to fruitful results. Literature has shown that direct comparative analysis is not as useful as approximate estimation. So far, there is no compact solution available that could outperform for a robust translation from DNA to RNA.

It is a common phenomenon that nucleotide sequences in DNA perform a period three property [3, 11] due to codon composition and structure in the strand. This fundamental characteristic can be exploited to predict the codon regions that help in determination of RNA sequences in DNA. This finding is of immense importance as cell growth and function is determined by the type of protein the cell produces and helps in drug design and revealing genetic disorders as a result of mutation in structure of nucleotide bases (order in which they appear over chain). Many approaches have been proposed in literature that addresses this open optimization problem in computational biology.

Discrete Fourier Transforms [6, 7, and 8] normally result in spectral leakage that doesn't preview the optimal power spectral density estimation. On the other hand, the Short Time Fourier Transforms [2, 4] minimize the leakage but are considered useful when we desire to have the frequency contents with location information. It can plot the components for time, amplitude and frequency of a genetic signal.

Digital Filters [5, 7, and 13] present the spectral contents of signal around the periodicity property of coding regions but don't specify the frequency time relationship with amplitude.

Dosay-Akbulut [14] emphasized the classification of introns in two groups based on RNA secondary structure and self splicing ability in variant species using PCR.

A. Parent et al., [15] describe the importance of coordination between transcription and RNA processing that carboxy-terminal domain of RNA polymerase II acts as a common link in both.

Al Wadi et al. [16] used wavelet transforms for forecasting volatility in experimental results. M. Hashemi et al. [17] provided Identification of *Escherichia coli* O157:H7 Isolated from Cattle Carcasses in Mashhad Abattoir by Multiplex PCR.

A. Ali et al. [18] have presented a Histopathological Study for development of a model for Tumor Lung Cancer Assessing Anti-neoplastic Effect of PMF in Rodents.

J. Singh et al. [19] proposed a technique for Prediction of in vitro Drug Release Mechanisms from Extended Release Matrix Tablets.

2. Proposed approach

The proposed approach consists of a series of components that analyze the DNA signal and enhances the prediction accuracy of genic regions over DNA sequence. The major steps of proposed approach are,

- Conversion of target DNA stretch to a digital pattern employing an indicator sequence
- Decomposition of signal using wavelet transforms
- Calculations of approximate coefficients of signal at level three
- Calculations of detail coefficients of signal
- Density estimation of signal
- Signal analysis for denosing
- Depiction of original and synthesized signal at level three
- Histogram estimations of signal
- Signal extension to a desired length

- Shannon entropy calculation of signal
- Magnitude and power estimation of signal
- Calculation of discrimination measure for PSD analysis
- Exon and intron boundaries' estimation

As an elaboration, the DNA sequence is passed through a filter that transforms it into a digital pattern. This phase is accomplished employing an indicator sequence with the following weights for nucleotides,

Adenine (A) = X (A) = 0.260
Thymine (T) = X (T) = 0.375
Guanine (G) = X (G) = 0.125
Cytosine (C) = X (C) = 0.370

The corresponding transform becomes

$$X_{\text{IndSeq}}[k] = \sum_{n=1}^{N} x_{\text{IndSeq}}[n] e^{-j2\pi kn/N} \quad k = 1,\ 2, \ldots, N \tag{1}$$

Indicator sequence

The signal is decomposed employing the wavelet transforms of order three at level three

$$\begin{aligned} y(t) &= A_1(t) + D_1(t) \\ &= \sum_k cA_2(k)\phi_{j-2,k}(t) + \sum_k cD_2(k)w_{j-2,k}(t) + \sum_k cD_1(k)w_{j-1,k}(t) \\ &= A_2(t) + D_2(t) + D_1(t) \\ &= A_3(t) + D_3(t) + D_2(t) + D_1(t) \\ &= \sum_k cA_3(k)\phi_{j-3,k}(t) + \sum_k cD_3(k)w_{j-3,k}(t) + \sum_k cD_2(k)w_{j-2,k}(t) + \sum_k cD_1(k)w_{j-1,k}(t) \end{aligned} \tag{2}$$

3rd order wavelet decomposition

The wavelet decomposition passes the signal into a series of low and high pass filters that decompose and synthesize the signal for reducing flicker noise (pink noise).

The signal is then convoluted with a window function (Kaiser Window) defined below,

$$w(n) = \begin{cases} I_0\left(\beta\left(1-((n-\alpha)/\alpha)^2\right)^{1/2}\right) / I_0(\beta) & 0 \le n \le M-1 \\ 0 & otherwise \end{cases} \tag{3}$$

Kaiser window of length 351 bp

Each section of the signal is traversed for calculation of absolute and power values. Each segment is plotted over the power spectral graph keeping the period three property maintained at each step. The exon boundaries appear as sharp peaks. The final discrimination measure depicts the degree of relevance in exon and introns.

3. Results and discussions

A specimen gene pattern S.cerevisiae chromosome III (AF099922) has been taken for experiments over proposed approach. The gene is passed through the series of steps defined,

At processing stage, the dataset is passed through two kinds of filters. First filter refines the data and outputs a data file that purely contains nucleotide characters. Second filter operates on output file obtained from first filter application and generates a file that contains numeric data. This data is fed into central engine for further processing.

Figure 1 Shows dataset that contains nucleotide characters and some other characters. This is first necessary step because this input when fed into our engine will badly degrade the performance and brings false results.

```
gttaatgtag cttaataaca aagcaaagca ctgaaaatgc ttagatg
gtaaaat tacacatgca121 aacctccata gaccggtgta aaatccc
gacacct tgcctagcca cacccccacg ggactcagca 241 gtgata
01 atttcgtgcc agccaccgcg gtcatacgat taacccaaac taat
atccaac ttatatgtga aaattcattg  421 ttaggaccta aactc
ttagata ccccactatg cttagccata aacctaaata attaaattta
ggcggta601 ctttatatcc atctagagga gcctgttcta taatcga
cagcaaa ccctaaaaag gtattaaagt 721 aagcaaaaga atcaaa
tcttata aaagaacatt actataccct ttatgaaact aaaggactaa
agtacgc 901 acacaccgcc cgtcaccctc ctcaaattaa attaaa
cgtaaca aggtaagcat actggaaagt gtgcttggaa1021 taatca
gacactc tgaactaatc ctagccctag ccctacacaa atataattat
tcgtaca tctaggagct1201 atagaactag taccgcaagg gaaaga
cttttgc ataatgaact aactagaaaa cttctaacta aaagaattac
ctcgtct1381 atgtggcaaa atagtgagaa gatttttagg tagagg
```

Figure 1. Preprocessed dataset

Figure 2 represents a data glimpse that contains pure nucleotide characters.

```
gttaatgtagcttaataacaaagcaaagcactgaaaatgcttagatgga
atgcaaacctccatagaccggtgtaaaatcccttaaacatttgcttaaa
ggactcagcagtgataaatattaagcaataaacgaaagtttgactaagt
attttcggcgtaaaacgtgtcaactataaataaataaatagaattaaaa
tataatacacgacagctaagacccaaactgggattagataccccactat
ttaaaactcaaaggacttggcggtactttatatccatctagaggagcct
cttcagcaaaccctaaaaaggtatttaagtaagcaaaagaatcaaacat
acattactatatcctttatgaaactaaaggactaaggaggatttagtag
ctcaaattaaattaaacttaacataattaatttctagacatccgtttat
ttaatattaaagcatctggcctacacccagaagatttcatgaccaatgg
atttatcctactaaaagtattggagaaagaaattcgtacatctaggagc
ccttgtaccttttgcataatgaactaactagaaaacttctaactaaaag
atgtggcaaaatagtgagaagatttttaggtagaggtgaaaagcctaac
aacaaaatcaaaaagtaagtttaaattatagccaaaagagggacagctc
aaaagcagccaccaacaaagaaagcgttcaagctcaacataaaatttca
caacactgttagtatgagtaacaagaattctaattctccaggcatacac
```

Figure 2. Refined dataset

The EIIP indicator sequence transforms the nucleotides in numeric values as per its definition. A part of signal is described in Figure 3 below using EIIP indicator sequence as,

0.1335	0.1335	0.0806	0.1260	0.1260	0.1335	0.1335	0.1340	0.1260	0.1260	0.1335
0.1340	0.1260	0.1335	0.0806	0.1340	0.1335	0.1335	0.1335	0.1335	0.1335	0.1335
0.1335	0.1260	0.1260	0.1260	0.1260	0.1260	0.0806	0.1260	0.0806	0.1340	0.1260
0.1260	0.1260	0.1260	0.1335	0.1335	0.1335	0.1335	0.1335	0.1335	0.1340	0.1260
0.0806	0.0806	0.0806	0.1260	0.1260	0.1260	0.1335	0.1340	0.1340	0.0806	0.1335
0.0806	0.1340	0.1335	0.1340	0.1260	0.1260	0.1335	0.1335	0.1335	0.1335	0.0806
0.1260	0.1260	0.1340	0.1335	0.1335	0.1260	0.0806	0.1335	0.0806	0.1340	0.1340
0.1335	0.1335	0.0806	0.1340	0.1335	0.1340	0.1340	0.1260	0.1340	0.1340	0.0806
0.1340	0.1335	0.1260	0.1260	0.1260	0.1335	0.1260	0.1335	0.1335	0.1335	0.1340
0.0806	0.0806	0.1260	0.1335	0.1335	0.1335	0.1340	0.1260	0.1260	0.1260	0.1335
0.1260	0.1260	0.1340	0.1260	0.1335	0.0806	0.1260	0.1260	0.0806	0.1335	0.1340
0.0806	0.1340	0.1335	0.1340	0.1335	0.0806	0.0806	0.1260	0.0806	0.1335	0.1335
0.1340	0.0806	0.1260	0.1260	0.1260	0.1260	0.1340	0.0806	0.1260	0.1260	0.1260
0.1340	0.1260	0.1335	0.1335	0.1335	0.1335	0.1335	0.1335	0.1340	0.0806	0.1340
0.1260	0.1260	0.1260	0.1260	0.1260	0.1260	0.1260	0.1260	0.1340	0.0806	0.1340
0.1260	0.1260	0.1340	0.0806	0.1340	0.0806	0.1260	0.1260	0.1260	0.1260	0.1260
0.0806	0.0806	0.1260	0.0806	0.1335	0.1340	0.0806	0.1340	0.1340	0.1260	0.1340
0.0806	0.1340	0.1260	0.1335	0.1335	0.1335	0.1260	0.1335	0.1335	0.1260	0.0806

Figure 3. Numeric translation of gene F56F11.5 (AF099922)

The binary indicator sequence is formed by replacing the individual nucleotides with values either 0 or 1. 1 stands for presence and 0 for absence of a particular nucleotide in specified location in DNA signal,

Figure 4 describes the glimpse of binary indicator sequence which is the one of four parts of translation of gene file. Only 1's and 0's are visible in this sequence.

```
0 0 0 0 0 0 0 1 0 0 0 0 0 0 0 0 1 0 0 0 1 0 0 0 0 0 0 0 0 0 0 0 0 0 0
0 0 0 0 0 1 0 0 1 0 0 0 0 0 0 0 0 0 0 0 0 1 0 0 0 1 0 0 0 0 0 0 0 0 0
1 0 0 1 0 0 0 0 0 1 0 1 0 0 0 0 0 0 0 1 0 1 1 0 0 0 1 0 0 0 0 0 0 1 1
0 0 0 0 0 0 0 1 0 1 1 0 1 1 0 0 0 0 0 0 1 0 0 0 0 0 0 0 0 0 1 0 0 0 0
0 0 0 0 0 0 1 0 0 0 0 0 0 0 0 0 0 0 1 0 0 0 0 0 0 0 1 0 1 0 1 0 0 1 0
0 0 0 0 0 0 0 0 0 0 0 0 1 0 0 0 0 0 1 0 0 0 0 0 0 0 0 0 1 0 0 0 0 0 0
1 0 1 0 0 0 1 1 0 0 0 0 0 0 0 0 1 0 1 0 0 0 0 0 0 0 1 0 1 0 0 0 0 0 0
0 0 0 1 0 0 0 0 0 1 0 1 1 0 1 0 1 0 1 0 0 1 0 0 0 0 0 0 0 0 0 0 0 0 1
0 1 0 0 0 0 1 0 0 0 0 0 0 1 1 0 0 0 0 0 0 0 0 0 1 1 0 0 0 0 0 0 0 0 0
0 0 0 0 0 0 0 0 0 0 0 1 0 0 0 0 0 1 0 0 0 0 0 0 0 0 0 1 1 0 0 0 0 0 0
0 0 0 1 0 0 0 0 0 0 0 0 0 0 0 0 0 0 0 0 0 0 0 0 0 0 0 0 0 1 0 0 0 0 0
0 0 0 0 1 0 1 1 0 1 0 0 1 0 1 0 0 0 0 0 0 0 1 0 0 0 1 1 0 0 1 0 0 0 0
0 0 0 0 0 0 0 0 0 0 0 0 0 0 0 0 0 0 0 0 1 0 0 0 0 0 0 0 0 0 0 0 0 0 0
0 0 0 0 1 0 0 0 0 0 0 0 0 0 0 0 0 0 0 0 0 0 0 0 0 0 0 0 0 0 0 0 0 1 1
1 0 0 0 0 0 0 0 1 0 1 0 0 0 0 0 0 1 0 0 0 1 0 0 0 0 0 0 0 0 1 0 0 0 0
```

Figure 4. Binary indicator sequence

The complex indicator sequence is defined by replacing the nucleotide with 1, -1, iota and -iota values.

Figure 5 shows a portion of gene AF099922 after application of complex indicator sequence.

```
0+1i;0+1i;-1;1;1;0+1i;0+1i;0-1i;1;1;0+1i;0+1i;1;1;1;1;0-1i;1;0+1i;-1;0-1i
1i;1;1;1;1;1;1;0+1i;0+1i;0+1i;0+1i;0+1i;0+1i;0-1i;1;1;1;0-1i;0+1i;-1;-1;-1
1i;0+1i;0+1i;0+1i;-1;0-1i;0+1i;0-1i;0-1i;-1;1;1;0-1i;0+1i;0+1i;1;-1;0+1i;-
1;0+1i;-1;-1;-1;-1;0-1i;0+1i;1;1;1;0+1i;1;0+1i;0+1i;0+1i;0-1i;0+1i;1;-1;0+
1;1;1;-1;0+1i;0-1i;1;0-1i;1;0-1i;-1;-1;0-1i;0+1i;0-1i;0+1i;-1;-1;1;-1;0+1i
1i;0+1i;0+1i;0+1i;0+1i;0-1i;-1;0-1i;1;1;-1;0-1i;0-1i;1;1;1;1;1;1;1;1;0-1i;
1i;0-1i;-1;0-1i;0-1i;1;0-1i;1;0-1i;0+1i;0-1i;-1;-1;0-1i;1;0+1i;0+1i;0+1i;1
1i;0+1i;0+1i;0-1i;0-1i;0+1i;0+1i;-1;-1;1;1;0+1i;-1;0+1i;0-1i;0-1i;1;-1;0+1
1i;1;0+1i;0+1i;0+1i;0+1i;0-1i;1;1;-1;1;0+1i;0+1i;0+1i;0+1i;0+1i;0-1i;0-1i;
1i;1;1;1;1;1;0+1i;0+1i;1;0+1i;0+1i;0+1i;0+1i;0+1i;1;0-1i;0+1i;-1;0+1i;1;-1
1i;-1;0+1i;1;0-1i;0-1i;1;-1;0-1i;-1;-1;-1;1;0+1i;0+1i;0+1i;0+1i;0+1i;0+1i;
1i;1;1;0+1i;1;1;0+1i;0+1i;0+1i;0+1i;0+1i;0+1i;0+1i;-1;0-1i;0+1i;-1;0+1i;0+
1;1;0+1i;0+1i;1;-1;0+1i;0-1i;0-1i;-1;0-1i;1;1;0+1i;0+1i;0+1i;0+1i;-1;0-1i;
1i;0+1i;1;0-1i;-1;1;1;1;1;0+1i;0-1i;0+1i;1;-1;0+1i;0+1i;1;1;0+1i;0+1i;0+1i
1i;1;0+1i;1;0+1i;0+1i;0+1i;0+1i;0+1i;1;-1;0-1i;0-1i;0+1i;0-1i;0-1i;0+1i;0+
1;1;1;1;0-1i;1;1;1;1;1;1;0+1i;0+1i;-1;0+1i;-1;0+1i;0+1i;0-1i;1;0-1i;0+1i;0
+1i;0+1i;0+1i;0+1i;0-1i;1;0-1i;1;0+1i;0+1i;1;0+1i;0+1i;0+1i;0+1i;1;0+1i;-1
1i;1;0+1i;0+1i;0+1i;0+1i;0-1i;0+1i;-1;-1;1;1;0+1i;0+1i;0+1i;0-1i;0-1i;0-1i
1i;-1;1;-1;-1;-1;0+1i;0+1i;1;0-1i;0+1i;-1;0+1i;1;-1;1;1;-1;0+1i;1;0-1i;1;0
1i;0-1i;-1;0-1i;-1;-1;-1;1;0+1i;0+1i;0+1i;0+1i;0+1i;0+1i;-1;1;0+1i;0+1i;0+
1i;0+1i;0+1i;0+1i;0+1i;-1;0-1i;-1;-1;0+1i;0+1i;-1;0+1i;-1;0+1i;0-1i;0-1i;0
```

Figure 5. Complex indicator sequence applied to gene

The complex indicator sequence transforms the sequence into four digital patterns with associated weights. It is worth mentioning that this indicator sequence provided close range estimation for nucleotides in the literature.

This signal is then passed through the steps of windowed STFT for exonic prediction spectral analysis. This helps to extend the length of the signal to a target length so that perfect analysis could be performed over the signal.

Figure 6 shows that signal has been extended to a desired length. The length of signal was 8000 patterns. The convolution method suggests that to perform a better approximation, the signal should be extended to 8192 patterns. The signal should be mapped employing Kaiser Window of length 351 base pairs. The previous power of two shows a numerical value 4096 which truncates the signal from its original length. Truncation phenomenon can degrade the results and may bring faulty approximation that would lead to differ from the standard range of exons.

Figure 7 depicts the wavelet sketch for db3 wavelet. Scaling and wavelet functions have been described. Decomposition of low pass filter and high pass filters have been identified, similarly signal synthesis for low and high pass filters have been shown. This sketch demonstrates that signal should be passed through these defined filters to further analyze it for denoising and enhancement. The upward and downward curves self explains the convolution of signal with the window function at desired location of nucleotides.

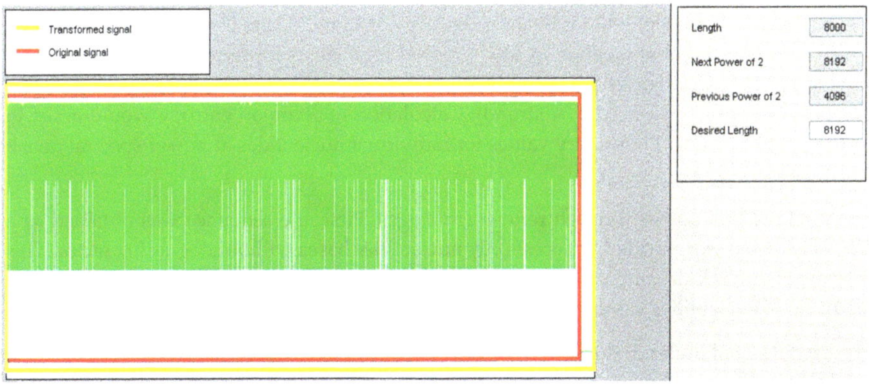

Figure 6. Signal extension to desired length

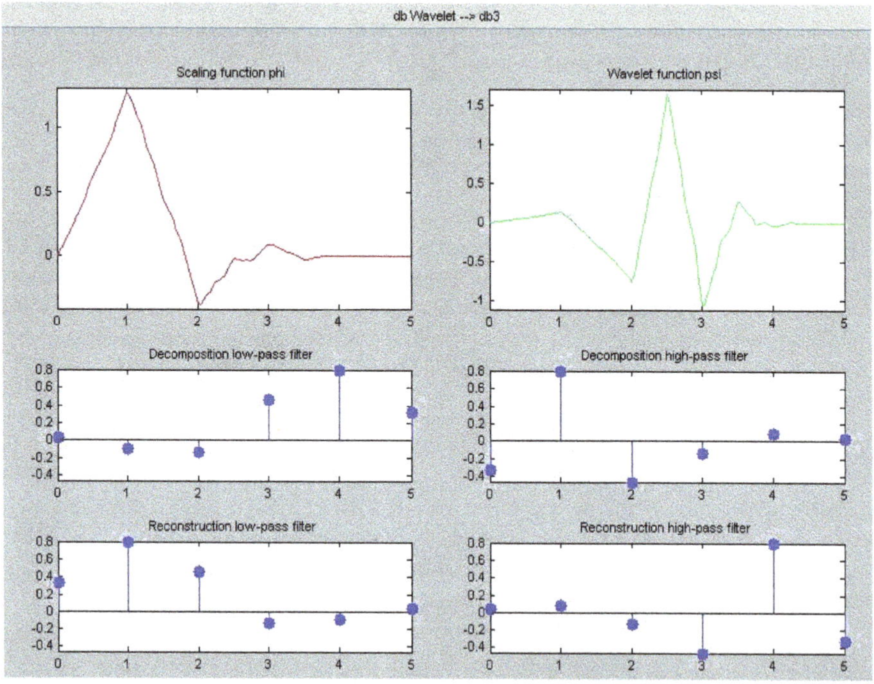

Figure 7. Wavelet of db3 sketch

Figure 8 shows a wavelet tree for Shannon entropy of signal. There is a tree structure for the nodes depicting different position factors. Colored coefficients for terminal nodes can be observed. The first rectangle at the right top shows the analyzed signal at different nucleotides places (diffusion of bases at DNA strand). Calculation of Shannon entropy would assist in further identification of boundary values for individual nucleotides at power spectral density estimation graphs.

The digital signal passes through refinement stages. First, the sequence was obtained as a raw data which was purified to access only nucleotides bases without degrading factor. This is termed as an important process because any kind of unwanted characters may lead to different set of nucleotides values that would be away from actual results.

The digital signal under discussions contains 8000 base pairs. The same dataset was used extensively in literature by other researchers and it is being used as a bench mark. The spectral estimation graph reveals that it contains five exonic regions at different nucleotides ranges. Identification of these ranges close to standard range demands to denoise the signal and selection of an appropriate window function that could be used for perfect convolution. The standard convolution requires to multiply the signal with a portion of window function, this is the reason that signal was extended to a power of two to make it to desired length. Each frame of the signal is calculated numerically equal sized so that power spectral graph is uniform in all characteristics.

For discrete wavelet transforms of order three, the signal is decomposed and synthesized. These db3 performs the quick vanishing of coefficients for approximate and detail patterns.

Figure 9 shows a glimpse of original signal. There are 8000 base pairs shown in the form of a digital pattern. Cumulative histogram of signal shows different range of weight values assigned to nucleotides base pairs. It can be seen that nucleotides with numerals higher than 0.25 have high frequency while those between 0.1 and less than 0.25 have lower frequency. The individual histogram also shows three separate characterizations of nucleotide weight values. The standard deviation has been found to be 0.09037, median of absolute deviation is 0.11 while mean absolute deviation is 0.07843. The maximum range is 0.375 while minimum range is 0.125 and the average range is depicted as 0.25.

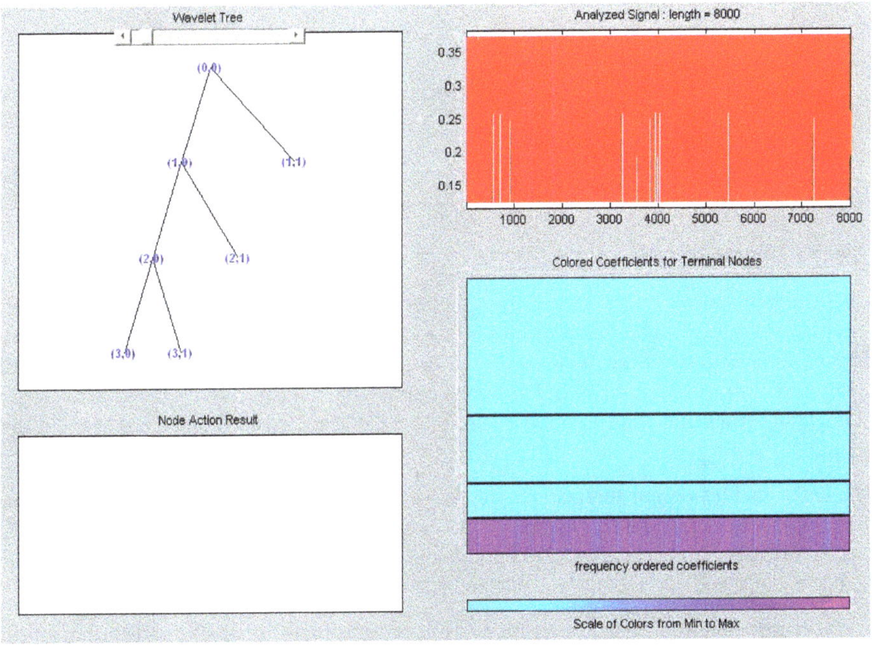

Figure 8. Wavelet tree for Shannon entropy

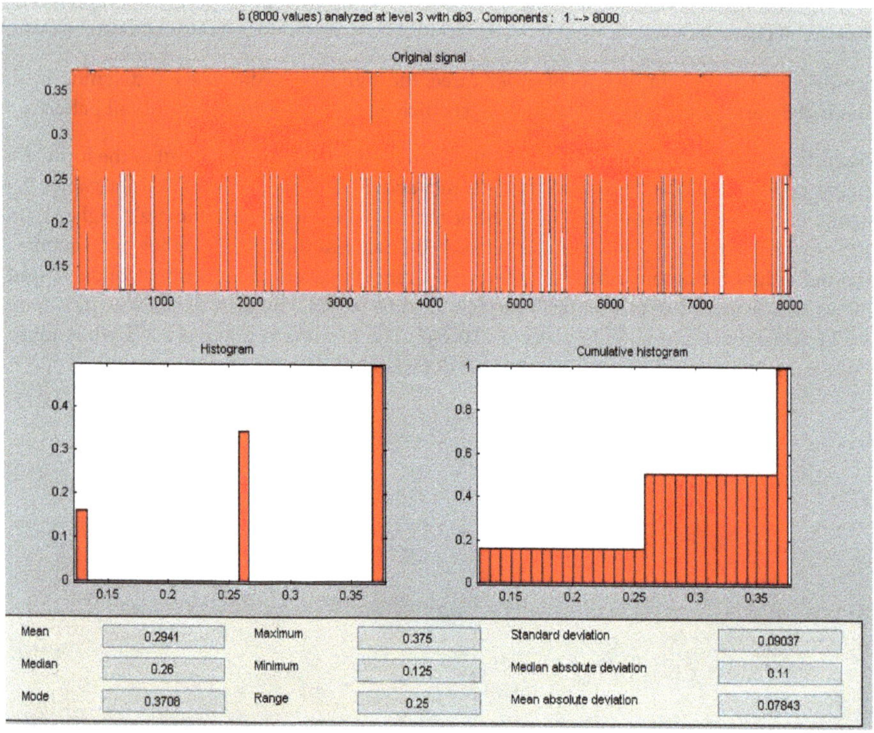

Figure 9. Original signal at level 3

The histogram calculates the frequency of nucleotides bases in the signal. Since the signal was mapped with an enhanced indicator sequence which assigns perfect weights to nucleotides bases, histogram of such a signal is uniform. It is observed that almost half of the signal is diffused in first band and half in the second band. First half band shows the smaller histogram values (frequency components) while second half band depicts some larger histogram values.

The individual histogram components dependant over the individual nucleotide bases, for instance, the numeric value of Adenine is 0.260, which is plotted against the other numeric values for bases in individual histogram. Depending over the weights assigned to Thymine and cytosine, the histogram shape may change.

It is also important to note that histogram of frequency components present the redundancy of bases in the digital pattern. This repetition depends over the order of nucleotides in DNA sequence which defines the habits, traits and other characteristics of species.

Figure 10 shows the synthesized signal at level three. Like the original signal, the synthesized signal owns the same histogram characteristics. There are 8000 base pairs shown in the form of a digital pattern. Cumulative histogram of signal shows different range of weight values assigned to nucleotides base pairs. It can be seen that nucleotides with numerals higher than 0.25 have high frequency while those between 0.1 and less than 0.25 have lower frequency. The individual histogram also shows three separate characterizations of nucleotide weight values. The standard deviation has been found to be 0.09037, median of absolute deviation is 0.11 while mean absolute deviation is 0.07843. The maximum range is 0.375 while minimum range is 0.125 and the average range is depicted as 0.25.

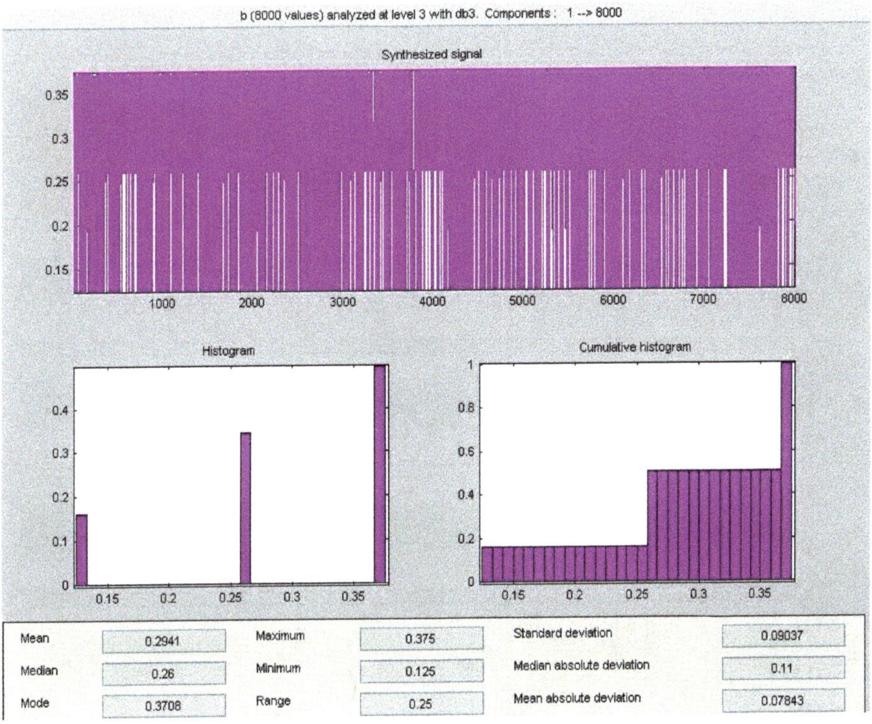

Figure 10. Synthesized signal at level 3

The synthesized signal shows the same histogram even after its decomposition. The synthesized signal is perfectly reconstructed by employing discrete wavelet transforms. The approximate and detail coefficients of signal are obtained in passing through a series of filters. These digital filters have been defined and constructed using Matlab. The decomposed signal is addition of approximate and detail coefficients at level three along with detail coefficients at level two and level one.

As for as, we decompose the signal, the components are loosely packed.

Figure 11 depicts the signal decomposition into approximate and detail coefficients. Symbol s represents the original signal. Approximate and detail coefficient at level three show the reduced complexity in the signal.

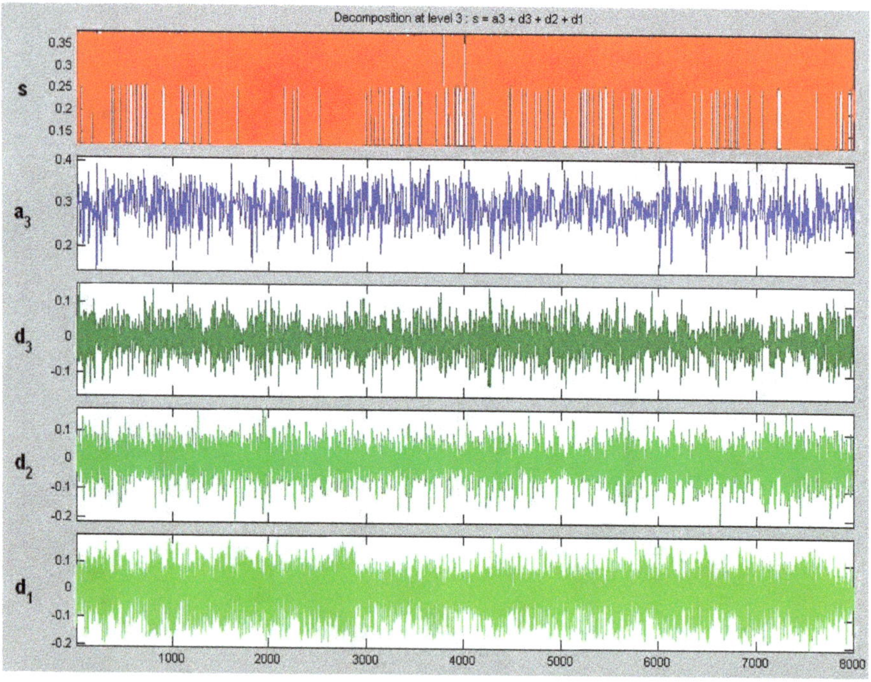

Figure 11. Signal decomposition

Figure 12 presents the histogram for approximate and detail coefficients. At level one, the concentration of components is less than other levels. Level two shows that signal components are more concentrated. At level three, the signal components are more closely packed. Likewise, the histogram for approximate coefficients presents the same phenomenon. At level one, the concentration of components is less than other levels. Level two shows that signal components are more concentrated. At level three, the signal components are more closely packed. It can be observed that original signal and synthesized signal contain the same number of components. The concentrations of signal components are uniform over these histograms' plots, which depict the perfect reconstruction of digital signal.

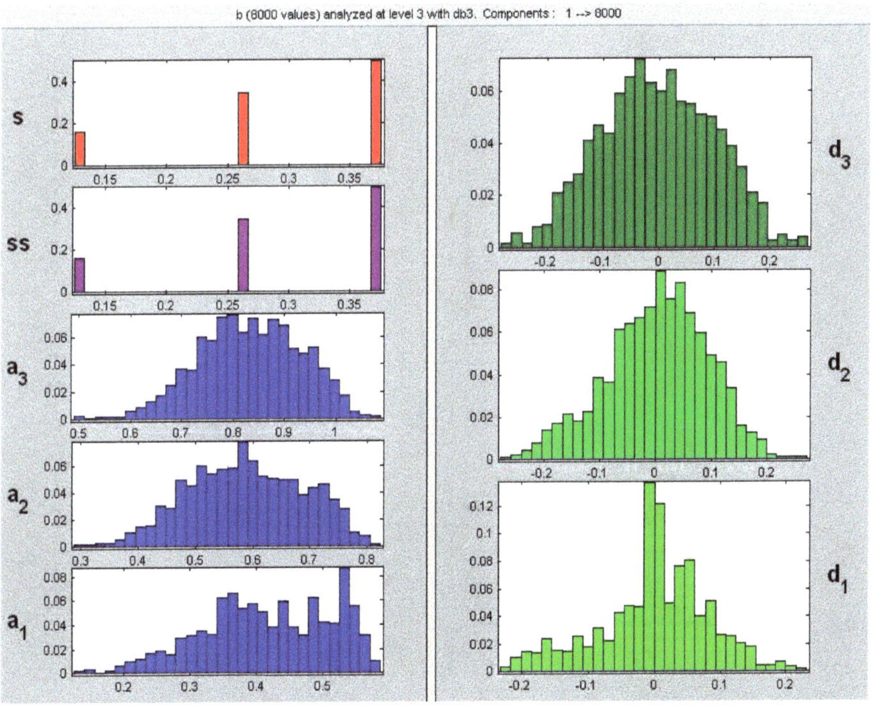

Figure 12. Histogram of signal

Figure 13 shows the density estimation of approximate and detail coefficients. The density estimate of original signal shows the numerals for nucleotides present in the signal in digital format as a general. The approximate coefficients at level three presents a sharp peak at some 0.25 points. The signal remains uniform through the course except at another peak value ranging from 0.37 to the end of the signal. The density estimation for detail coefficients at level one shows the same sharp peak around 0.27 points. The same peak can be observed around 0.40 at level two. At level three, the phenomenon is same but the signal components are loosely packed than level two. At granular level, the components are more packed at level one than other levels.

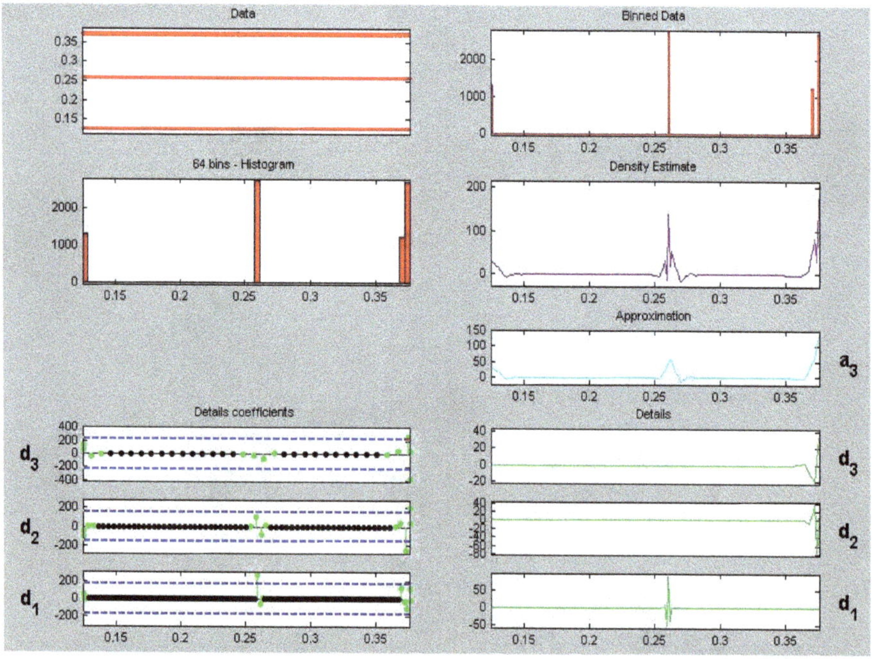

Figure 13. Density estimation of signal

Figure 14 shows the resultant denoised signal. It is obvious that preview of detail coefficients at level three shows the loosely packed signal components. The original signal is represented in red color. The threshold coefficients are shown in vertical bars for all nucleotide range (8000 base pairs). The coefficients at detail level depicts a hierarchy of packed, loosely packed and more loosely packed components, which shows a gradual improvement in the signal for denosing.

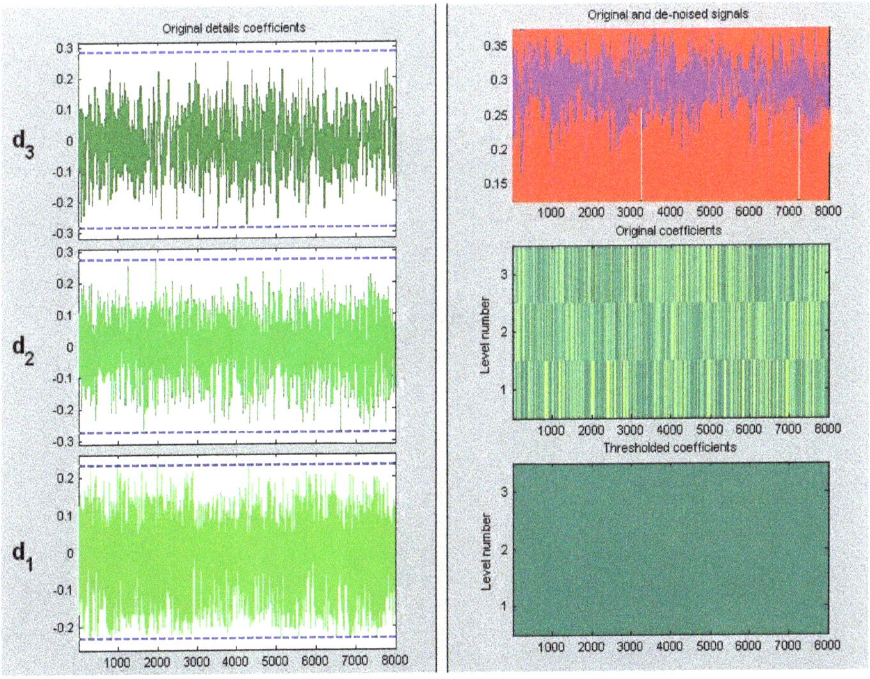

Figure 14. Denoised signal

Figure 15 shows the approximate coefficients at level three. A sharp gradual change can be observed in a commutative histogram. The peaks are more pronounced at from point one to onwards. In another histogram, the peaks are not much visible around first 0.6 points, there is a sharp gradual increment in the bars reaching the maximum of 0.07 points then a gradual decrement is observed leading it to point one. The peaks are less pronounced after this point. The coefficients of approximation at level three show the signal as loosely packed components.

It can be observed that detail coefficient at level one are packed showing more concentration of nucleotides while detail coefficients at level two are loosely packed. The coefficients at level three are more significant than other levels, which represents that the signal is filtered for refinement. The signal was passed through a series of filters for the wavelet db3 which denoised the signal as a result of reconstruction of signal.

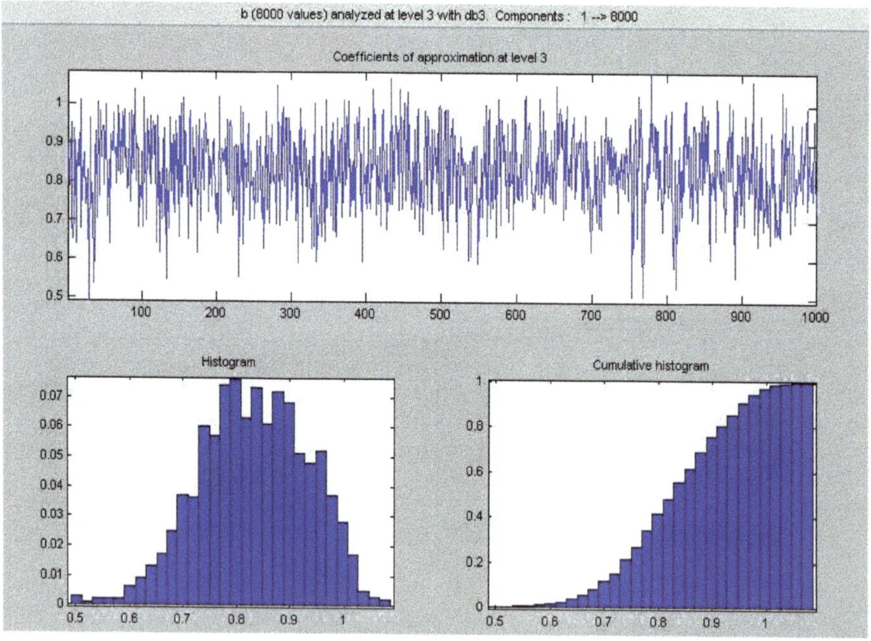

Figure 15. Approximation coefficients at level 3

Figure 16 shows the detail coefficients at level three. A sharp gradual change can be observed in a commutative histogram. The peaks are more pronounced at from point one to onwards. In another histogram, the peaks are not much visible around first 0.6 points, there is a sharp gradual increment in the bars reaching the maximum of 0.07 points then a gradual decrement is observed leading it to point one. The peaks are less pronounced after this point. The coefficients of detail at level three show the signal as loosely packed components.

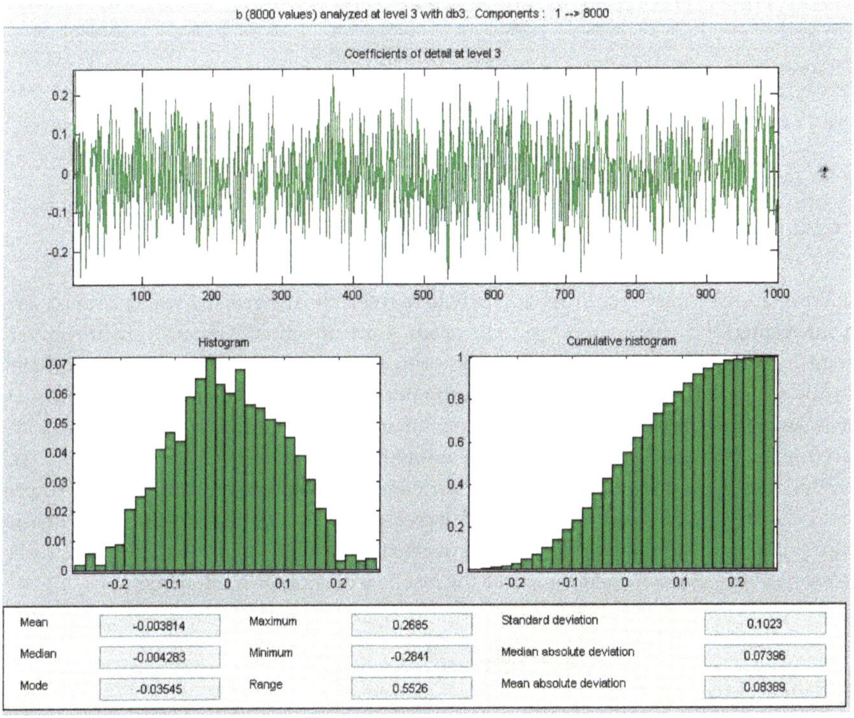

Figure 16. Coefficients of detail at level 3

It can be observed that detail coefficient at level one are packed showing more concentration of nucleotides while detail coefficients at level two are loosely packed. The coefficients at level three are more significant than other levels, which represents that the signal is filtered for refinement. The signal was passed through a series of filters for the wavelet db3 which denoised the signal as a result of reconstruction of signal.

Table 1 presents the nucleotide range for exons. Clear differences can be observed as a comparative analysis of various approaches. Binary and EIIP methods show a wide range difference compared with the standard range. Complex method results are better than the first two approaches. Digital filters behave accordingly. The proposed approach has more significant results than other prevailing approaches.

Method	E_1	E_2	E_3	E_4	E_5
Binary Method	656-1206	2406-3106	3806-4406	5306-5806	7106-7706
EIIP Method	706-1206	2206-2906	3906-4406	5206-5806	7206-7706
Complex Method	750-1100	2600-2906	3600-4406	5206-5706	7106-7600
Filter 1 (Anti-notch)	656-1206	2450-3106	3806-4450	5306-5850	7106-7750
Filter 2 (Multistage)	706-1250	2206-2950	3906-4450	5206-5850	7206-7706
proposed Method	750-1050	2450-2906	3950-4380	5206-5600	7220-7680
NCBI Range	928-1039	2528-2857	4114-4377	5465-5644	7255-7605

Table 1. Range of exons for different methods

4. Conclusion

Bioinformatics is a very rapidly emerging field of research. The genome sequence analysis is an interesting and challenging task that needs great attention. The analysis brings very promising relevance between species. The proposed approach provides a way to better identify the genetic regions in mixture of exon-intron noise. The focus directed to minimize the leakage of frequency contents by adoption of an optimal indicator sequence. We also reduced the signal noise by using Kaiser Window function with length 351 base pairs. The spectral density estimation was enhanced with application of wavelet transforms. The proposed dimensions reduced the noise and increased the sharp peaks of exons in density graphs. We have observed significant improvement in results as a comparative analysis between existing techniques and compared the results with strands NCBI range.

Author details

Noor Zaman[1*], Ahmed Muneer[1] and Fausto Pedro García Márquez[2]

*Address all correspondence to: nzaman@kfu.edu.sa

*Address all correspondence to: mmalik@kfu.edu.sa

*Address all correspondence to: FaustoPedro.Garcia@uclm.es

1 College of Computer Sciences & Information Technology,King Faisal University, Saudi Arabia

2 ETSI Industriales, Universidad Castilla-La Mancha, Ciudad Real, Spain

References

[1] Tina P George and Tessamma Thomas, "Discrete wavelet transform de-noising in eukaryotic gene splicing", BMC Bioinformatics, 11(Suppl 1):S50, doi: 10.1186/1471-2105-11-S1-S50, 2010

[2] Roy, M., Biswas, S. and Barman, S., "Identification and Analysis of Coding and Noncoding Regions of a DNA Sequence by Positional Frequency Distribution of Nucleotides (PFDN) Algorithm", 4th International Conference on Computers and Devices for Communication. CODEC 2009, Page(s): 1 - 4, 2009

[3] GuoShuo and Zhu Yi-sheng, "Prediction of Protein Coding Regions by Support Vector Machine", International Symposium on Intelligent Ubiquitous Computing and Education, Digital Object Identifier: 10.1109/IUCE.2009.141, Page(s): 185 - 188, 2009

[4] J. Quintanilla-Domínguez, B. Ojeda-Magaña, J. Seijas, A. Vega-Corona, D. Andina, "Edges Detection of Clusters of Microcalcifications with SOM and Coordinate Logic Filters", Proceedings of the 10th International Work-Conference on Artificial Neural Networks, Pages: 1029 - 1036, ISBN:978-3-642-02477-1, 2009

[5] Ruchira. Ajay Jadhav, Roopa Ashok Thorat, "Computer aided breast cancer analysis and detection using statistical features and neural networks", Proceedings of the International Conference on Advances in Computing, Communication and Control, Mumbai, India, ISBN:978-1-60558-351-8, 2009

[6] Muneer Ahmad and Hassan Mathkour, "Multiple Sequence Alignment with GAP consideration by pattern matching technique", International Conference on signal acquisition and processing, Malaysia, 2009

[7] Muneer Ahmad and Hassan Mathkour, "Genome sequence analysis by Matlab histogram comparison between Image-sets of genetic data", WCSET, Hong Kong, 2009

[8] Muneer Ahmad and Hassan Mathkour, "A pattern matching approach for redundancy detection in bi-lingual and mono-lingual Corpora", IAENG 2009

[9] HazrinaYusofHamdani and SitiRohkmahMohdShukri, "Gene prediction system", International Symposium on Information Technology, Volume: 2, Digital Object Identifier: 10.1109/ITSIM.2008.4631728, Page(s): 1 - 7, 2008

[10] ShuoGuo and Yi-Sheng Zhu, "An integrative algorithm for predicting protein coding regions", IEEE Asia Pacific Conference on Circuits and Systems,Digital Object Identifier: 10.1109/APCCAS.2008.4746054, Page(s): 438 - 441, 2008

[11] Kakumani, R., Devabhaktuni, V., and Ahmad, M.O., "Prediction of protein-coding regions in DNA sequences using a model-based approach", IEEE International Symposium on Circuits and Systems, Digital Object Identifier: 10.1109/ISCAS.2008.4541818, Page(s): 1918 - 1921, 2008

[12] Akhtar, M., Ambikairajah, E. and Epps, J., "Optimizing period-3 methods for eukaryotic gene prediction", IEEE International Conference on Acoustics, Speech and Signal

Processing, Digital Object Identifier: 10.1109/ICASSP.2008.4517686, Page(s): 621 - 624, 2008

[13] Hota, M.K. and Srivastava, V.K., "DSP technique for gene and exon prediction taking complex indicator sequence", IEEE Region 10 Conference, Digital Object Identifier: 10.1109/TENCON.2008.4766667, Page(s): 1 - 6, 2008

[14] Dosay-Akbulut, M., 2006. Group I introns and splicing mechanism and their present possibilities in elasmobranches. J. Boil. Sci., 6: 921-925. DOI: 10.3923/jbs.2006.921.925

[15] Parent, A., I. Benzaghou, I. Bougie and M. Bisaillon, 2004. Transcription and mRNA processing events: The importance of coordination. J. Biological Sci., 4: 624-627. DOI: 10.3923/jbs.2004.624.627

[16] S. Al Wadi, MohdTahir Ismail, M.H. Alkhahazaleh and SamsulAriffinAddulKarim, "Orthogonal Wavelet Transforms in Forecasting Volatility: An Expermintal Results", World Applied Sciences Journal, Volume 10 number 3, 2010

[17] M. Hashemi, S. Khanzadi and A. Jamshidi, " Identification of Escherichia coli O157:H7 Isolated from Cattle Carcasses in Mashhad Abattoir by Multiplex PCR", World Applied Sciences Journal, Volume 10 number 6, 2010

[18] A. Ali, F. Khorshid, H. Abu-araki and A.M. Osman, ,"Tumor Lung Cancer Model for Assessing Anti-neoplastic Effect of PMF in Rodents: Histopathological Study", Trends in Applied Sciences Research Volume 6, Number 10, 1214-1221, 2011

[19] J. Singh, S. Gupta and H. Kaur , "Prediction of in vitro Drug Release Mechanisms from Extended Release Matrix Tablets using SSR/R2 Technique", Trends in Applied Sciences Research Volume 6, Number 4, 400-409, 2011

Chapter 2

Maintenance Management Based on Signal Processing

Fausto Pedro García Márquez,
Raúl Ruiz de la Hermosa González-Carrato,
Jesús María Pinar Perez and Noor Zaman

Additional information is available at the end of the chapter

1. Wind Turbines

Most of the wind turbines are three-blade units (Figure 1.) [55]. Once the wind drives the blades, the energy is transmitted via the main shaft through the gearbox (supported by the bearings) to the generator. The generator speed must be as near as possible to the optimal for the generation of electricity. At the top of the tower, assembled on a base or foundation, the housing or nacelle is mounted and the alignment with the direction of the wind is controlled by a yaw system. There is also a pitch system in each blade. This mechanism controls the wind power and sometimes is employed as an aerodynamic brake. The wind turbine features a hydraulic brake to stop itself when it is needed. Finally, there is a meteorological unit that provides information about the wind (speed and direction) to the control system.

1.1. Maintenance in Wind Turbines

Maintenance is a key tool to ensure the operation of all components of a set. One of the objectives is to use available resources efficiently. The classical theory of maintenance was focused on the corrective and preventive maintenance [9] but alternatives to corrective and preventive maintenance have appeared in recent years. One of them is Condition Based Maintenance, which ensures the continuous monitoring and inspection of the wind turbine detecting emerging faults and organizing maintenance tasks that anticipate the failure [59]. Condition Based Maintenance implies acquisition, processing, analysis and interpretation of data and the selection of proper maintenance actions. This is achieved using condition monitoring systems [27, 28]. Thereby, CBM is presented as a useful technique to improve not only the maintenance but the safety of the equipments. Byon and Ding [14] or McMillan and Ault [50] have demonstrated its successful application in wind turbines, making the CBM one of the most employed strategies in this industry. Another example of the maintenance evolution is the Reliability Centred Maintenance. It is defined as a process to determine what must be done to ensure that any physical asset works in its operating context [71]. Nowadays it is the most common type of maintenance for many industrial fields [25, 26] and it involves maintenance system functions or identifying failure modes among others maintenance tasks [52].

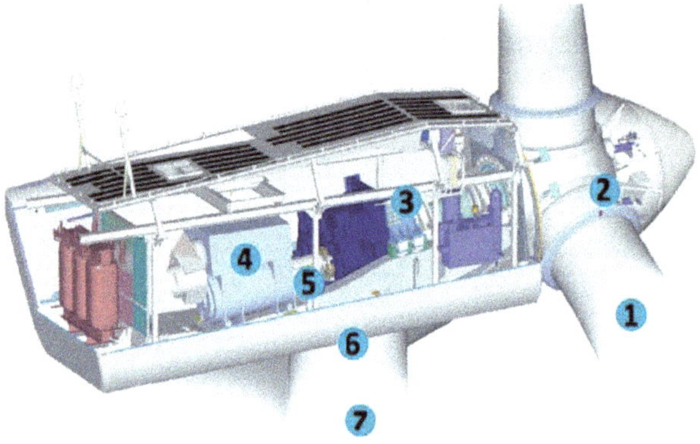

Figure 1. Main parts of a turbine: (1) blades, (2) rotor, (3) gearbox, (4) generator, (5) bearings, (6) yaw system and (7) tower [36].

1.2. Condition Monitoring applied to Wind Turbines

Condition Monitoring systems operate from different types of sensors and signal processing equipments. They are capable of monitoring components ranging from blades, gearboxes, generators to bearings or towers. Monitoring can be processed in real time or in packages of time intervals. The procurement of data will be critical to determine the occurrence of a problem and determine a solution to apply. Therefore, the success of a Condition Monitoring system will be supported by the number and type of sensors used and the signal collection and processing.

Any element that performs a rotation is susceptible of being analysed by vibration. In the case of the wind turbines, vibration analysis is mainly specialized in the study of gearboxes [48, 49] and bearings [81] [85]. Different types of sensors will be required depending on the operating frequency: position transducers, velocity sensors, accelerometers or spectral energy emitted sensors.

Acoustic emissions (AE) describe the sound waves produced when a material undergoes stress as a result of an external force [35]. They can detect the occurrence of cracks in bearings [84] and blades [91] in earlier stages.

Ultrasonic tests evaluate the structural surface of towers and blades in wind turbines [22] [24]. Consistent with some other techniques, it is capable of locating faults safely.

Oil analysis may determine the occurrence of problems in early stages of deterioration. It is usually a clear indicator of the wearing of certain components. The technique is widely used in the field of maintenance, being important for gearboxes in wind turbines [47].

Thermographic technique is established for monitoring mainly electrical components [72]; although its use is extended to the search of abnormal temperatures on the surfaces of the blades [64]. Using thermography, hot spots can be found due to bad contacts or a system failure. It is common the introduction of online monitoring systems based on the infrared spectrum.

There are techniques that not being so extended, are also used in the maintenance of wind turbines. In many cases, their performance is heavily influenced by the costs or their excessive specialization, making them not always feasible. Some examples are strain measurements in blades [68]; voltage and current analysis in engines, generators and accumulators [67]; shock pulse methods detecting mechanical shocks for bearings [13] or radiographic inspections to observe the structural conditions of the [61].

1.3. Signal processing methods

Fast Fourier Transform (FFT)

The FFT converts a signal from the time domain to the frequency domain. The use of FFT also allows its spectral representation [56]. Each frequency range is framed into a particular failure state. It is very useful when periodic patterns are searched [5]. Vibration analysis also provides information about a particular reason of the fault origin and/or its severity [43]. There is extensive literature demonstrating the development of the method for rolling elements. The FFT of a function *f(x)* is defined as [12]:

$$\int_{-\infty}^{\infty} f(x)e^{-2i\pi xs}dx \tag{1}$$

This integral, which is a function of s, may be written as *F(s)*. Transforming *F(s)* by the same formula, equation (2) where *F(s)* is the Fourier transform of *f(x)* is obtained.

$$\int_{-\infty}^{\infty} f(s)e^{-2i\pi\omega s}ds \tag{2}$$

There are a considerable number of publications regarding the diagnosis of faults for rolling machinery that justifies the models and patterns based on the Fast Fourier Transform. Misalignment is one of the most commonly observed faults in rotating machines, being the second most common malfunction after unbalance. It may be present because of improper machine assembly, thermal distortion and asymmetry in the applied load. Misalignment causes reaction forces in couplings that are the major cause of machinery vibration. Some authors evaluated numerically the effect of coupling misalignment and suggested the occur rence of strong vibrations at twice the natural frequency [70] [95], although rotating machinery can excite vibration harmonics from twice to ten harmonics depending on the signal pickup locations and directions [53].

Faults do not have a unique nature and most of the time, problems on a smaller scale are linked, e.g. in the case of misalignment, when an angular misalignment is studied, parallel misalignment (minor fault) needs to be take into account. Al-Hussain and Redmond reported vibrations for parallel misalignment at the natural frequency from experimental investigations [4].

To facilitate the diagnosis in rolling elements, some companies and researchers tabulate the most common failure modes in the frequency domain, so that the analysis can be carried out easier. Thus, the appearance of different frequency peaks determines the existence of developing problems such as gaps, unbalances or misalignments among other circumstances [31].The great advantage of these tables is that the value of the frequency peak is not a particular value and may be adapted to any situation where the natural frequency (or the rotational speed) is known.

Wavelet transform is a time-frequency technique similar to Short Time Fourier Transform although it is more effective when the signal is not stationary. Wavelet transform decompose an input signal into a set of levels at different frequencies [77]. Wavelet transforms have been applied to the fault detection and diagnosis in various wind turbine parts.

A hidden Markov model is a statistical model in which the system being modelled is assumed to be a Markov process with hidden states. A hidden Markov model can be considered as the simplest dynamic Bayesian network [8]. Ocak and Loparo presented the application for the bearing fault detection [57].

They are used when a statistical study is required. In these cases, common statistical, i.e. the root mean square or peak amplitude; to diagnose faults are employed. Other parameters can be maximum or minimum values, means, standard deviations to energy ratios or kurtosis. Moreover, trend analysis refers to the collection of information in order to find a trend.

There are many methods that, as happened with the techniques available for CM, are very specific and therefore they are used for very specific situations. Filtering methods, for example, are designed to remove any redundant information, eliminating unnecessary overloads in the process. Analysis in time domain will be a way of monitoring wind turbine faults as inductive imbalances o turn-to-turn faults. Other methodology, the power cepstrum, defined as the inverse Fourier Transform of the logarithmic power spectrum [92], reports the occurrence of deterioration through the study of the sidebands. Time synchronous averaging, amplitude demodulation and order analysis are other signal processing methodologies used in wind turbines.

2. Wavelet transform

The wavelet transform is a method of analysis capable of identifying the local characteristics of a signal in the time and frequency domain. It is suitable for large time intervals, where great accuracy is requested at low frequencies and vice versa, e.g. small regions where precision details for a deeper processing are required at higher frequencies [23]. The wavelet transform can be defined as a signal on a temporal base that is filtered successive times and whose average value is zero. These wavelets are irregular and asymmetrical [51]. The transform has many applications in control process and detection of anomalies. It enables to analyse the signal structures that depend on time and scale, being a useful method to characterize and identify signals with spectral features, unusual temporary files and other properties related to the lack of stationary. When the frequency range corresponding to each signal is known, the data can be studied in terms of time, frequency and amplitude. Therefore it is possible to see which frequencies are in each time interval, and may even reverse the wavelet transform when it is necessary. Previously to the wavelet transform, the FFT was able to work with this type of signals in the frequency domain but without great resolution in the time domain [38].

The wavelet transform of a function *f(t)* is the decomposition of *f(t)* in a set of functions and $\psi_{s,\tau}(t)$, forming a base. It is defined as [88] [66]:

$$W_f(s,\tau)=\int f(t)\psi^*_{s,\tau}(t)dt \tag{3}$$

Wavelets transforms are generated from the translation and scale change from a same wavelet function $\psi(t)$, called *mother wavelet*, which is given by equation (4):

$$\psi_{s,\tau}(t)=\frac{1}{\sqrt{s}}\psi\left(\frac{t-\tau}{s}\right) \tag{4}$$

where s is the scale factor, and τ is the translational factor.

The wavelets $\psi_{s,\tau}(t)$ generated from the same mother wavelet function $\psi(t)$ have different scale s and location τ, but the same shape. Scale factors are always $s>0$. The wavelets are dilated when the scale $s>1$ and contracted when $s<1$. Thus, the changing of the value s can cover different ranges of frequencies. Large values for the parameter s correspond to lower frequencies ranges or a large scale for $\psi_{s,\tau}(t)$. Small values of s correspond to lower frequencies ranges or very small scales.

The wavelet transform can be continuous or discrete. The difference between them is that the continuous transform provides more detailed information but consuming more computation time while the discrete signal is efficient with fewer parameters and less computation time [17]. The Discrete Wavelet Transform coefficients are a group of discrete intervals of time and scales. These coefficients are used to formalize a set of features that characterize different types of signals. Any signal can be divided into low frequency approximations (*A*) and high frequency details (*D*). The sum of *A* and *D* is always equal to the original signal. The division is done using filters (Figure 2).

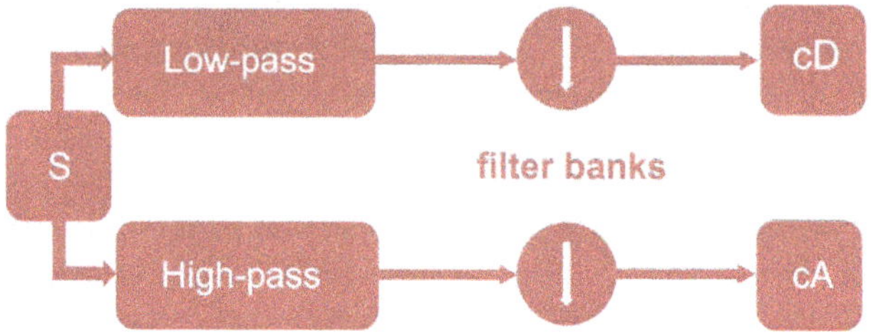

Figure 2. Decomposition diagram.

To reduce the computational and mathematical costs due to duplication of data, a sub-sampling is usually performed, containing the half of the collected information from *A* and *D* but without losing information. It is common to accompany this information with a graphical representation where the original signal is divided in low pass filters and high pass filters [15]. When the signals are complex, the decomposition must be to further levels and it is not sufficient with two frequency bands. From this need, multilevel filters appear. Multilevel filters repeat the filtering process iteratively with the output signals from the previous level. This leads to the so called wavelet decomposition trees (Figure 3.) [2]. By decomposing a signal in more frequency bands, additional information is obtained. A suitable branch to each signal is highly recommended as more decompositions do not always mean higher quality results.

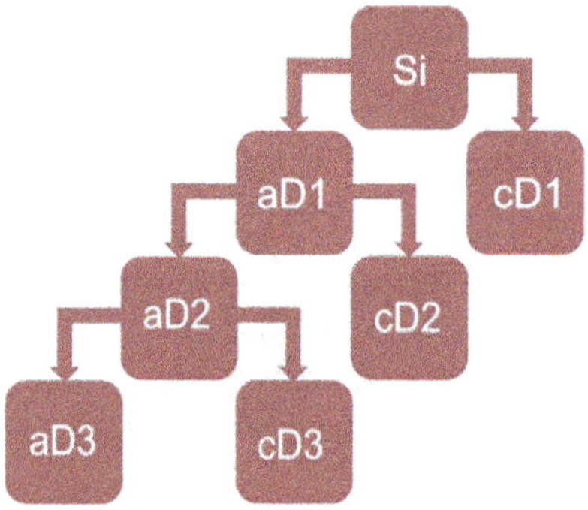

Figure 3. Wavelet decomposition tree.

The calculation of the Continuous Wavelet Transform starts for an initial time and a scale value. The result of multiplying the two signals is integrated into the whole space of time. Subsequently, this integral is multiplied by the inverse of the square root scale value, obtaining a transformed function with a normalized energy. This process is iterative until the end of the original signal is reached and must be repeated for all the values of scale that sweep the frequency range to be studied.

2.1. Wavelet families

The concept of wavelet has emerged and evolved during the last decades. Though new families of wavelet transforms are rapidly increasing, there are a number of them that have been established with more strength over time. In most situations, the use of a particular family is set by the application.

Daubechies wavelets are the most used wavelets, representing the foundations of wavelets signal processing and founding application in Discrete Wavelet Transform. They are defined as a family of orthogonal and smooth basis wavelets characterized by a maximum number of vanishing moments. The degree of smoothness increases as long as the order is higher. Daubechies wavelets lead to more accurate results in comparison to others wavelet types and also handle with boundary problems for finite length signals in an easier way [58] [29] [60] [94]. Wavelets have not an explicit expression except for order 1, which is the Haar wavelet. The inability to present a wavelet equation by a particular formula will be the general trend for almost all types of wavelet families [76].

As above mentioned, Haar wavelets are Daubechies wavelets when the order is 1. They are the simplest orthonormal wavelets. The main drawback for Haar wavelets is their discontinuity as a consequence of not solving breaking points problems for its derivates. The Haar transform is one of the earliest examples of a wavelet transform and it is supported by a function is an odd rectangular pulse pair [33]. Haar functions are widely used for applications as image coding, edge extraction and binary logic design and are defined as [46] [41] [34] [30]:

$$H(t) = \begin{cases} 1 & 0 \le t < \frac{1}{2} \\ -1 & \frac{1}{2} \le t < 1 \\ 0 & elsewhere \end{cases} \tag{5}$$

The main advantages of the Haar wavelet are its accuracy and fast implementation compared with others methods, its simplicity and small computational costs, and its capacity for solving boundaries problems [87].

Symlet wavelet transform is an orthogonal wavelet defined by a scaling filter (a low-pass finite impulse response filter of length $2N$ and sum 1). Symlet wavelet transform is sometimes called SymletN, where N is the order. Symlet wavelets are near symmetric. Furthermore, they have highest number of vanishing moments for a given width [7].

Coiflet wavelets are a family of wavelets whose main characteristics are similar to the Symlet ones: a high number of vanishing moments and symmetry. Coiflet family is also compactly supported, orthogonal and capable to give a good accuracy when the original signal has a distortion. The Coiflet wavelets are defined for 5 orders [18].

Biorthogonal wavelets have become very popular because of its versatility, being capable of supporting symmetric or antisymmetric signals. They perform very well under certain boundaries conditions [97]. Moreover the Biorthogonal wavelet transform is an invertible transform. They have two sets of lowpass filters for reconstruction, and highpass filters for decomposition [32].

Along with the Haar wavelets, the Meyer family is one of the exceptions that can be represented by an equation. The Meyer wavelets have numerous applications in the theory of functions, solving differential equations, signal processing, etc. [39]. Meyer family has not compact support being this one of its drawbacks. It is defined by equation (6) [44]:

$$\lambda(\omega)\Big|_{[0,\infty]} = \begin{cases} \dfrac{\pi}{4} + \theta(\omega - \pi), & \omega \in \left[\dfrac{2\pi}{3}, \dfrac{4\pi}{3}\right], \\ \dfrac{\pi}{4} + \theta(\dfrac{\omega}{2} - \pi), & \omega \in \left[\dfrac{4\pi}{3}, \dfrac{8\pi}{3}\right], \\ 0, & \omega \in \left[0, \dfrac{2\pi}{3}\right] \cup \left[\dfrac{8\pi}{3}, +\infty\right], \end{cases} \tag{6}$$

where $\theta(\omega)$ is a continuously and differentiable function equal to $\frac{\pi}{4}$ for $\omega \geq \frac{\pi}{3}$.

2.2. Wavelet transform applications

The use of the wavelet transform has been developed over the past two decades focused on the process diagnosis and instrumentation. In 1990, Leducq introduces them in the analysis of hydraulic noise for a centrifugal pump [45]. Later other authors demonstrates its usefulness for the detection of mechanical failures and the health monitoring control in gears [74] [11] [90] [21] [82] [80]. Cracks in rotors [1], structures [73] [63] [89] [10] or composite plates [75] has been another exploitation source for wavelet transforms. In 1994, Newland researches on their properties and applications, and coins the term harmonic wavelet. Harmonic wavelets are used for ridge and phase identification in signals [54]. The results showed that the cracks found reduced the rotor speed. The effectiveness of wavelets has also been compared with the envelope detection methodology in the diagnosis of faults in the bearings, obtaining results in shorter time analysis [85].

Due to its good analytical skills in time regarding the frequency, wavelet transform is a guarantee of success in the study of transient processes. Chancey and Flowers [16] managed to discover a relation between vibration patterns and the coefficients of a wavelet. Kang and Birtwhistle [40] or Subramanian, Badrilal and Henry [78] developed techniques to find problems in power transformers. Yacamini [96] proposed a method to detect torsional vibrations in engines and generators from the stator currents.

At present, the development of techniques associated to the scopes mentioned previously are still being implemented but others wavelet transforms purposes are emerging, such as classification of linear frequency modulation signals for radar emitter recognition [83] or applications to damages caused by corrosion in chemical process installations [86]. As follow there is an explanation for some of the most examined in the scientific literature.

The application of wavelets transforms in wind turbines focuses on the implementation of adaptive controllers for wind energy conversion systems. Wavelet transform is capable of providing a good and quick approximation. The drivers studied under different noise levels achieved higher performances [69]. Other works study the monitoring and diagnosis of faults in induced generators with satisfactory results. In these cases a combination of DWTs, accompanied by statistical data and energy is proposed. The use of decomposed signals spectral components is other highly interesting technique of study. Its harmonic content has suitable characteristics to be employed in fault diagnosis as an alternative to conventional methods [3].

Rolling bearing plays an important role in rotating machines. The choice of a particular wavelet family is crucial for the maintenance and fault diagnosis. The location of peaks on the vibration spectrum can identify a particular fault. Wavelet decomposition trees are a useful tool for this identification. The mean square error extracted from the terminal nodes of a tree reports the failure and its size [17]. There are also studies focused on determining what type of wavelet is suitable for bearing maintenance [79].

The wavelet transform is a good signal analysis method when a variation of time but not of space exists. The analysis provides information about the frequency of the signal, being a solution for the engine failure detection. There are detection algorithms that identify the presence of a fault in working condition and are ahead of the shutdown of the system, reducing costs and downtimes [19] [20]. These algorithms are independent of the type of engine used. Other studies in this field, present methods to detect imbalances in the stator voltage of a three phase induction motor. The wavelet transform of the stator current is analysed. Computationally, these methods are less expensive than other existing and can detect faults in an early stage. In the same vein, monitoring fatigue damage has been studied [65].

3. Condition Monitoring for engine-generator mechanism

A novel approach for Condition Monitoring based on wavelet transforms is introduced. A system for a mechanism based on an engine and a generator will be shown. It has been designed to represent any similar mechanism located in a wind turbine, generally in the nacelle. These mechanisms are used in cooling devices (generators, gearboxes), electric motors for service crane, yaw motors, pitch motors (depending on the configuration) or pumps (oil, water) according to the sub systems configurations, ventilators, etc (Figure 4).

A set of faults are induced in different experiments: ski-slope faults, misalignment faults, angular misalignment faults, parallel misalignment faults, rotating looseness faults and external noise faults. Pattern recognition is obtained from the extraction of vibration and acoustic signals. A Fault Detection and Diagnosis method is developed from the patterns of these signals. In order to recognize the patterns, three basic steps have been followed [37]:

1. The data acquisition on the testing bench (Figure 5).
2. The extraction of the features of the experiment using specific algorithms.
3. A decision-making.

A classification has been done to obtain the optimal pattern recognitions employing the data from Fast Fourier Transform and wavelet transforms applied to the vibrations and sounds signals respectively.

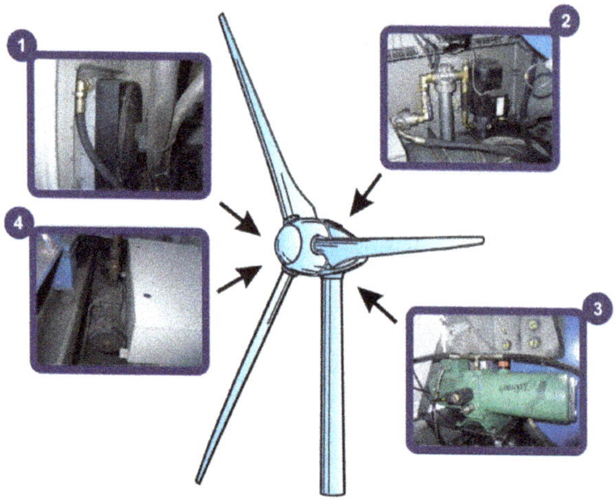

Figure 4. Different locations of a wind turbine where the CM can be used: (1) fans, (2) gear oil pump, (3) oil pump for brake and (4) water cooling pump.

3.1. Case study

The experiments were made on a mechanism consisting of an engine and a generator linked by an elastic coupling joint. The sensors employed were a current sensor, an ambient temperature sensor, another temperature sensor located in strategic points of the mechanism, a vibration sensor; and a sound sensor (microphone). The data obtained by these sensors are stored in a data acquisition board, except for the vibration which is collected directly with a vibrometer. The software employed was LabView and specific software for vibration provided by the manufacturer Kionix. The speed of the engine and its associated frequency were set by a frequency variator, and the energy is dispelled using a resistive element.

The allocation of the vibration measurements were: two points for the engine and two for the generator. Points of selection were located at the end of each machine and as close as possible to the axis which is the main rotational element of the mechanism (Figure 6).

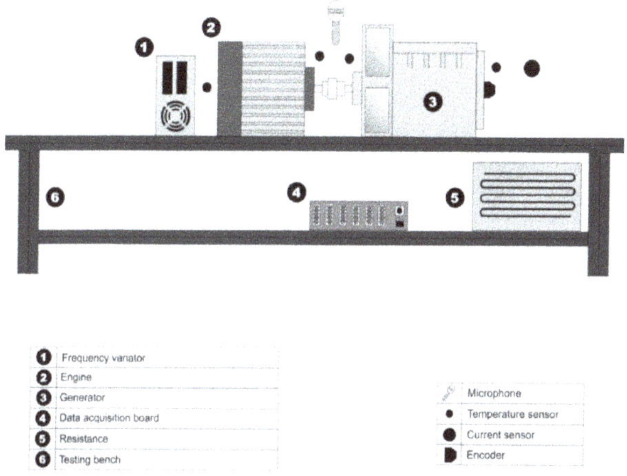

Figure 5. Experimental mechanism.

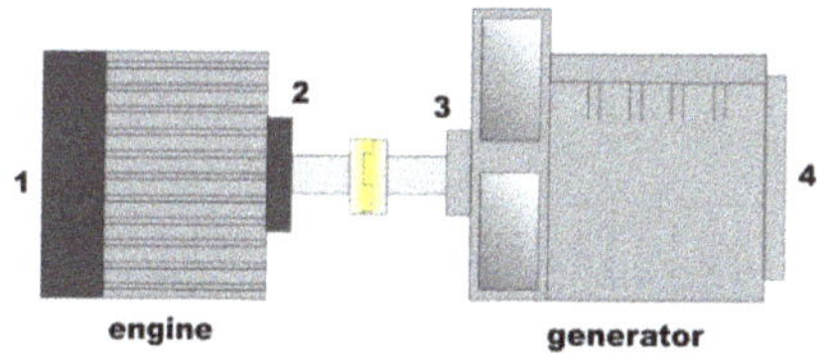

Figure 6. Measuring points.

The experiments were completed for an average time of 10 seconds each one, and every experiment was repeated 3 times. Therefore, for each experiment 12 measurements of temperatures, currents, sound, velocities and vibrations were taken (Figure 7). In the case of vibration, the vibrometer is capable of storing samples for the 'x', 'y' and 'z' axis, in addition to a total measurement for the point studied (Figure 8).

The experiments were carried out in order to identify couplings and misalignments in different degrees. The engine has 4 rubber clamping (silemblocks), while the generator has 3 rubbers clamping. The silemblocks were located at the ends, having two on the right side of the engine and two on the left side. The generator has them placed in a triangle, two in the area closest to the coupling and one at the end. The first experiment recorded under free fault conditions, and the rest of experiments were performed when the silemblocks were removed from the engine and the generator in order to create the different degrees of decoupling (Figure 9).

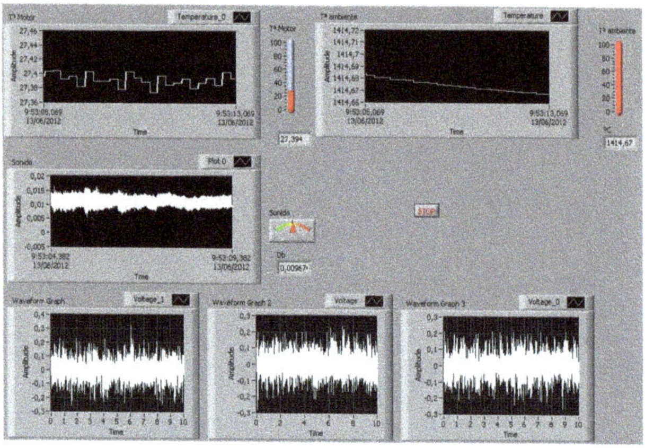

Figure 7. Data collection in LabView.

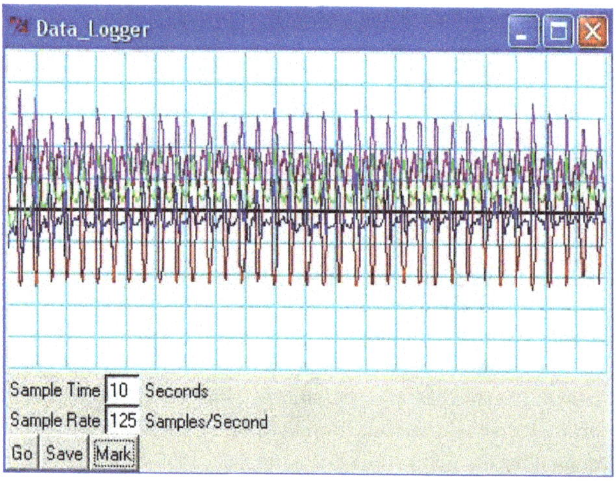

Figure 8. Data collection with Kionix software (vibration).

The rotational speed is 1500 rpm, i.e. 25 Hz. In order to do an analysis above the natural frequency, the number of samples was increased from 25 Hz to 125 Hz, being 25 Hz the default samples. This guarantees a range 5 times bigger than the natural frequency of the engine.

Experiment	Type of experiment	Data set
1	Free fault conditions	From 1 to 12
2	Misalignment removing silemblocks from the right side of the engine	From 13 to 24
3	Misalignment removing silemblocks from the right side and the front left one of the engine	From 25 to 36
4	Generation of resistance in the coupling	From 37 to 48
5	Misalignment removing the silemblock from the right side of the generator	From 49 to 60
6	Misalignment removing 2 silemblocks near to the coupling in the generator	From 61 to 72
7	Misalignment removing the silemblock from the right side of the generator and one from the left side of the engine	From 73 to 84
8	Use of a rigid coupling	From 85 to 96

Table 1. Experiments (1500 rpm).

The FFT of each signal has been developed in Matlab. An algorithm that allows the comparison of two signals for a given frequency was created. The main purpose is to compare pattern conditions with the signals of the rest of experiments that represent a fault and to analyse the peaks found in the natural frequency and its multiples. In some cases it is important to analyse the area located below the natural frequency. Another advantage of the program is that it is possible to obtain the amplitude values for a certain frequency range (Figure 10). With a click on a particular peak, the program provides the data.

Figure 9. Misalignments induced removing silemblocks from the engine and the generator and experimentation with a rigid coupling.

Values for 25 Hz (natural frequency or 1X), 50 Hz (2X), 75 Hz (3X) and 100 Hz (4X) have been taken into account. Frequencies above these values have been discarded.

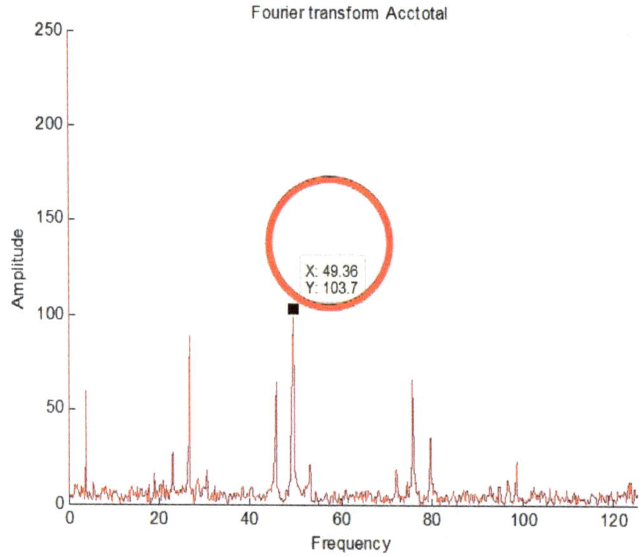

Figure 10. FFT of a vibration signal.

3.2. Vibration diagnosis and results

The most common spectrums for engine-generator mechanisms are presented. Examples based on the experiments held are shown.

Ski-slope fault

A ski-slope fault appears when the spectrum begins at a high level and then it goes down slowly (Figure 11). A ski-slope shows a problem with the quality of the sensor. Sometimes it happens because the sensor has experienced a transient during the measurement process. The transient may be mechanical, thermal or electrical.

Misalignment faults

Misalignment fault appears when the centrelines of coupled shafts do not coincide. If the misaligned shaft centrelines are parallel but not coincident, then the misalignment is a parallel misalignment. If the misaligned shafts meet at a point but they are not parallel, the misalignment is angular. Most of the cases are a combination of them. The diagnosis is based on dominant vibration from the natural frequency (1X) at twice the rotational rate (2X), with increased rotational rate levels (3X, 4X, etc.) acting in the axial, vertical or horizontal directions.

Angular misalignment fault

Angular misalignment fault produces a bending moment on both shafts and this generates a strong vibration at 1X, and some others at 2X and 3X for the axial direction. There will also be strong radial components for vertical and horizontal directions (Figure 11).

Parallel misalignment fault

Parallel misalignment fault produces a shear force and a bending moment on the coupled end of each shaft. High vibration levels at 2X as well as 1X are produced in the radial direction. Most often the 2X component is higher than 1X. Depending on the coupling, there can be 3X or 4X, even reaching 8X when the misalignment is severe (Figure 11).

Rotating looseness fault

Rotating looseness fault will create harmonics or sub-harmonics every 0.5X. Even 1/3 order harmonics are possible (Figure 11).

External noise fault

It is very common to find a peak in a spectrum that is difficult to analyse. This happens because of the vibration from another machine or process. The peak will typically be at a non-synchronous frequency (Figure 11). External noise can be verified stopping the machine (or varying the speed) and seeing if the vibration is still present or checking local machines for the same frequency source.

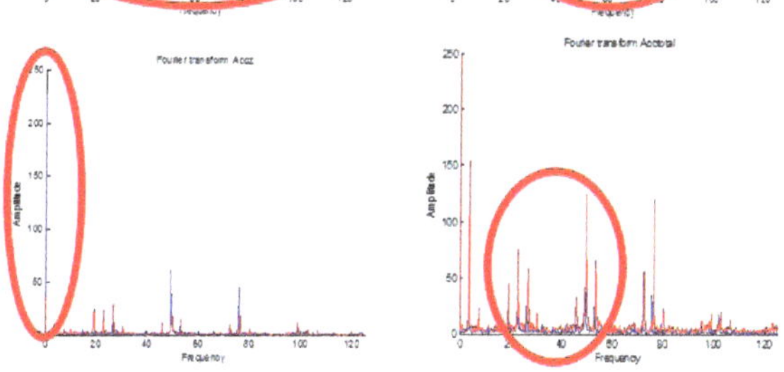

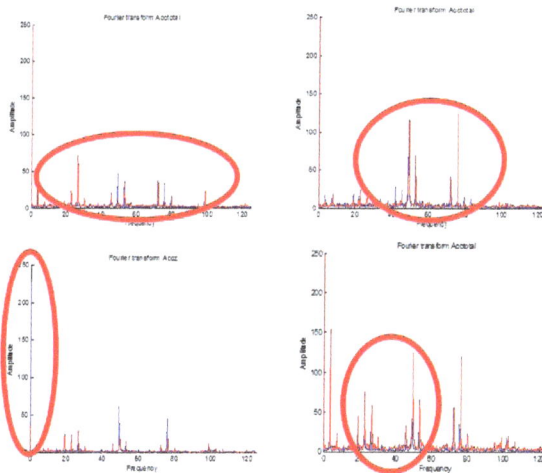

Figure 11. a) Angular misalignment fault (red) and pattern condition (blue), (b) parallel misalignment fault (red) and pattern condition (blue), (c) ski-slope fault (blue) and pattern condition (red) and (d) rotating looseness (blue); and external noise (red).

3.3. Vibration results

As a rule, the natural frequency (1X) has been kept as the reference. Following the same nomenclature, the peaks at 50 Hz, 75 Hz and 100 Hz have been named 2X, 3X and 4X.

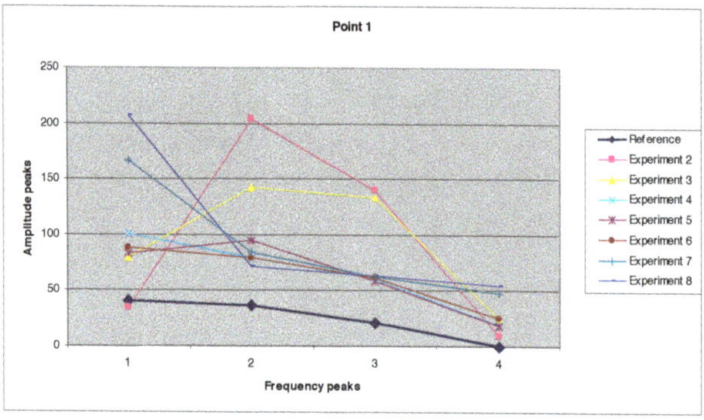

Figure 12. Vibration for point 1.

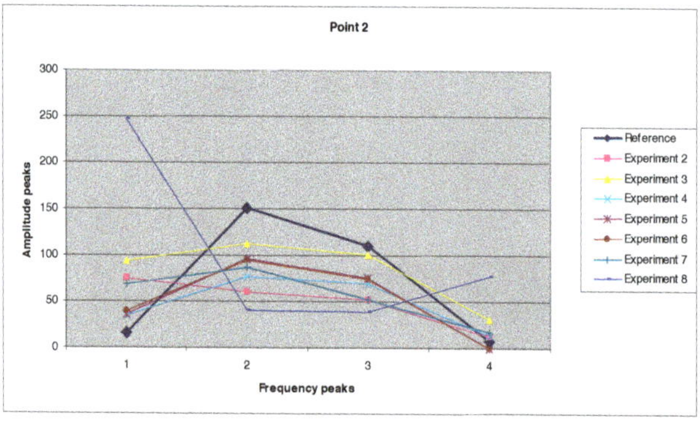

Figure 13. Vibration for point 2.

Vibration patterns are different for the four operating points. It has been detected that the natural frequency, regardless of its amplitude, tends to predominate in the experiments associated with the end points of the set (Figures 12 and 15). Additionally, the generator's closest point to the coupling also has a similar pattern (Figure 14). The second point differs from the rest, yielding most predominant peaks from the frequency at 50 Hz (Figure 13). To make the vibration analysis, it must be taken into consideration not only the appearance of peaks, but also the amplitude. The same diagnosis for two experiments can vary its amplitude depending on the severity of the faults found. The main symptoms appear when peaks at 0.5X,

1X, 2X and 3X, sidebands and noise sources are detected. When a failure is studied at an advanced stage, peaks at 4X are noticeable (case of rigid coupling).

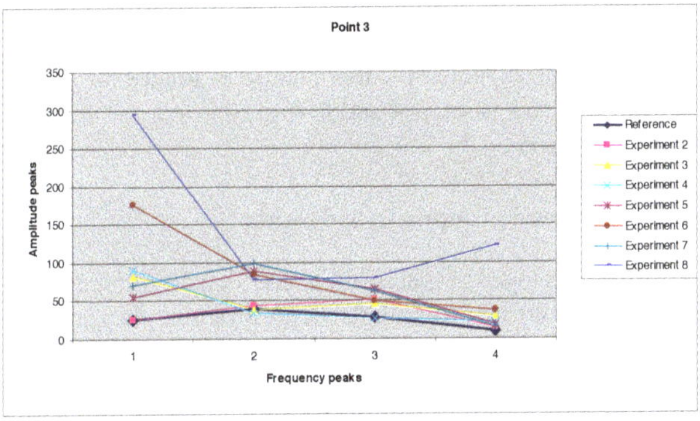

Figure 14. Vibration for point 3.

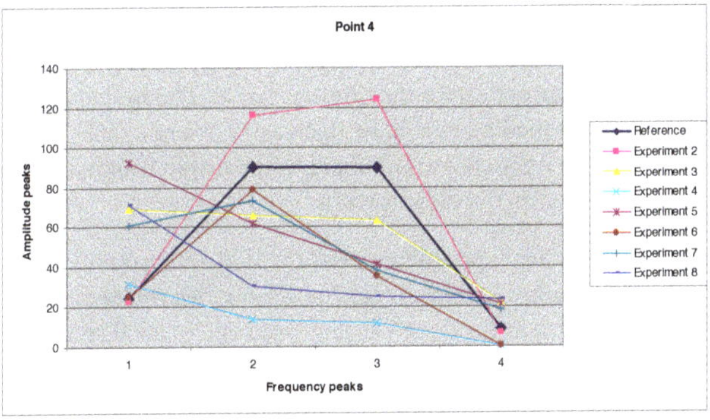

Figure 15. Vibration for point 4.

The diagnosis of the experiments reveals that the mechanism has a minor looseness which causes the appearance of a high peak at the natural frequency in some cases, even under free fault conditions. This looseness appears because the engine and the generator are not anchored directly to the test bench. The assembly was done on a surface that has facilitated the removal of the silemblocks when the experiments required it, e.g. to create different degrees of misalignment. On the other hand, this action expands the vibration intentionally because in this way it is closer to the actual behavior of the nacelle. These frequency peaks change their trend in 1X as long as the study advances from the end of the engine to the generator.

From point 2, the peak at frequencies as 2X and 3X becomes more significant and sometimes exceed the amplitude of the natural frequency.

The results for experiment 8 are also remarkable. The rigid coupling added causes a severe looseness and vibration. The growth of a frequency at 4X and a constant noise over the spectrum is observed. Although it is usual to find sidebands, peaks below 1X and high frequency peaks for all this type of experiments, this feature is unique to this last experiment. Initially, a similar diagnosis for cases 1, 4 and 8 was expected, but the behavior has been slightly different for this reason.

3.4. Wavelet transform processing approach and results

Wavelet transforms were employed to analyse the sound signals. As for the Fast Fourier Transform, an algorithm has been written with Matlab. This program plots and compares two signals. Data has been transformed in 5 decompositions named a_4, d_4, d_3, d_2 and d_1, where each of them has an energy rate associated from the original signal (Figure 16). The algorithm also returns a percentage value per decomposition. These values of energy, the decomposition levels attached and the peak amplitudes are examined in order to look for patterns.

Functions in the time domain can be represented as a linear combination of all frequency components present in a signal, where the coefficients are the amount of energy provided by each frequency component to the original signal. The main decomposition is associated with a_4 (*main* or *mother wavelet*) that usually has the highest energy, though it is not always necessarily the case. It has a similar pattern to the original signal. The first (d_4), second (d_3), third (d_2) and fourth (d_1) transformed signals have decreasing energy rates, being s the original signal. Usually a_4 is the low frequency component of the original signal while d_i is the high frequency component, having d_1 the biggest value.

It is necessary to verify that the experiments performed at 1500 rpm can be extrapolated to other speeds. In the case of wind turbines, most of the engines rotate at speeds close to 3000 rpm. A certain number of tests were done varying from 500 to 3000 rpm (at intervals of 500 rpm) in order to ensure the existence of the proportional pattern.

The results showed that regardless of the speeds or the points of study, all the graphical representations for the different decompositions of energy had the same patterns. Figure 17 indicates the existence of a similar behavior where only changes the numerical value. The biggest ones will correspond to the *main* signals, while the results for decompositions d_1 and d_2 are similar.

Data can be studied according to the evolution of a single point along the different experiments or analysing the evolution of the set points for all the experiment. Each row in Figure 18 contains two graphics, one with the amplitude peaks (left) and the other one with the energy distribution of the sound signal (right). The first two graphics correspond to the engine end (point 1). The following two graphics are the closest to the coupling (point 2). The third row belongs to the points of the generator next to the coupling (point 3), and finally, the last two graphics are for the end of the generator (point 4).

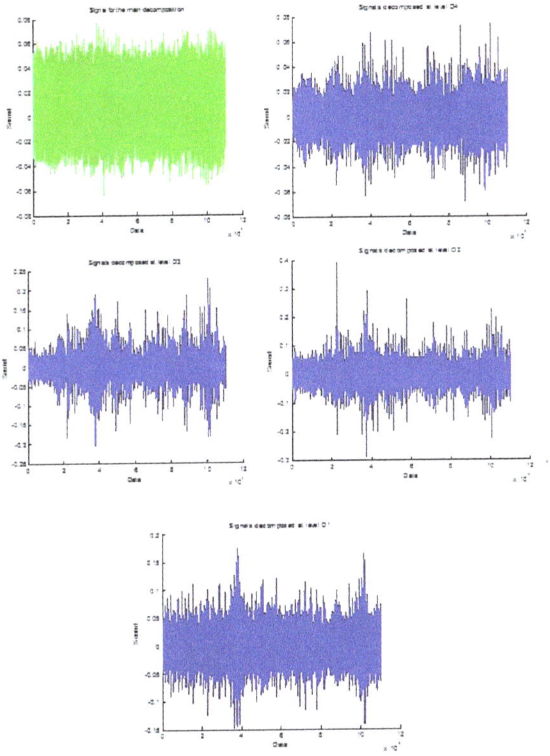

Figure 16. Wavelet decompositions.

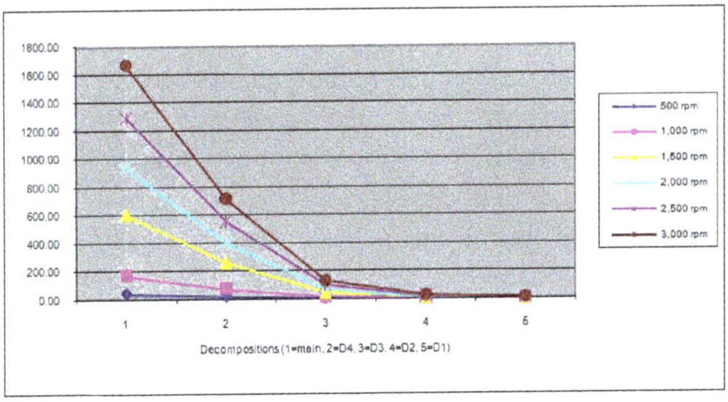

Figure 17. Energies at different rotational speeds.

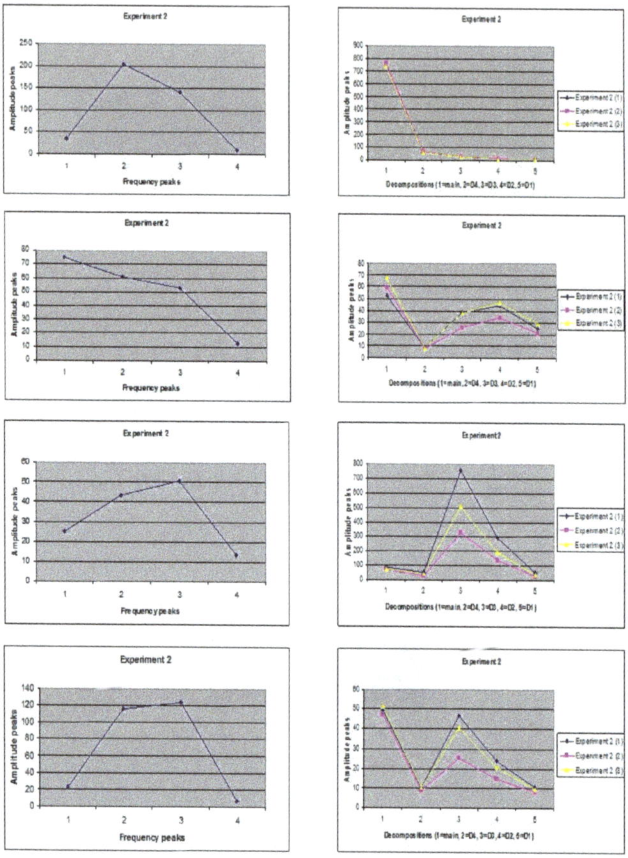

Figure 18. Evolution of the frequency peaks and wavelet energy decompositions for each point in experiment 2.

Based on the distribution of the energy among the 5 different decompositions of every signal, the energy distribution for point 1, end of the engine-generator set is ruled by an almost similar pattern where each experiment has a maximum of energy in the main signal and a minimum for decomposition d_1 or d_2. It means that by performing a decomposition of the signal, the energy has a closest resemblance to the original value, often exceeding 85% of the total energy, remaining a residual percentage for d_1 or d_2. When the experiments are closer to the generator (points 2, 3 and 4), the energy is distributed among the 5 decompositions and not concentrated in the *mother wavelet*, as it is for point 1.

All the decompositions have been registered with their energy maximum and minimum values and their patterns distribution. An example for 2 experiments is shown in Table 2.

Experiment	Main	d_4	d_3	d_2	d_1	Energy
A	17,19%	9,10%	22,12%	24,95%	26,63%	167,9
B	82,32%	9,90%	5,17%	1,77%	0,84%	311,8

Table 2. Energy distribution for experiments A and B.

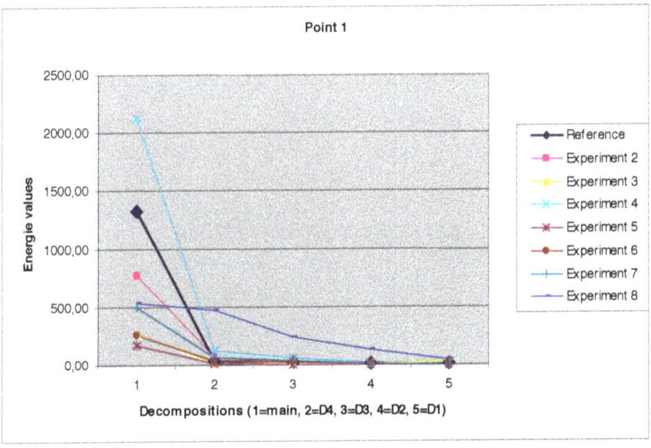

Figure 19. Energy values for point 1.

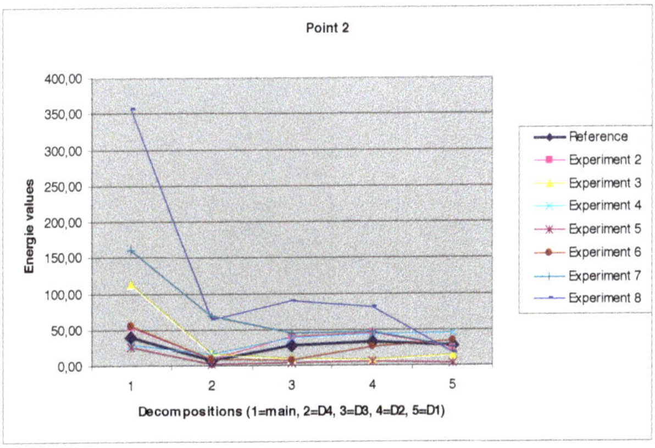

Figure 20. Energy values for point 2.

Experiment A is associated to point 2, belonging to the engine and situated close to the coupling. Experiment B, however, is related to point 1, left end of the assembly. Experiment A has the maximum percentage of energy in d_1 and the minimum in d_4. Furthermore, the experiment B has its maximum in the *main signal* and the minimum located in d_1. The maximum-minimum patterns are d_1 -d_4 and *main*-d_1 respectively. Numerically, the most compensated distribution of energy is close to the coupling (experiment A – point 2) above mentioned. The patterns *main*-d_1 and *main*-d_2 appear for all the cases in point 1. However, the same maximum-minimum distribution is smaller for the points 2, 3 and 4. Unlike in point 1, there are different patterns for the 8 experiments in these points. Figures 19, 20, 21 and 22 represent the numerical values of the energy per point and experiment. It must be noted that the numerical values are higher or lower, depending on the type of experiment.

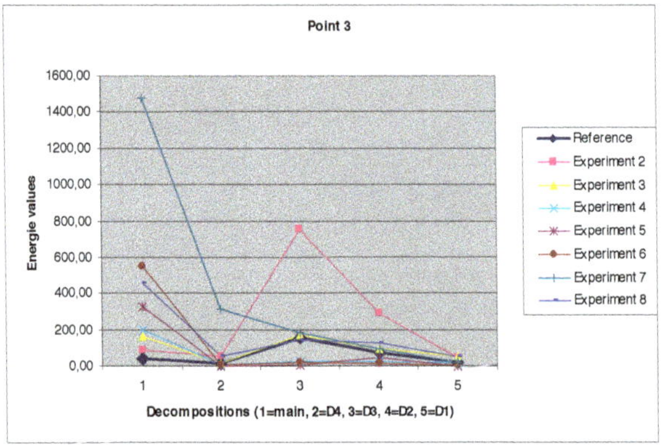

Figure 21. Energy values for point 3.

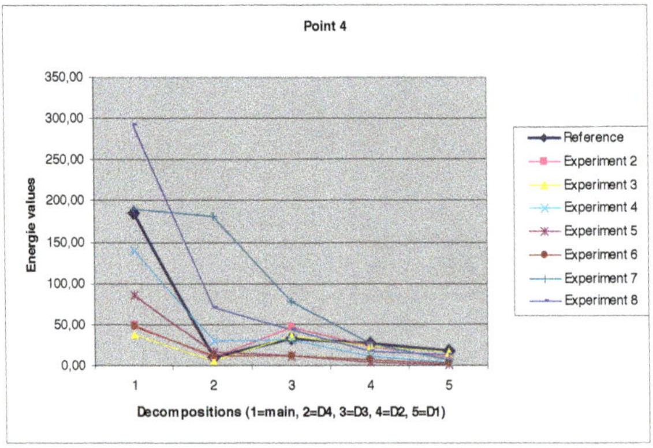

Figure 22. Energy values for point 4.

4. Conclusions

Wind turbines are complex systems that require a high level of reliability, availability, maintainability and safety. This chapter is focused on to guarantee these correct levels for mechanisms used in cooling devices for generators and gearboxes, electric motors for service crane, yaw motors, pitch motors, pumps, ventilators, etc.

The mechanism brake of the engine has been simulated linking a generator by a coupling joint. The signals collected have been:

- Vibration.
- Sound.
- Current.
- Temperature.
- Velocity.

The experiments have been done in working conditions for different points of the mechanism and considering the following failures:

- Misalignment removing silemblocks from the right side of the engine.
- Misalignment removing silemblocks from the right side and the front left one of the engine.
- Induction of resistance in the coupling.
- Misalignment removing the silemblock from the right side of the generator.
- Misalignment removing 2 silemblocks near to the coupling in the generator.
- Misalignment removing the silemblock from the right side of the generator and one from the left side of the engine.
- Using a rigid coupling.

A fault detection and diagnosis model based on the Fast Fourier Transform applied to the vibration signals; together with the wavelet transform applied to sound signals has been developed. The model detects and diagnoses correctly 100% of the failures considered.

It has been observed that for the outer ends of the engine and the generator, the appearance of a pronounced peak amplitude at the natural frequency or *2X* (vibration) was associated to the maximum energy values for the *main* signal, the most suitable with the original, and minimum values for decomposed signals d_1 and d_2 (sound). In contrast, the results obtained close to the coupling did not follow a clear trend as the results were conditioned by the type of experiment. The numerical values of each peak were also taken into account in the establishment of the pattern recognitions, being different for each experiment. The same conclusion was reached for the energy values. Different models and results were expected because the objective was not to find similar patterns between different experiments, and the tests were never performed under identical conditions. The objective was to have different vibration patterns and their associated sound models in order to create a catalogue of possible scenarios for predictive maintenance in the mechanisms. Thus, it is possible to extend the range of possibilities to relate the result of an acoustic signal with the frequency domain using the Fast Fourier Transform.

Author details

Fausto Pedro García Márquez[1], Raúl Ruiz de la Hermosa González-Carrato[1], Jesús María Pinar Perez[1] and Noor Zaman[2]

1 University of Castilla-La Mancha, Spain

2 CCSIT, King Faisal University, Saudi Arabia

References

[1] Adewusi, S. A., & Al-Bedoor, B. O. (2001). Wavelet analysis of vibration signals of an overhang rotor with a propagating transverse crack. *Journal of Sound and Vibration.*, 246(5), 777-793.

[2] Aktas, M., & Turkmenoglu, V. (2010). Wavelet-based switching faults detection in direct torque control induction motor drives. *Science, Measurement and Technology.*, 4(6), 303-310.

[3] Al-Ahmar, E., Benbouzid, M. E. H., & Turri, S. (2008). Wind energy conversion systems fault diagnosis using Wavelet analysis. *International Review of Electrical Engineering.*, 3(4), 646-652.

[4] Al-Hussain, K. M., & Redmond, I. (2002). Dynamic response of two rotors connected by rigid mechanical coupling with parallel misalignment. *Journal of Sound and Vibration.*, 249(3), 483-498.

[5] Amidror, I., & Hersch, R. (2009). The role of Fourier theory and of modulation in the prediction of visible moiré effects. *Journal of Modern Optics.*, 56(9), 1103-1118.

[6] Anon. (2005). Managing the wind: Reducing kilowatt-hour costs with condition monitoring. *Refocus.*, 6(3), 48-51.

[7] Arora, R., Sharma, L., Birla, N., & Bala, A. (2011). An algorithm for image compression using 2D wavelet transforms. *International Journal of Engineering Science & Technology.*, 3(4), 2758-2764.

[8] Baum, L. E., & Petrie, T. (1966). Statistical inference for probabilistic functions of finite state Markov chains. *The Annals of Mathematical Statistics.*, 37(6), 1554-1563.

[9] Ben-Daya, M. S., & Duffuaa, A. R. (2009). Handbook of maintenance management and engineering. *Springer Verlag London Limited.*

[10] Bieman, C., Staszewski, W. J., Boller, C., & Tomlinson, G. R. (1999). Crack detection in metallic structures using piezoceramic sensors. *Key Engineering Materials.*, 167, 112-121.

[11] Boulahbal, D., Farid, G. M., & Ismail, F. (1999). Amplitude and phase wavelet maps for the detection of cracks in geared systems. *Mechanical Systems and Signal Processing.*, 13(3), 423-436.

[12] Bracewell, R. (2000). The Fast Fourier Transform and its applications. *McGraw Hill Higher Education.*

[13] Butler, D. E. (1973). The shock pulse method for the detection of damaged rolling bearings. *NDT International.*, 6(2), 92-95.

[14] Byon, E., & Ding, Y. (2010). Season-dependent condition-based maintenance for a wind turbine using a partially observed Markov decision process. *IEEE Transactions on Power Systems.*, 25(4), 1823-1834.

[15] Canal, M. R. (2010). Comparison of wavelet and short time Fourier transform methods in the analysis of EMG signals. *Journal of Medical Systems.*, 34(1), 91-94.

[16] Chancey, V. C., & Flowers, G. T. (2001). Identification of transient vibration characteristics using absolute harmonic wavelet coefficients. *Journal of Vibration and Control.*, 7(8), 1175-1193.

[17] Chebil, J., Noel, G., Mesbah, M., & Deriche, M. (2010). Wavelet decomposition for the detection and diagnosis of faults in rolling element bearings. *Jordan Journal of Mechanical & Industrial Engineering.*, 4(5), 260-266.

[18] Chourasia, V. S., & Mittra, A. K. (2009). Selection of mother wavelet and denoising algorithm for analysis of foetal phonocardiographic signals. *Journal of Medical Engineering & Technology.*, 33(6), 442-448.

[19] Combastel, C., Lesecq, S., Petropol, S., & Gentil, S. (2002). Model-based and wavelet approaches to induction motor on-line fault detection. *Control Engineering Practice.*, 10(5), 493-509.

[20] Cusidó, J., Romeral, L., Ortega, J. A., García, A., & Riba, J. R. (2010). Wavelet and PDD as fault detection techniques. *Electric Power Systems Research.*, 80(8), 915-924.

[21] Dalpiaz, G., Rivola, A., & Rubini, R. (2000). Effectiveness and sensitivity of vibration processing techniques for local fault detection in gears. *Mechanical Systems and Signal Processing.*, 14(3), 387-412.

[22] Deshpande, V. S., & Modak, J. P. (2002). Application of RCM for safety considerations in a steel plant. *Reliability Engineering and System Safety.*, 3(78), 325-334.

[23] Dong, Y., Shi, H., Luo, J., Fan, G., & Zhang, C. (2010). Application of wavelet transform in MCG-signal denoising. *Modern Applied Science.*, 4(6), 20-24.

[24] Endrenyi, J., Mc Cauley, J., & Singh, C. (2001). The present status of maintenance strategies and the impact of maintenance on reliability. *IEEE Transaction Power System.*, 16(4), 638-646.

[25] García, F. P., Schmid, F., & Collado, J. C. (2003). A reliability centered approach to remote condition monitoring. A railway points case study. *Reliability Engineering & System Safety.*, 80(1), 33-40.

[26] García, F. P., Schmid, F., & Collado, J. C. (2003). Wear assessment employing remote condition monitoring: A case study. *Wear.*, 2(255), 1209-1220.

[27] García, F. P., Pedregal, D. J., & Roberts, C. (2010). Time series methods applied to failure prediction and detection. *Reliability Engineering & System Safety.*, 95(6), 698-703.

[28] García, F. P., Roberts, C., & Tobias, A. (2010). Railway point mechanisms: Condition monitoring and fault detection. *Proceedings of the Institution of Mechanical Engineers, Part F, Journal of Rail and Rapid Transit. Professional Engineering Publishing,* 224(1), 35-44.

[29] Genovese, L., Neelov, A., Goedecker, S., Deutsch, T., Ghasemi, S., Willand, A., Caliste, D., Zilberberg, O., Rayson, M., Bergman, A., & Schneider, R. (2008). Daubechies wavelets as a basis set for density functional pseudopotential calculations. *Journal of Chemical Physics.*, 129(1), 104-109.

[30] Ghanbari, M., Askaripour, M., & Khezrimotlagh, D. (2010). Numerical solution of singular integral equations using Haar wavelet. *Australian Journal of Basic & Applied Sciences.*, 4(12), 5852-5855.

[31] Goldman, P., & Muszynska, A. (1999). Application of full spectrum to rotating machinery diagnostics. *Orbit First Quarter.*, 20(1), 17-21.

[32] Hajjara, S., Abdallah, M., & Hudaib, A. (2009). Digital image watermarking using localized biorthogonal wavelets. *European Journal of Scientific Research.*, 26(4), 594-608.

[33] Hariharan, G., & Kannan, K. (2010). Haar wavelet method for solving FitzHugh-Nagumo equation. *International Journal of Computational & Mathematical Sciences.*, 4(6), 281-285.

[34] Hassan, S., Al-Saegh, M., Mohamed, A., & Batarfi, H. (2011). Haar wavelet spectrum of a pulsed-driven qubit. *Nonlinear Optics, Quantum Optics: Concepts in Modern Optics.*, 42(1), 37-50.

[35] Huang, M., Jiang, L., Liaw, P. K., Brooks, C. R., Seeley, R., & Klarstrom, D. L. (1998). Using acoustic emission in fatigue and fracture materials research. *JOM Nondestructive Evaluation: Overview.*, 50(11), 1-12.

[36] Igarashi, T., & Hamada, H. (1982). Studies on the vibration and sound of defective roller bearings (First report: vibration of ball bearing with one defect). *Bulletin of the Japan Society of Mechanical Engineers.*, 25(204), 994-1001.

[37] Jardine, A., Lin, D., & Banjevic, D. (2006). A review on machinery diagnostics and prognostics implementing condition-based maintenance. *Mechanical Systems and Processing.*, 20(7), 1483-1510.

[38] Jia, M., & Wang, Y. (2003). Application of wavelet transformation in signal processing for vibrating platform. *Journal of Shenyang Institute of Technology.*, 22(3), 53-55.

[39] Johnstone, I. M., Kerkyacharian, G., Picard, D., & Raimondo, M. (2004). Wavelet deconvolution in a periodic setting. *Journal of the Royal Statistical Society: Series B (Statistical Methodology).*, 66(3), 547-573.

[40] Kang, P., & Birtwhistle, D. (2003). Condition assessment of power transformer on-load tap-changers using wavelet analysis. *IEEE Transactions on Power Delivery.*, 18(1), 78-84.

[41] Karimi, H., & Robbersmyr, K. (2011). Signal analysis and performance evaluation of a crash test with a fixed safety barrier based on Haar waveletes. *International Journal of Wavelets, Multiresolution & Information Processing.*, 9(1), 131-149.

[42] Knezevic, J. (1993). Reliability, maintainability and supportability engineering: A probabilistic approach. *McGraw Hill.*

[43] Lahdelma, S., & Juuso, E. (2007). Advanced signal processing and fault diagnosis in condition monitoring. *Non-destructive Testing and Condition Monitoring.*, 49(12), 719-725.

[44] Lebedeva, E., & Protasov, V. (2008). Meyer wavelets with least uncertainty constant. *Mathematical Notes.*, 84(5), 680-687.

[45] Leducq, D. (1990). Hydraulic noise diagnostics using wavelet analysis. *Proceedings of the International Conference on Noise Control Engineering.*, 997-1000.

[46] Lepik, Ü. (2011). Buckling of elastic beams by the Haar wavelet method. *Estonian Journal of Engineering.*, 17(3), 271-284.

[47] Leske, S., & Kitaljevich, D. (2006). Managing gearbox failure. *Dewi Magazine.*, 29.

[48] Mc Fadden, P. D. (1987). Examination of a technique for the early detection of failure in gears by signal processing of the time domain average of the meshing vibration. *Mechanical Systems and Signal Processing.*, 1(2), 173-183.

[49] Mc Fadden, P. D. (1996). Detecting fatigue cracks in gears by amplitude and phase demodulation of the meshing vibration. *Journal of Vibration, Acoustics, Stress and Reliability in Design.*, 108, 165-170.

[50] Mc Millan, D., & Ault, G. W. (2008). Condition monitoring benefit for onshore wind turbines: Sensitivity to operational parameters. *IET Renewable Power Generation.*, 2(1), 60-72.

[51] Misrikhanov, A. M. (2006). Wavelet transform methods: Application in electroenergetics. *Automation and Remote Control.*, 67(5), 682-697.

[52] Moubray, J. (1997). Reliability-centered maintenance. *New York: Industrial Press.*

[53] Nakhaeinejad, M., & Ganeriwala, S. Observations on dynamic responses of misalignments. *Technologial Notes. SpectraQuest Inc.*, http://spectraquest.com/.

[54] Newland, D. E. (1999). Ridge and phase identification in the frequency analysis of transient signals by harmonic wavelets. *Journal of Vibration and Acoustics, Transactions of the ASME.*, 121(2), 149-155.

[55] Novaes de, G., Alencar, E., & Kraj, A. Remote conditioning monitoring system for a hybrid wind diesel system-application at Fernando de Naronha Island. http://www.ontario-sea.org.

[56] Oberst, U. (2007). The Fast Fourier Transform. *SIAM Journal on Control & Optimization.*, 46(2), 1-45.

[57] Ocak, H., & Loparo, K. A. (2001). A new bearing fault detection and diagnosis Scheme based on Hidden Markov modeling of vibration signals. *Acoustics, Speech and Signal Processing.*, 5, 3141-4144.

[58] Patil, S., Kasturiwala, S., Dahad, S., & Jadhav, C. (2011). Wavelet tool: Application for human face recognition. *International Journal of Engineering Science & Technology.*, 3(3), 2392-2398.

[59] Pedregal, D. J., García, F. P., & Roberts, C. (2009). An algorithmic approach for maintenance management. *Annals of Operations Research.*, 166, 109-124.

[60] Peng, Z., & Chu, F. (2004). Application of the wavelet transform in machine condition monitoring and fault diagnostics: a review with bibliography. *Mechanical Systems and Signal Processing.*, 18, 199-221.

[61] Peters, S. T. (1998). Handbook of composites. *Chapman & Hall. London.*, 839-855.

[62] PROTEST (PROcedures for TESTing and measuring wind energy systems). (2009). Deliverable D1: State of the art report. *FP7-ENERGY-2007-1-RTD.*

[63] Quek, S. T., Wang, Q., Zhang, L., & Ang, K. K. (2001). Sensitivity analysis of crack detection in beams by wavelet technique. *International Journal of Mechanical Sciences.*, 43(12), 2899-2910.

[64] Rumsey, M. A., & Musial, W. (2001). Application of infrared thermography nondestructive testing during wind turbine blade tests. *Journal of Solar Energy Engineering.*, 123(4), 271.

[65] Samsi, R., Ray, A., & Mayer, J. (2009). Early detection of stator voltage imbalance in three-phase induction motors. *Electric Power Systems Research.*, 79(1), 239-245.

[66] Schmitt, E., Idowu, P., & Morales, A. (2010). Applications of wavelets in induction machine fault detection. *Ingeniare. Revista chilena de ingeniería.*, 18(2), 158-164.

[67] Schoen, R. R., Lin, B. K., Habetler, T. G., Schlag, J. H., & Farag, S. (1995). An unsupervised, on-line system for induction motor fault detection using stator current monitoring. *IEEE Transactions on Industry Applications.*, 31(6), 1280-1286.

[68] Schroeder, K., Ecke, W., Apitz, J., Lembke, E., & Lenschow, G. (2006). A fibre Bragg grating sensor system monitors operational load in a wind turbine rotor blade. *Measurement Science and Technology.*, 17(5), 1167-1172.

[69] Sedighizadeh, M., & Rezazadeh, A. (2008). Nonlinear model identification and PI control of wind turbine using neural network adaptive frame wavelets. *International Journal of Applied Engineering Research.*, 3(7), 861-877.

[70] Sekhar, A. S., & Prabhu, B. S. (1995). Effects of coupling misalignment on vibration of machines. *Journal of Sound and Vibration.*, 185(4), 655-671.

[71] Smith, A. M. (1993). Reliability-centred maintenance. *New York: McGraw-Hill, Inc.*

[72] Smith, B. M. (1978). Condition monitoring by thermography. *NDT International*, 11(3), 121-122.

[73] Srinivas, H. K., Srinivasan, K. S., & Umesh, K. N. (2010). Application of artificial neural network and wavelet transform for vibration analysis of combined faults of unbalances and shaft bow. *Advances in Theoretical and Applied Mechanics.*, 3(4), 159-176.

[74] Staszewski, W. J., & Tomlinson, G. R. (1994). Application of the wavelet transform to fault detection in a spur gear. *Mechanical Systems and Signal Processing.*, 8(3), 289-307.

[75] Staszewski, W. J., Pierce, S. G., Worden, K., & Culshaw, B. (1999). Cross-wavelet analysis for lamb wave damage detection in composite materials using optical fibres. *Key Engineering Materials.*, 167, 373-380.

[76] Strang, G. (1992). The optimal coefficients in Daubechies wavelets. *Physica D: Nonlinear Phenomena.*, 60, 239-244.

[77] Strang, G., & Nguyen, T. (1997). Wavelets and filter banks. *Wellesley-Cambridge Press.*

[78] Subramanian, S., Badrilal, M., & Henry, J. (2010). Wavelet transform based differential protection for power transformer and classification of faults using SVM and PNN. *International Review of Electrical Engineering.*, 5(5), 2186-2198.

[79] Sugumaran, V., & Ramachandran, K. I. (2009). Wavelet selection using Decision tree for fault diagnosis of roller bearings. *International Journal of Applied Engineering Research.*, 4(2), 201-225.

[80] Suh, J. H., Kumara, S. R. T., & Mysore, S. P. (1999). Machinery fault diagnosis and prognosis: Application of advanced signal processing techniques. *CIRP Annals- Manufacturing Technology.*, 48(1), 317-320.

[81] Sun, Q., & Tang, Y. (2002). Singularity analysis using continuous wavelet transform for bearing fault diagnosis. *Mechanical Systems and Signal Processing.*, 16(6), 1025-1041.

[82] Sung, C. K., Tai, H. M., & Chen, C. W. (2000). Locating defects of a gear system by the technique of wavelet transform. *Mechanism and Machine Theory.*, 35(8), 1169-1182.

[83] Swiercz, E. (2011). Automatic classification of LFM signals for radar emitter recognition using wavelet decomposition and LVQ classifier. *Physical Aspects of Microwave and Radar Applications.*, 119(4), 488-494.

[84] Tan, C. C. (1990). Application of acoustic emission to the detection of bearing failures. *Proceedings Tribology Conference. Brisbane.*, 110-114.

[85] Tse, P. W., Peng, Y. H., & Yam, R. (2001). Wavelet analysis and envelope detection for Rolling element bearing fault diagnosis. Their effectiveness and flexibilities. *ASME Journal of Vibration and Acoustics.*, 123, 303-310.

[86] Van Dijck, G., & Van Hulle, M. M. (2011). Information theory filters for wavelet packet coefficient selection with application to corrosion type identification from acoustic emission signals. *Sensors.*, 11(6), 5695-5715.

[87] Venkatesh, S., Ayyaswamy, S., & Hariharan, G. (2010). Haar wavelet method for solving initial and boundary value problems of Bratu-type. *International Journal of Computational & Mathematical Sciences.*, 4(6), 286-289.

[88] Wang, D., Miao, Q., Fan, X., & Huang, H. Z. (2009). Rolling element bearing fault detection using an improved combination of Hilbert and wavelet transform. *Journal of Mechanical Science and Technology.*, 23, 3292-3301.

[89] Wang, Q., & Deng, X. M. (1999). Damage detection with spatial wavelets. *International Journal of Solids and Structures.*, 36(23), 3443-3468.

[90] Wang, W. Q., Ismail, F., & Golnaragh, M. F. (2001). Assessment of gear damage monitoring techniques using vibration measurements. *Mechanical Systems and Signal Processing.*, 15(5), 905-922.

[91] Wei, J., & Mc Carty, J. (1993). Acoustic emission evaluation of composite wind turbine blades during fatigue testing. *Wind Engineering.*, 17(6), 266-274.

[92] Wismer, N. J. (1994). Gearbox analysis using cepstrum analysis and comb liftering. *Application Note. Brüel & Kjaer. Denmark.*

[93] World Wind Energy Association. World wind energy report 2009. http://www.wwindea.org.

[94] Wu, J., & Liu, C. (2008). Investigation of engine fault diagnosis using discrete wavelet transform and neural network. *Expert Systems with Applications.*, 35, 1200-1213.

[95] Xu, M., & Marangoni, R. (1994). Vibration analysis of a motor-flexible coupling-rotor system subjected to misalignment and unbalance Part I: Theoretical model and analysis. *Journal of Sound and Vibration.*, 176(5), 663-679.

[96] Yacamini, R., Smith, K. S., & Ran, L. (1998). Monitoring torsional vibrations of electro-mechanical systems using stator currents. *Journal of Vibration and Acoustics, Transactions of the ASME.*, 120(1), 72-79.

[97] Yang, X., Shi, Y., & Yang, B. (2011). General framework of the construction of biorthogonal wavelets based on Bernstein bases: theory analysis and application in image compression. 5(1), 50-67.

Chapter 3

Deterministic Sampling for Quantification of Modeling Uncertainty of Signals

Jan Peter Hessling

Additional information is available at the end of the chapter

1. Introduction

Statistical signal processing [1] traditionally focuses on extraction of information from noisy measurements. Typically, parameters or states are estimated by various filtering operations. Here, the quality of signal processing operations will be assessed by evaluating the statistical uncertainty of the result [2]. The processing could for instance simulate, correct, modulate, evaluate, or control the response of a physical system. Depending on the addressed task and the system, this can often be formulated in terms of a differential or difference *signal processing model equation* in time, with uncertain parameters and driven by an exciting input signal corrupted by noise. The quantity of primary interest may not be the output signal but can be extracted from it. If this uncertain dynamic model is linear-in-response it can be translated into a linear digital filter for highly efficient and standardized evaluation [3]. A statistical model of the parameters describing to which degree the dynamic model is known and accurate will be assumed given, instead of being the target of investigation as in system identification [4]. *Model uncertainty* (of parameters) is then *propagated* to *model-ing uncertainty* (of the result). The two are to be clearly distinguished – the former relate to the input while the latter relate to the output of the model.

Quantification of uncertainty of complex computations is an emerging topic, driven by the general need for quality assessment and rapid development of modern computers. Applications include e.g. various mechanical and electrical applications [5-7] using uncertain differential equations, and statistical signal processing. The so-called brute force Monte Carlo method [8-9] is the indisputable reference method to propagate model uncertainty. Its main disadvantage is its slow convergence, or requirement of using many samples of the model (large ensembles). Thus, it cannot be used for demanding complex models. The ensemble size is a key aspect which motivates deterministic sampling. Small ensembles are found by substituting the random generator with a customized deterministic sampling rule. Since any computerized random generator produces a pseudo-random rather than a truly random sequence, this is equivalent of modifying the random generator to be *accurate* for *small* ensembles of *definite* size, rather than being *asymptotically exact* (infinite ensembles). Correctness of very large ensembles is of theoretical but hardly practical interest for complex models, if the convergence to the asymptotic result is very slow.

2. Modeling uncertainty of signals

2.1. Problem definition

Suppose the (output) signal $y(x, t) \in \mathbb{R}$ of interest is generated from the (input) signal $x(t) \in \mathbb{R}$ passing through a dynamic system H, with parameters $a_k \in \mathbb{R}$, $b_k \in \mathbb{R}$,

$$\left[\sum_{k=0}^{u} a_k D^k\right] y = \left[\sum_{k=0}^{v} b_k D^k\right] x, \quad a_0 = 1. \tag{1}$$

The model is given in $n = u + v + 1$ uncertain parameters, which can be arranged in a column vector $q = (b_0 \ \cdots \ b_v \ a_1 \ \cdots \ a_u)^T$. For systems continuous-in-time (CT), $D = \partial_t$ is the differential operator in time while for systems discrete-in-time (DT), $D = \Delta^{-1}$ is the negative unit displacement operator, $\Delta^{-1} x_k = x_{k-1}$. There are several approximate methods to sample CT systems to DT systems, see [3] and references therein. The discretization techniques are beyond the scope of this presentation and DT systems will be assumed. If $u \geq 1$, there is feedback in the system which results in an impulse response $h(q, t)$ of infinite duration. For finite accuracy however, the duration is finite. The system is linear-in-response, $y(x = \alpha x_1 + \beta x_2, t) = \alpha y(x_1, t) + \beta y(x_2, t)$. Most importantly, the system is non-linear-in-parameters if $u \geq 1$. This is the typical situation addressed here.

Systems of the form in Eq. 1 may be directly realized as digital filters, $y(q, x, t) = h(q) * x(t)$, where $*$ denotes the filtering operation. The coefficients b_k and a_k are the numerator and denominator coefficients of the filter with impulse response $h(q)$, respectively. Its z-transform $H(q, z)$ is obtained with the substitution $\Delta \to z$. The parameterization can be changed to for instance gain K, poles p_k and zeros z_k, or poles p_k and residues r_k,

$$Y(z) = H(z)X(z): \quad H(q,z) = \frac{\sum_{k=0}^{v} b_k z^{-k}}{\sum_{k=0}^{u} a_k z^{-k}} = K \frac{\prod_k (z - z_k)/(1 - z_k)}{\prod_k (z - p_k)/(1 - p_k)} = \sum_k \frac{r_k}{(z - p_k)} \tag{2}$$

The parameterization should be carefully chosen as it affects the convergence rate of Taylor expansions (section 3.1) as well as the physical interpretation. The parameters and their statistics are preferably extracted from measurements using system identification techniques [4]. Note that complex-valued poles and zeros are conjugated in pairs [10].

The problem to be addressed is the statistical evaluation of any function $g(y(t)=h(q, t)*x(t))$, given statistical models of q and x. It will here consist of evaluating its time-dependent mean $\langle g(y)\rangle$ and standard deviation $\sqrt{\langle[g(y)-\langle g(y)\rangle]^2\rangle}$. Without loss of generality, the analysis will be made for $g(y)=y$. Digital filtering will be utilized for evaluating samples of the model, i.e. filtering with definite sets of q and signals x.

2.2. Nomenclature

Statistical expectations of any signal, model or function $g(q)$ over finite discrete E as well as continuous ensembles or probability distributions (no subscript) are defined as,

$$\begin{aligned} \langle g\rangle_E &= \frac{1}{m}\sum_{k=1}^{m} g\left(\hat{q}^{(k)}\right), \\ \langle g\rangle &= \int_Q g(q) f_q(q)dq. \end{aligned} \quad (3)$$

Samples of q are labeled $\hat{q}$, with their components organized in columns. Sample indices will be given as superscripts in parenthesis, eg. $\hat{q}^{(k)}$ is a column vector denoting the k-th sample of parameter q. Variations from the mean are written as $\delta q_{(E)} \equiv q-\langle q\rangle_{(E)}$.

Only uniform (UNI) and normal distributions (NRM) will be utilized. Either the mean and standard deviation, or the interval in brackets will be given in parenthesis, e.g. $q\sim \mathrm{UNI}(0.5, 1/2\sqrt{3})=\mathrm{UNI}([0, 1])$. Statistical moments $M_i^{(k)}=\sqrt[k]{\langle(\delta q_i)^k\rangle}$ carry the information contained in the marginalized probability density functions (pdf) $f_i(\delta q_i)\equiv\int_Q f_q(\delta q)dq_1\cdots dq_{i-1}dq_{i+1}\cdots dq_n$, where Q denotes the sample space. While $M_i^{(2)}$ describes the width of $f_i(\delta q_i)$, $M_i^{(3)}$ is related but different to its *skewness* [11]. Further, the shape is reflected in $M_i^{(4)}$, similarly to the *curtosis* [11]. Since UNI(0,1) and NRM(0,1) are normalized and symmetric $f_i(\delta q_i)=f_i(-\delta q_i)$, $M_i^{(2)}=1$ and $M_i^{(3)}=0$. Their differences are first reflected in their fourth moment, $M_i^{(4)}=1/4\sqrt{5}, 1/2\sqrt{3}\approx(0.11, 0.29)$ for UNI(0,1) and NRM(0,1), respectively. The maximum variation of the parameter q_i is expressed by the range $M_i^{(\infty)}\equiv\lim_{k\to\infty}|M_i^{(k)}|=\max(|\delta q_i|)$. Dependencies are expressed in mixed moments $\langle(\delta q_{i1})^{k_1}(\delta q_{i2})^{k_2}\cdots\rangle_{(E)}$. The discussion will be limited to correlations described by the covariance matrix $\mathrm{cov}(q)=\langle\delta q\delta q^T\rangle$, where the vector multiplication is an outer product.

Matrix size will be indicated with subscripts, e.g. $V_{n\times m}$ is a matrix of n rows and m columns with elements V_{jk}, $j=1, \ldots n$ and $k=1, \ldots m$. The identity matrix will be denoted I, while matrices with equal elements (i) will have their size attached, $(i_{n\times n})_{jk}=i$. For a matrix (vector) D, diag(D) is a vector (diagonal matrix) with components (diagonal elements) equal to the diagonal elements (components) of D. The trace of a matrix is denoted Tr.

A method will be stated intrusive if manipulations of the model are required. For the targeted highly complex models, it will be assumed that the computational cost for their evaluation dominates all other calculations. The efficiency ρ of any method will accordingly be defined by the least required number of evaluations of the original model.

2.3. Fundamentals of non-linear propagation of uncertainty

Linearity in parameters (LP) is to be distinguished from linearity in response (LR),

$$\begin{aligned} LR: \quad y(q,a_1x_1+a_2x_2,t) &= a_1y(q,x_1,t)+a_2y(q,x_2,t), \quad \forall x_1,x_2 \\ LP: \quad y(q_1+q_2,x,t) &= y(q_1,x,t)+C(x,t)^T(q_2-q_1) \quad \forall q_1,q_2 \end{aligned}, \tag{4}$$

for some vector $C_{n\times 1}$. Different concepts of linearity are used, $y(q_1+q_2, x, t)\neq y(q_1, x, t)+y(q_2, x, t)$ for LP models. Strictly speaking, LP denotes models that are affine, i.e. written as linear combinations of their parameters. Most constructed systems are designed to be as close to LR as possible while most models are not LP. There is hence no contradiction in non-linear (LP) propagation of uncertainty with linear (LR) digital filters, as here.

For non-linear propagation of uncertainty, the asymmetry of the resulting pdf is central. It can be expressed as a lack of commutation of non-linear propagation and statistical evaluation of a center value ($\cdot_C$), as measured with the *scent* [12],

$$\varsigma \equiv y_C(q)-y(q_C). \tag{5}$$

The method for evaluating the center is left unspecified, as there are several alternatives. The most common choice is to use the mean, $\cdot_C=\langle\cdot\rangle$. The lowest order approximation of the scent can then be obtained by calculating the expectation of a Taylor expansion (section 3.1), $\varsigma=\mathrm{Tr}[\mathrm{cov}(q)\cdot H(y)]/2$, where $H(y)_{jk}=\partial^2 y/\partial q_j\partial q_k$ is the Hessian matrix signal of y, evaluated at $\langle q\rangle$. The scent is related to the skewness $\gamma=\langle\delta y^3\rangle/\langle\delta y^2\rangle^{3/2}$. The *additional* asymmetry caused by the non-linearity of the model is measured with the scent but differently. The scent addresses how parametric uncertainties are propagated and not how the result is distributed, e.g. $\varsigma=0$ for all LP models for which γ may attain any value. A finite scent thus implies the model is not LP, but not the reverse. The scent should not be confused with bias. Bias is a property of an estimator, while scent is a property of a model. For every model, such as the

REF (section 6.1), many different estimators of $y_C(q)$ can be used, e.g. the different ensembles in section 5.6, see result in Fig. 5 (left). Consequently, an unbiased estimator of $y_C(q)$ correctly accounts for rather than ignores its finite scent, or deviation from $y(q_C)$.

The scent is important since y_C and not $y(q_c)$ is the main result utilized in applications. The corresponding difference [13] in the standard deviation $M_y^{(2)}$ from its linearized approximation $\sqrt{\nabla y^T \text{cov}(q) \nabla y}$, with $(\nabla y)_{jk} = \partial_j y(t_k)$, affects the confidence in the result. Its accuracy is usually less critical. An accurate evaluation of the scent is perhaps the strongest feature of the unscented Kalman filter, which provides the foundation for the presented approach as well as the origin of the term 'scent'.

3. Conventional methods

A brief resume of the most traditional related methods of uncertainty propagation, applicable to signal processing models, is here given together with their pros and cons. Advanced intrusive methods like e.g. polynomial chaos expansions [14-15] not directly related to the proposed method are omitted.

3.1. Taylor expansions

The indisputable default methods of uncertainty propagation are based on Taylor expansions. These methods are intrusive if the differentiations are made analytically. Convergent series require regular differentiable models and numerical or analytical complexity make them error prone. Their applicability is therefore limited for complex models.

The transfer function $H(q, z)$ of the digital filter can be expanded in a Taylor series,

$$\begin{aligned} \delta H(q,z) &= H(q,z) - H(\langle q \rangle, z) = \sum_{k=1}^{+\infty} \frac{1}{k!} \left(\delta q^T \nabla_q \right)^k H(q,z) = \delta q^T \nabla_q H + \frac{1}{2} \sum_{k,l}^{n} \delta q_k \delta q_l \frac{\partial^2 H}{\partial q_l \partial q_k} + \dots \\ &= \delta q^T E^{(1)}(\langle q \rangle, z) + \text{Tr}\left\{ \left[\delta q \delta q^T \right] \cdot E^{(2)}(\langle q \rangle, z) \right\} + \dots \end{aligned} \quad (6)$$

This defines n sensitivity systems (column vector) $E_k^{(1)}(\langle q \rangle, z)$, $n(n+1)/2$ unique quadratic variation systems (matrix) $E^{(2)}(\langle q \rangle, z)$, and so on. These variation systems differ (intrusive) from $H(q, z)$ but may nevertheless be realized as digital filters [3,7,10], just as $H(q, z)$. The corresponding variation of $y(q, x, t) = h(q, t) * x(t)$ is given by,

$$\delta y(q,x,t) = \delta q^T \left[e^{(1)}(\langle q \rangle, t) * x(t) \right] + \frac{1}{2} \text{Tr}\left\{ \left[\delta q \delta q^T \right] \cdot \left[e^{(2)}(\langle q \rangle, t) * x(t) \right] \right\} + \dots, \quad (7)$$

where $e^{(k)}(\langle q\rangle, t)$ are the impulse responses of the systems $E^{(k)}(\langle q\rangle, z)$. Utilizing digital filters with impulse responses $e^{(k)}(\langle q\rangle, t)$, the differentiations are conveniently done *once*, and not repeatedly for every signal $x(t)$. The linearity in parameters of the model can easily be studied for many different input signals $x(t)$, by evaluating $e^{(k)}(\langle q\rangle, t) * x(t)$. Due to the large number of variation systems, higher order perturbation analyses rapidly become intractable though. The established method is limited to linearization (LIN) [16] $(e^{(1)})$. It will always incorrectly yield vanishing scent, $\zeta=0$. A first order estimate of ζ is instead given by the expectation of second term in Eq. 7, $\zeta \approx \mathrm{Tr}[\mathrm{cov}(q) H(\langle q\rangle, t)]/2$, where the matrix of Hessian signals $H(\langle q\rangle, t) = e^{(2)}(\langle q\rangle, t) * x(t)$ is obtained with repeated digital filtering.

3.2. Brute force Monte Carlo

Monte Carlo (MC) methods [8-9], or *random sampling* of uncertain models was originally introduced and phrased 'statistical sampling' by Enrico Fermi already in the 1930's [17]. The MC methods *realize* uncertain signal processing models in finite *ensembles*. Every ensemble consists of a possible set of well-defined model systems, all (usually) having the same structure but slightly different parameter values. In the original so-called brute force Monte Carlo method, each set of parameters is assigned to the output of random generators with appropriate statistics. The convergence to the assigned statistics is very slow [5] but it is asymptotically exact and the required number of samples is essentially independent of the number of parameters. Hence it does not suffer from the curse-of-dimensionality of many other methods. The outstanding simplicity in application is likely the cause of its popularity, just as the slow convergence or low efficiency is the main reason for its failures.

In MC, arbitrary distributions and dependencies are usually obtained by means of transformations of samples of elementary distributions. Independent samples $\hat{q}^{(k)}$ of any probability density function (pdf) $\phi(x)$ can be constructed with the inverse transform method [9]. It consists of a calculation of the inverse of its cumulative distribution function (cdf) $\Phi(y)$ and generation of a uniformly distributed random sequence $\hat{z}^{(k)}$,

$$\hat{q}^{(k)} = \Phi^{-1}\left(\hat{z}^{(k)}\right), \quad \Phi(y) = \int_{-\infty}^{y} \phi(x)dx, \quad z \sim \mathrm{UNI}(0,1), \quad k = 1,2,\ldots m. \tag{8}$$

Covariance may be included with an appropriate transformation of samples of *canonical* parameters $\tilde{q}$: $q = U^T S \tilde{q}$ with $\mathrm{cov}(\tilde{q}) = I$,

$$\mathrm{cov}(q) = \left\langle \delta q \delta q^T \right\rangle = \left\langle U^T S \delta\tilde{q} \left(U^T S \delta\tilde{q}\right)^T \right\rangle = U^T S \left\langle \delta\tilde{q}\delta\tilde{q}^T \right\rangle S U = U^T S^2 U, \quad \begin{cases} U^T U = U U^T = I \\ S_{jk}^2 = 0, \, j \neq k \end{cases}, \tag{9}$$

The matrices S, U are found by calculating the eigenvalues (S^2) and eigenvectors (U) [11] of $\mathrm{cov}(q)$. This transformation makes the marginal pdfs $f_k(q_k)$ to differ substantially from the univariate pdfs ϕ_k of the independent but scaled parameters $S\tilde{q}_k$,

$$f_k(q_k) = \int \phi_1([Uq]_1)\phi_2([Uq]_2)\cdots\phi_k([Uq]_k)\cdots\phi_n([Uq]_n)dq_1\cdots dq_{k-1}dq_{k+1}\cdots dq_n \neq \phi_k(q_k), \text{ if } U \neq I. \quad (10)$$

All ϕ_k are hence mixed according to U. Dependencies are thus difficult to account for. One rare exception is provided by the multinomial distribution [9]. It is often better to assign the pdfs to the canonical parameters in the original instead of the canonical basis. The transformation then reads $\tilde{q}: q = U^T SU\tilde{q}$. As required, it leaves cov(q) invariant. The marginalization in Eq. 10 changes accordingly, $U \to SU^T S^{-1}U$. Since the transformation $U^T SUS^{-1}$ of $S\tilde{q}_k$ contains cancelling operations U, U^T and S, S^{-1}, it is generally less distorting than U^T. Indeed, if the commutator $[S, U^T] \equiv SU^T - U^T S$ vanishes, $U^T SUS^{-1} = I$. The transformation U^T must satisfy the stronger criterion $U = I$ to avoid mixing. For any transformation $q \to Wq$, an indicator of mixing of the components of q is given by,

$$\Psi(W) \equiv \frac{1}{n}\sum_{r=1}^{n}\left(1 - \frac{\max_c |W_{rc}| - \min_c |W_{rc}|}{\|W_{r,:}\|}\right) \in [0,1], \quad \|W_{r,:}\| \equiv \sqrt{\sum_{c=1}^{n} |W_{rc}|^2}. \quad (11)$$

A simple example illustrates that the mixing effect can be considerable, even for minute correlations. Assume a model has two parameters with a covariance matrix,

$$\mathrm{cov}(q) = \begin{pmatrix} 0.90 & 0.10 \\ 0.10 & 0.90 \end{pmatrix} \Leftrightarrow U = \frac{1}{\sqrt{2}}\begin{pmatrix} 1 & 1 \\ 1 & -1 \end{pmatrix}, S^2 = \begin{pmatrix} 1 & 0 \\ 0 & 0.8 \end{pmatrix} \Rightarrow \begin{cases} \phi_1(S\tilde{q}_1) & = & \mathrm{UNI}([0,1]) \\ \phi_2(S\tilde{q}_2) & = & \mathrm{UNI}([0,\sqrt{0.8}]) \end{cases}. \quad (12)$$

Large rotations are required because the canonical variances S_{jj}^2 are similar, i.e. cov(q) is almost *degenerate*. As shown in Fig. 1, the large rotations mix the assigned pdfs $\phi_k(S\tilde{q}_k)$ to marginal
6 to
pdfs $f_k(q_k)$ beyond recognition for the transformation U^T but not for $U^T SUS^{-1}$.
marginal pdfs $f_k(q_k)$ beyond recognition for the transformation U^T but not for $U^T SUS^{-1}$.

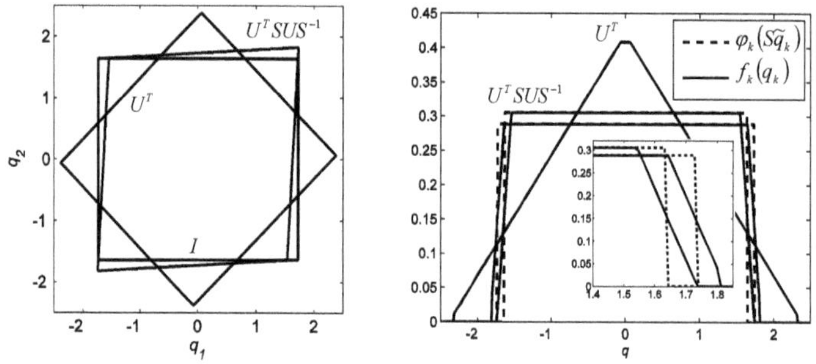

8

Figure 1. Left: The sample space of independent scaled parameters

Figure 1. Left: The sample space of independent scaled parameters ($I: q_k = S\tilde{q}_k$) (Eq. 12), and of the two transformations (U^T) (rotated) and ($U^T SUS^{-1}$) (skewed and tilted). Right: Assigned pdfs $\phi_k(S\tilde{q}_k)$ (dashed) and obtained marginal pdfs $f_k(q_k)$ (solid)) with mixing $\psi(U^T)=1.00$ and $\psi(U^TSUS^{-1})=0.058$, and magnified upper transition region (inset).

Specifying both marginal probability distributions and covariance is either redundant or inconsistent, as the latter is uniquely determined by the former. Nevertheless, this reflects the typical available information for signal processing applications. The moments can be accurately determined [4] for sufficiently large data sets but the joint distribution $f(q)$ is hardly ever known with any precision. Some of its properties are usually assigned, with varying degree of confidence. For instance, the allowed maximal range $M^{(\infty)}$ of the parameters of digital filters is given by stability constraints. The transformation technique above is well adapted to these facts, since the covariance is prioritised. The transformation $q=U^TSU\tilde{q}$ will be utilized in section 5.2 to include correlations with limited mixing of the statistics assigned to independent normalized canonical parameters $\tilde{q}$.

3.3. Refinements of Monte Carlo

To increase the efficiency of MC, the original brute force sampling technique has been further developed in mainly two directions: model simplification and sample distribution improvement. In response surface methodology (RSM) [18], the model is replaced by a simple approximate surrogate model. A model of order v may be found by applying linear (with respect to C) regression at *collocation points* [15] $\hat{q}^{(k)}=\mu^{(k)}$,

$$H(\mu)\approx R(\mu)C,\quad \begin{cases} R_{kj} &= R_j\left(\mu^{(k)}\right),\quad j=1,2,\ldots,v \\ H_k &= H\left(\mu^{(k)}\right) \end{cases},\quad \begin{cases} C &= \left(C_1 \;\cdots\; C_v\right)^T \\ \mu &= \left(\mu^{(1)} \;\cdots\; \mu^{(m)}\right)^T \end{cases},\quad m\geq v, \tag{13}$$

where $R_j(q)$ is basis function j. Since it may be non-linear, RSM allows for non-linear propagation of uncertainty and may give a substantially different and more accurate result than LIN. If only linear basis functions are used $R_j(q)=q_j$, RSM becomes equivalent to LIN. The best least square approximation is directly obtained from Eq. 13 [19],

$$C=\left(R^TR\right)^{-1}R^TH \tag{14}$$

Let RSM(r) utilize a complete set of mixed polynomial basis functions up to order r. Its least number (v) of collocation points grows rapidly with both the number of parameters (n) and polynomial order (r) [12],

$$v=\sum_{k=0}^{r} w(n,k): \quad w(n,k)=\sum_{j=1}^{\min(n,k)} \binom{n}{j}\cdot w(j,k-j), \quad w(j,0)=1. \tag{15}$$

In practice, $r>3$ often yields an unacceptable number of samples, see table 1.

	$n=2$	$n=5$	$n=10$	$n=20$
$r=1$	3	6	11	21
$r=2$	6	21	66	231
$r=3$	10	56	286	1771

Table 1. Efficiency $\rho=v$ for RSM(r), for selected polynomial orders r and numbers n of parameters.

The distribution of samples may be improved with stratification, as in Latin Hypercube sampling (LHS) [18]. By dividing the sample space into intervals, or stratas representing equal probability the need for large ensembles is reduced. In LHS, each parameter is sampled exactly once in each of its stratas giving a generalized *latin square* [20]. This selection pushes the samples away from each other and distributes them more evenly. To illustrate the improvement with stratification, sample one parameter q~NRM(0, 1). After division into m intervals of equal probability, samples are found with the inverse transform method described in section 3.2 (Eq. 8). As seen in Fig. 2, the convergence improves dramatically. Still, even for $m=100$ samples the second moment (left) varies noticeably. The convergence is generally poorer for higher order moments $M^{(k)}$, as shown for $k=4$ (right).

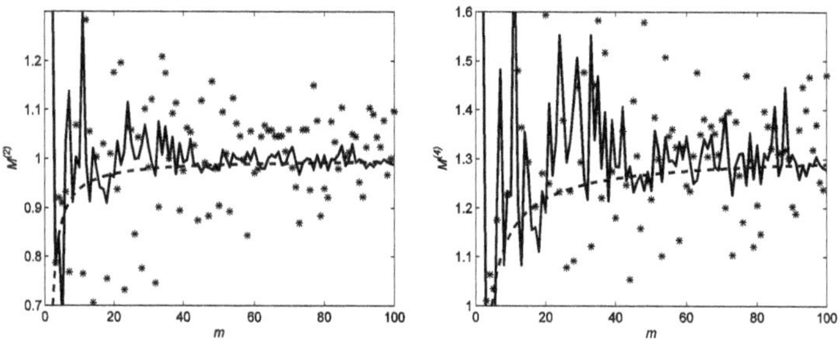

Figure 2. The second $M^{(2)}$ (left) and fourth $M^{(4)}$ (right) moments for stratified (solid) and brute force sampling ($*$) of q~NRM(0, 1), compared to a fixed grid (dashed).

In this case, it is questionable if 100 samples are sufficient to represent as few as four moments $M^{(1)}-M^{(4)}$. The probabilistically evenly distributed fixed grid (dashed) converges more rapidly to the proper statistics. Despite the prevailing tradition, there is no absolute requirement of using a random generator to represent statistical information. Fixed grids are examples of deterministic sampling. Stratification provides an interesting intermediate type of sampling since it is partially deterministic – the strata are constructed deterministically but the samples within each stratum are generated randomly. The construction of a fixed grid requires focus on the most relevant features. To reproduce $M^{(1)} - M^{(4)}$ *exactly*, a very sparse grid or few deterministic samples are needed,

$$\hat{q} = \begin{cases} \left(\pm 1.376 \quad \pm 0.325\right) & q \sim \mathrm{UNI}(0,1) \\ \left(\pm 1.732 \quad 0(\times 4)\right) & q \sim \mathrm{NRM}(0,1) \end{cases}. \tag{16}$$

If the problem at hand only depends on these moments, the exact solution will be obtained. The size of such small ensembles must be fixed, no matter how they are generated. Adding, or perturbing a single sample would modify the statistics substantially.

4. Deterministic sampling

Deterministic sampling (DS) of uncertain systems is a viable alternative to random sampling (RS). Instead of using random generators, specific DS rules are devised to generate appropriate, but still statistical (Fermi's notation, see section 3.2) ensembles. A rudimentary example illustrates the principle: Assume a model $y(q)$ depends on one parameter q with mean $\langle q \rangle$ and variance $\langle \delta q^2 \rangle$. To estimate the mean $\langle y \rangle$ and the variance $\langle \delta y^2 \rangle$ of the model, the samples (filter parameters) $\hat{q}^{(1,2)} = \langle q \rangle \pm \sqrt{\langle \delta q^2 \rangle}$ are appropriate since they satisfy the desired statistics, $\langle \hat{q} \rangle_E = \langle q \rangle$ and $\langle \delta \hat{q}^2 \rangle_E = \langle \delta q^2 \rangle$. The formula for $\hat{q}^{(1,2)}$ constitutes the sampling rule and $\hat{q}^{(1,2)}$ is the statistical ensemble containing only two model samples. By paying the computational cost of using more samples and improving the sampling rule, additional moments $\langle \delta q^k \rangle$, $k > 2$ or other statistical features can be accounted for.

In deterministic sampling the model evaluations involve no approximations and are *non-invasive*. In many respects, deterministic sampling is constructed and optimized for quantification of modeling uncertainty: Minimal ensembles allow for evaluation of the most numerically demanding models. The model evaluations are exact and non-invasive to fully respect non-linear deeply hidden parameter dependences. Only vaguely known statistics of the model is approximated.

4.1. Concepts of deterministic sampling

DS does not per se specify the goal of sampling, e.g. given mean and covariance of the parameters. In the example at the end of section 3.3, the primary target was the joint pdf of the parameters. In section 4.2, the target is $M^{(2)}(q)$. In section 5, this will be complemented with additional requirements. DS can also be utilized for direct evaluation of confidence intervals [12]. The targets of various DS methods may differ but the focus on the most influential statistical aspect and customization is shared. In stark contrast, almost without exception RS targets the joint pdf of the parameters and ignores the final utilization. Adaptation and fixed ensemble sizes provides the principal means to improve the efficiency of sampling.

4.2. Propagation of covariance in the standard unscented Kalman filter

The reference will be the specific variant of DS used for propagating covariance in what will be referred to as the standard unscented Kalman filter (UKF) [21-23]. The ensemble consists of $2n$ samples, or *sigma-points*,

$$\hat{q}^{(s,k)} \equiv \langle q \rangle + s \cdot \sqrt{n} \cdot \Delta_{:k}, \quad \Delta\Delta^T = \text{cov}(q), \quad k = 1,2,\ldots n, \quad s = \pm \tag{17}$$

where $\Delta_{:k}$ denotes the k-th column of Δ. The sampling rule is manifested in the square root calculation of the covariance matrix (Δ). As suggested [23] it may be found with a Cholesky factorization [19]. The square root matrix is not unique though – the Cholesky root is upper triangular and thus asymmetric. A more symmetric standard alternative is to evaluate the matrix square root in a canonical basis [24] Uq where cov(Uq) is diagonal. The canonical variations $U\delta\hat{q}^{(s,v)}$ will be unit vectors in the n positive and negative directions of the *principal* axes of the covariance matrix, amplified by the marginal standard deviations and most importantly, $\sqrt{n}$. For many parameters with large covariance, the scaling with $\sqrt{n}$ may cause the UKF to fail since the scaling is not related to the variability of the parameters, only their total number. A possible solution to the scaling problem is provided by the scaled unscented transformation [25]. However, it is based on Taylor expansions and thus suffers from an approximation problem of the model.

5. Sampling with conservation of moments

One class of methods of deterministic sampling conserves a limited number of statistical moments. The model parameters are sampled to satisfy these moments and collected in ensembles, similar to how parameters are sampled to fulfill probability distributions in RS.

5.1. Principle

The constraints of satisfying statistical moments constitute an infinite system of equations for the samples $\delta\hat{q}_i^{(v)}$. It can formally be viewed as sampling ($\hat{=}$) of the joint pdf $f(q)$,

$$\begin{array}{lllll}
0=\langle \delta q_i \rangle & \equiv \int \delta q_i f(q)dq & \hat{=} \frac{1}{m}\sum_{v=1}^{m} \delta \hat{q}_i^{(v)} & \equiv \langle \delta \hat{q}_i \rangle_E \\
\langle \delta q_{i1} \delta q_{i2} \rangle & \equiv \int \delta q_{i1} \delta q_{i2} f(q)dq & \hat{=} \frac{1}{m}\sum_{v=1}^{m} \delta \hat{q}_{i1}^{(v)} \delta \hat{q}_{i2}^{(v)} & \equiv \langle \delta \hat{q}_{i1} \delta \hat{q}_{i2} \rangle_E \\
\langle \delta q_{i1} \delta q_{i2} \delta q_{i3} \rangle & \equiv \int \delta q_{i1} \delta q_{i2} \delta q_{i3} f(q)dq & \hat{=} \frac{1}{m}\sum_{v=1}^{m} \delta \hat{q}_{i1}^{(v)} \delta \hat{q}_{i2}^{(v)} \delta \hat{q}_{i3}^{(v)} & \equiv \langle \delta \hat{q}_{i1} \delta \hat{q}_{i2} \delta \hat{q}_{i3} \rangle_E \\
\vdots & \equiv \vdots & \hat{=} \vdots & \equiv \vdots
\end{array} \quad (18)$$

The infinite number of equations requires an infinite number of samples. However, it is implicitly assumed that relatively few moments are known and significantly influence the result of interest. Only a few moments then needs to be accurately represented by $\{\delta \hat{q}^{(v)}\}$. Typically, $\langle \delta q_i \rangle$ and $\langle \delta q_{i1} \delta q_{i2} \rangle$ are estimated when models are identified [4,7]. In addition, the range $M^{(\infty)}$ or another higher diagonal moment can generally be determined from underlying physical constraints like stability. Clearly, any sampling rule must generate a fixed number of samples and create them simultaneously. The samples are consequently strongly dependent. One obvious sampling method is to solve Eq. 18 numerically for a sufficiently large number of samples $\delta \hat{q}$, as in Eq. 16. Due to the strong non-linearities, this is quite difficult for a large number of moments but may be feasible for a few moments.

5.2. The excitation matrix

The UKF (section 4.2) utilizes DS with conservation of all first $(\langle \delta q_i \rangle)$ and second $(\langle \delta q_{i1} \delta q_{i2} \rangle)$ statistical moments. The invariance in its *formulation* allows for any additional 'half' unitary transformation $\Delta \to \Delta V$, $V: VV^T = I$. This results in another equally valid matrix $\tilde{\Delta}$, since $\tilde{\Delta}\tilde{\Delta}^T = \Delta V[\Delta V]^T = \Delta V V^T \Delta^T = \Delta \Delta^T$. Since the transformation V is allowed and influences the result, the result of applying the UKF is not unique. The matrix V condenses this invariance and provides practical means to manipulate the UKF ensemble. A key feature of V is the absence of constraints on $V^T V$. That makes it possible to stretch V 'horizontally' (as long as $VV^T = I$). That corresponds to adding samples (sigma-points). The improved transformation $U^T SU$ (section 3.2) can be applied by also combining U with V. The square root of the covariance matrix will then read $\Delta = U^T SUV$ instead of $\Delta = U^T SV$,

$$\Sigma_{n\times m} \equiv \langle q \rangle \cdot 1_{1\times m} + U^T SU\hat{V}, \quad \hat{V} \equiv \sqrt{m} \cdot V_{n\times m}, \quad VV^T = I, \quad V \cdot 1_{m\times 1} = 0. \quad (19)$$

The samples $\hat{q}^{(k)}$ are here collected in columns of the ensemble matrix Σ. The matrix $\mathrm{cov}(q) = \Delta \Delta^T = U^T S^2 U$ is diagonalized, $S^2 = \mathrm{eig}(\mathrm{cov}(q))$, with the unitary transformation U [24]. The normalization factor $\sqrt{m}$ is included in the *excitation matrix* $\hat{V}$ to satisfy the correct covariance, $\langle \Sigma \Sigma^T \rangle = \Delta \Delta^T$, just as the factor $\sqrt{n}$ was included in Eq. 17 (the ensemble is here expanded from n to m samples). The excitation matrix controls the sampling *beyond* the first and second moments, e.g. the range of the samples. Row k of the matrix $S\hat{V}$ can be interpreted as deterministic samples of the pdf $\phi_k(S\tilde{q}_k)$, assigned to canonical parameters in RS, see section 3.2. All ensembles will be described with a unique excitation matrix $\hat{V}$.

The adopted transformation $U^T SUS^{-1}$ of $S\hat{V}$ distorts all higher moments than the second. This mixing effect is indicated by the index $\Psi(U^T SUS^{-1})$ defined in Eq. 11. To diagonalize large matrices cov(q) many efficient techniques have been developed. This should not cause any difficulties even for $n \sim 1000$, especially since cov(q) usually is either very sparse, or rank deficient for models with many parameters.

5.3. Elimination of singular values

The rationale for applying the reduction to be presented is that any model is derived from a limited set of experiments, resulting in a usually moderate rank of cov(q). If the number of parameters is large, it is thus often (nearly) rank deficient. The widely practiced singular value decomposition (SVD) [19] may then be used to reduce the excitation matrix and hence the number of samples. The most general form of SVD cannot be used here since it renders an asymmetric decomposition $\text{cov}(q) = U^T S^2 W$, where $UU^T = U^T U = WW^T = W^T W = I$ and $S^2_{jk} = \delta_{jk} S_{kk} \geq 0$. Different matrices U, W allow for decomposition of an arbitrary matrix. For the symmetric matrix cov(q), a symmetric SVD $U = W$ can be found with the less general eigenvalue decomposition [24], according to the spectral theorem [11]. As cov(q) is positive definite, all its eigenvalues S^2_{kk} fulfill the requirement of being positive. This is required to directly obtain a real-valued matrix square root $\Delta = U^T SUV$.

The ensemble may now be reduced by elimination of singular values (ESV). Choose a threshold α and remove row r and column r from S and row r from U for all r such that,

$$|S_{rr}| < \alpha \cdot \max_k |S_{kk}|, \quad \alpha \ll 1. \tag{20}$$

Proceeding as in many applications of SVD, this reduction (indicated by tilde below) will not change the result significantly, if α is small enough. Accordingly, samples are eliminated using the alternative decomposition of the square root of cov(q),

$$\Delta_{n\times m} = \left(U^T\right)_{n\times n} S_{n\times n} V_{n\times m} \approx \left(\tilde{U}^T\right)_{n\times r} \tilde{S}_{r\times r} \tilde{V}_{r\times \mu} = \tilde{\Delta}_{n\times \mu}. \tag{21}$$

Unfortunately, the less distorting transformation $U^T SUS^{-1}$ of $S\hat{V}$ advocated in section 3.2 do not allow for $r < n$ rows of the matrix V. The increase in distortion of $M^{(k>2)}$ indicated by $\Psi(U^T SUS^{-1}) \rightarrow \Psi(U^T)$ is less important for the intended use though. Signal processing models with large numbers of parameters are typically non-parametric and usually describe samples of signals like impulse responses, or noise signals. The required LR property of the system then implies LP. The propagation of covariance is then linear and only the undistorted first and second moments need to be encoded.

5.4. Correlated sampling of non-parametric models

A major difference between parametric and non-parametric models is the dimensionality. A conceptual dissimilarity is that non-parametric models usually refer to correlated signals, rather than abstract model structures. The parameters may describe discrete samples of input noise [7], or an impulse response [6]. A common parametric pole-zero model may contain 20 parameters, while a non-parametric model can be expressed in perhaps 1000 parameters. The ensembles of non-parametric models often need to be reduced drastically.

Due to limited resolution, the correlation times of any signal or impulse response is finite. Their 'memory' is thus finite so sample variations may be regenerated or repeated, as long as the time between repetitions exceeds the correlation time. This *correlated sampling* (CRS) provides efficient and accurate reduction of the ensembles. The minimal number of parameters n is then set by the correlation time of the model. Most importantly, the size of the ensemble becomes independent of the size of the model (the length of the signal).

A finite correlation length $\tau \in N$ of any model $\delta x(t)$ is normally inferred from the decay of its autocorrelation function$C(t, T)$, where t denotes the lag and T refers to a non-stationary variation. Here, a global τ will be defined through its l^2-norm and determined for a relative truncation threshold β (argmin returns the minimizing argument),

$$\tau = \max_T \left[\arg\min_\tau \left| \sum_{t=\tau+1}^{\infty} \left|C(t,T)\right|^2 - \beta^2 \sum_{t=0}^{\infty} \left|C(t,T)\right|^2 \right| \right], \quad C(t,T) \equiv \left\langle \delta x\left(T+\frac{t}{2}\right) \delta x\left(T-\frac{t}{2}\right) \right\rangle, \quad \beta << 1. \tag{22}$$

If the model is expressed as a convolution $\delta x(t) = h(t) * w(t)$ of an impulse response $h(t)$ and time-dependent white noise $w(t)$ as in section 6.2,

$$C(t,T) = \sum_{u=0}^{\infty} \eta^2\left(T - \left(u+\frac{t}{2}\right)\right) h(u) h(u+t) \approx \eta^2(T) \sum_{u=0}^{\infty} h(u) h(u+t), \quad \eta(t) = \sqrt{\operatorname{var}(w)}, \quad \langle w \rangle = 0. \tag{23}$$

By padding the model to an integer multiple of $\gamma \geq 2\tau$ samples, it is always possible to choose an excitation matrix partitioned to block-diagonal form,

$$\hat{V}_{n\times m} = \begin{pmatrix} c\tilde{V}_{\gamma\times\gamma} & 0_{\gamma\times\gamma} & 0_{\gamma\times\gamma} & \cdots \\ 0_{\gamma\times\gamma} & c\tilde{V}_{\gamma\times\gamma} & 0_{\gamma\times\gamma} & \cdots \\ 0_{\gamma\times\gamma} & 0_{\gamma\times\gamma} & c\tilde{V}_{\gamma\times\gamma} & \cdots \\ \vdots & \vdots & \vdots & \ddots \end{pmatrix}, \quad \tilde{V}_{\gamma\times\gamma} \tilde{V}_{\gamma\times\gamma}^T = \gamma \cdot I, \quad c = \sqrt{\frac{m}{\gamma}}, \tag{24}$$

where $\widetilde{V}_{\gamma\times\gamma}$ is any allowed deterministic sub-ensemble. The factor c accounts for the change from γ samples of $\widetilde{V}_{\gamma\times\gamma}$ to the $m>\gamma$ samples of $\hat{V}_{n\times m}$. By violating the normalization constraint $\hat{V}_{n\times m}\hat{V}^T_{n\times m}=m$, the size of the ensemble can be 'compressed' from m to γ samples by moving all sub-matrices $c\widetilde{V}_{\gamma\times\gamma}$ to the first block-column and skipping all zeros. The introduced constant c drops out as $m\to\gamma$,

$$\hat{V}_{n\times\gamma} \equiv \begin{pmatrix} \tilde{V}_{\gamma\times\gamma} \\ \tilde{V}_{\gamma\times\gamma} \\ \tilde{V}_{\gamma\times\gamma} \\ \vdots \end{pmatrix}, \quad \hat{V}_{n\times\gamma}\left(\hat{V}_{n\times\gamma}\right)^T = \gamma \begin{pmatrix} I_{\gamma\times\gamma} & I_{\gamma\times\gamma} & I_{\gamma\times\gamma} & \cdots \\ I_{\gamma\times\gamma} & I_{\gamma\times\gamma} & I_{\gamma\times\gamma} & \cdots \\ I_{\gamma\times\gamma} & I_{\gamma\times\gamma} & I_{\gamma\times\gamma} & \cdots \\ \vdots & \vdots & \vdots & \ddots \end{pmatrix} \neq \gamma \cdot I_{n\times n}. \tag{25}$$

Accordingly,

$$\operatorname{cov}(x)_{jk} \to \left(U^T S U V V^T U^T S U\right)_{jk} = \begin{cases} \operatorname{cov}(x)_{jk}, & |j-k| \le \gamma/2 \\ \operatorname{cov}(x)_{j,k+n\gamma}, & \begin{cases} |j-k| > \gamma/2 \\ n = \arg\min\limits_{l\in Z} |j-k-l\gamma| \end{cases} \end{cases} . \tag{26}$$

The consequence of violating the normalization constraint is that only a limited diagonal band of $\operatorname{cov}(x)$ is correctly reproduced. If a non-parametric model of a signal is propagated through a system model with impulse response h of correlation length $\sigma \le \gamma/2$ this will nevertheless *not* result in any error of $\operatorname{var}(h)$, as it is independent of all faulty elements $\operatorname{cov}(v)_{jk}$, $|j-k|>\gamma/2\ge\sigma$. To correctly evaluate $\operatorname{cov}(h)_{uv}$ though, the size of the sub-ensembles $\widetilde{V}_{\gamma\times\gamma}$ of correlated sampling must fulfill the stronger size constraint $\gamma\ge 2(\max(\tau,\sigma)+|u-v|)$. The symmetry of convolutions implies a corresponding result when the non-parametric model describes the impulse response h of the system, rather than a signal.

5.5. Combining covariance

A signal processing model generally includes both parametric and non-parametric sources of uncertainty. For instance, a device (parametric system model) may be fed with a signal corrupted with noise (non-parametric noise model). The question then arises how the two sources q_k, x_k of uncertainty can be combined. For propagation of uncertainty through LP models, the combined covariance is given by the Gauss approximation formula [16],

$$\operatorname{cov}(y) \overset{\text{LP}}{=} \sum_k \operatorname{cov}\left(y^{(k)}\right), \tag{27}$$

where $\mathrm{cov}(y^{(k)})$ is the propagated covariance of q_k, x_k. This will seize to apply for non-LP models. There exists no general non-linear summation rule for propagated covariance. A method of summation can be given though, if different ensembles are combined as in RS.

To combine ensembles of parametric (q) and non-parametric models (x), collect all parameters, $q \to \begin{pmatrix} q^T & x^T \end{pmatrix}^T$, and diagonalize the enlarged covariance matrix, $\mathrm{cov}(q) = U^T S^2 U$. Build $\hat{V}$ with two blocks and use CRS (section 5.4) for the non-parametric model,

$$\hat{V}_{(n+k\gamma)\times(m+v)} \equiv \begin{pmatrix} \sqrt{1+c^{-1}} \cdot \hat{V}_{n\times m} & 0 \\ 0 & \sqrt{1+c} \cdot \hat{V}_{\gamma\times v} \\ 0 & \sqrt{1+c} \cdot \hat{V}_{\gamma\times v} \\ \vdots & \vdots \\ 0 & \sqrt{1+c} \cdot \hat{V}_{\gamma\times v} \end{pmatrix}, \quad c = \frac{m}{v}, \quad \hat{V}_{(n+k\gamma)\times(m+v)} \left(\hat{V}_{(n+k\gamma)\times(m+v)} \right)^T = (m+v) \cdot I. \tag{28}$$

The scaling $\sqrt{1+c^{\pm 1}}$ may cause a similar scaling problem as the factor $\sqrt{n}$ in the UKF (section 4.2). Using extended excitation matrices these factors can be eliminated,

$$\hat{V}_{(n+k\gamma)\times c} \equiv \begin{pmatrix} \hat{A}_{n\times c} \\ \hat{B}_{\gamma\times c} \\ \hat{B}_{\gamma\times c} \\ \vdots \\ \hat{B}_{\gamma\times c} \end{pmatrix}, \quad \hat{E}_{(n+\gamma)\times c} = \begin{pmatrix} \hat{A}_{n\times c} \\ \hat{B}_{\gamma\times c} \end{pmatrix}, \quad \hat{E}_{(n+\gamma)\times c} \left(\hat{E}_{(n+\gamma)\times c} \right)^T = c \cdot I, \quad c = \max(m, v) > (n+\gamma). \tag{29}$$

A disadvantage of this summation is that the same type of ensemble must be used for all parameters. Both alternatives combine the statistics of the two models non-linearly. The uncertainties are propagated and combined by evaluating the model for all samples and calculating the desired statistics, just as if the combined ensemble described one model.

5.6. Selected ensembles

The standard (STD) ensemble employed in the UKF (as defined in section 4.2) utilizes the perhaps simplest possible excitation matrix,

$$\hat{V}_{\mathrm{STD}} = \sqrt{n} \cdot \begin{pmatrix} I_{n\times n} & -I_{n\times n} \end{pmatrix}, \quad m = 2n. \tag{30}$$

While the ultimate simplicity is its main advantage, the long maximal(!) range $M^{(\infty)}$ is its main disadvantage.

How far the reduction of samples might be driven is illustrated by the minimal simplex (SPX) ensemble,

$$\hat{V}_{\text{SPX}} = \sqrt{n+1} \cdot \perp \left\{ \begin{pmatrix} I_{n\times n} & -1_{n\times 1} \end{pmatrix} \right\}, \quad m = n+1, \tag{31}$$

where the operator $\perp$ performs classical Gram-Schmidt orthogonalization [11] and normalization of rows. The ensemble is constructed from half the STD ensemble, complemented with one sample $1_{n\times 1}$ to cancel the first moments. Since that violates the orthogonality of the rows of V, $\perp$ must be applied to satisfy $V V^T = I$. The high efficiency of the SPX ensemble is tarnished by its large skewness, or $M^{(3)}$. This may give considerable bias of propagated covariance for non-LP models, but is irrelevant for LP models.

The binary (BIN) ensemble has minimal range to guarantee allowed samples. By varying all parameters with an equal magnitude of one standard deviation in all samples, the diverging factor $\sqrt{n}$ of the STD is eliminated. Its excitation matrix $\hat{V}_{BIN}$ is fundamentally constructed from a standard binary array, with the difference that the allowed levels are ±1 instead of 0, 1 (see rows 1-3 in Eq. 32). It is then complemented with supplementary rows obtained in two ways, by cyclic shifting and mirror imaging,

$$\hat{V}_{BIN}^{(m)} = \begin{pmatrix} +1 & -1 & +1 & -1 & +1 & -1 & +1 & -1 & \cdots \\ +1 & +1 & -1 & -1 & +1 & +1 & -1 & -1 & \cdots \\ +1 & +1 & +1 & +1 & -1 & -1 & -1 & -1 & \cdots \\ -1 & +1 & +1 & -1 & -1 & +1 & +1 & -1 & \cdots \\ -1 & -1 & +1 & +1 & +1 & +1 & -1 & -1 & \cdots \\ +1 & -1 & +1 & -1 & -1 & +1 & -1 & +1 & \cdots \\ -1 & +1 & +1 & -1 & +1 & -1 & -1 & +1 & \cdots \\ \vdots & \vdots & \vdots & \vdots & \vdots & \vdots & \vdots & \vdots & \ddots \end{pmatrix}, \quad m = 2^{\operatorname{ceil}\left(\frac{n+5}{4}\right)}. \tag{32}$$

Cyclic shifts are applied to all original rows except the first, by a quarter of their periodicity. Mirror imaging of a row is defined to change the sign of its second half and is applied to all original rows except the last two, and all shifted rows except the last. For instance, in Eq. 32 row 4 and 5 are shifted versions of row 2 and 3, while rows 6 and 7 are the mirror images of rows 1 and the shifted row 4. The supplementary rows reduce the size of the ensemble drastically with a corresponding improvement of the efficiency. For $n = 20$ parameters, the size drops from roughly 10^6 to 128 samples. That size is acceptable in perspective of the $n + 1 = 21$ samples of the most efficient SPX. Eventually though, the number of samples will grow too large. The BIN can thus only be applied to moderately sized models.

By no means, this brief survey exhausts all possible ensembles. Many criteria for selecting the most appropriate ensemble can be formulated. Here, the first and second moments, parameter ranges and efficiency were in focus.

6. Application — Modeling uncertainty of a dynamic device

The task is to simulate the response of an electrical device such as an amplifier or oscilloscope, in the presence of non-stationary correlated noise on its input. An uncertain LR CT model of the device and its parametric covariance is usually found by applying system identification techniques [4] on calibration measurements [6]. Such a model of the system can be sampled into a digital filter and be described in the pole-zero form in Eq. 2. These standard steps will here be omitted. The system model will instead be assigned to a digital low-pass Butterworth filter, of order 10 and cross-over frequency $f_C=0.1f_N$, f_N being the Nyquist frequency and described by parameters K, p_1, p_2, $\dots p_{10}$, z_1, z_2, z_{10}. The complete correlations of complex-conjugated pole (p) and zero (z) pairs are eliminated by a transformation from $q=z$, p to $\mathrm{Re}(q)$, $\mathrm{Im}(q)\geq 0$, giving $n=21$ system model parameters,

$$q \equiv \begin{pmatrix} K & \mathrm{Re}(z_1) & \mathrm{Im}(z_1)\geq 0 & \cdots & \mathrm{Re}(p_1) & \mathrm{Im}(p_1)\geq 0 & \cdots & \mathrm{Re}(p_{10}) & \mathrm{Im}(p_{10})\geq 0 \end{pmatrix}^T. \tag{33}$$

To be most general, the non-parametric input noise model is chosen to be correlated/colored and non-stationary. The noise parameter δx_k represents the noise level at time sample k. Its *generating signal* [7] is a Dirac delta function δ_{jk}, centered at time k. The response of a system with impulse response h will be $\delta y_j = \delta x_k \cdot (h_j * \delta_{jk}) = h_{j-k} \cdot \delta x_k$. In matrix notation, $\delta y = \bar{h}^T \delta x$, where $\bar{h}_{kj} = h_{j-k}$. Hence,

$$\mathrm{cov}(y) = \left\langle \delta y \delta y^T \right\rangle = \bar{h}^T \left\langle \delta x \delta x^T \right\rangle \bar{h} = \bar{h}^T \mathrm{cov}(x) \bar{h} \tag{34}$$

Since the response is linear in noise parameters, it is sufficient to only capture $\mathrm{cov}(x)$.

6.1. Reference ensembles

Traditionally, any method for uncertainty propagation is evaluated by comparisons with the default method of linearization [10,16], and brute-force random sampling (MC) [9] as state-of-the-art. There are several drawbacks of this approach. Linearization is a coarse approximation for LP models and MC suffer from the difficulty of modeling dependencies and low efficiency. An alternative is to construct finite reference ensembles (REF) and by definition let them describe the truth. Their primary advantage is that the finite size of the REF makes it possible to propagate the uncertainty exactly, using all REF samples. A more or less arbitrary REF may be generated randomly, like any MC ensemble. All requirements are also automatically fulfilled since the REF is built of possible realizations. Also, the REF closes the loop as it makes it possible to compare 'true' and approximate samples directly on an equal footing (see Fig. 8). Even though the samples differ substantially, the resulting modeling uncertainties can be similar.

A plausible REF δq_j for the system model realized as a digital filter is created by randomly generating m samples of n parameters q_k from uniform distributionsUNI(0, σ_k), with σ_k listed in Fig. 3, top left. The joint pdf will have compact support [11], as required to guarantee stability. The mean is subtracted from all samples to remove the bias of the finite random ensemble, $\langle \delta q_j \rangle_E = \langle \delta q_j \rangle = 0,\ \forall\ j$. The covariance of the REF will have a desirable more or less random variation for small values of m. If the REF samples are arranged in columns of a matrix $\hat{\Lambda}_{n \times m}$, (as $\hat{V}$) $\mathrm{cov}(q)_{\mathrm{REF}} = 1/m \cdot \hat{\Lambda}\hat{\Lambda}^T$. For $m=31>n=21$, the strong correlations will expose the methods to severe tests with significant transformations U, S. The mixing Ψ (Eq. 13) using transformation $U^T S U S^{-1}$ was considerable, but less than for U^T, see caption Fig. 3. For the chosen REF, the resulting variations of poles and zeros are displayed in Fig. 3. The obtained variation of the parameters defined in Eq. 33 can be quantified with an averaged correlation index and standard deviation (Fig. 3, bottom left),

$$\xi_k \equiv \sqrt{\frac{1}{n-k}\sum_{j=1}^{n-k} \mathrm{cov}(q)_{j(j+k)}} \Bigg/ \sqrt{\frac{1}{n}\sum_{j=1}^{n} \sigma_j^2}, \quad \sigma_j^2 = \mathrm{var}(q_j). \tag{35}$$

A REF signal δx_j for non-stationary correlated noise may conveniently be generated from an autoregressive process (AR) acting on time-dependent zero mean white noise δw, $\delta x = \bar{g}^T \delta w$, where $\bar{g}_{kj} = g_{j-k}$ is the matrix of translated impulse responses g for the AR process defined by parameters α_k. Assigning a square wave time-dependence,

$$\begin{aligned} \sum_k \alpha_k \delta x_{j-k} &= \delta w_j, \quad \langle \delta x \delta x^T \rangle = \bar{g}^T \langle \delta w \delta w^T \rangle \bar{g} = \bar{g}^T \mathrm{cov}(w) \bar{g}, \quad \mathrm{cov}(w) = \mathrm{diag}[\eta(t)] \\ \eta(t) &= \mathrm{N}\left[1 + \frac{1}{1+\psi}\theta\left[\cos\left(\frac{2t\pi}{T} + \varphi\right)\right]\right], \quad \theta = \begin{cases} +(-)1 & x>(<)0 \\ 0, & x=0 \end{cases} \end{aligned} \tag{36}$$

The exact REF result of modelling covariance of noise is given by combining Eqs. 34 and 36,

$$\sigma_N \equiv \sqrt{\mathrm{Tr}(\mathrm{cov}(y))} = \sqrt{\mathrm{Tr}\left(\bar{h}^T \bar{g}^T \mathrm{diag}\left[\eta^2\right] \bar{g} \bar{h}\right)}. \tag{37}$$

An explicit realization of the REF for the noise model is hence not needed. Specifically, a second order system $\alpha = [1 \quad -0.4 \quad 0.6]$ with time parameters $\{N, \psi, T, \varphi\} = \{0.05, 0.3, 2f_C^{-1}, \pi/8\}$ was

Assigned variation σ		
K	$\mathrm{Re}(z_k)$, $\mathrm{Im}(z_k)$	$\mathrm{Re}(p_k)$, $\mathrm{Im}(p_k)$
2%	0.2%	0.9%

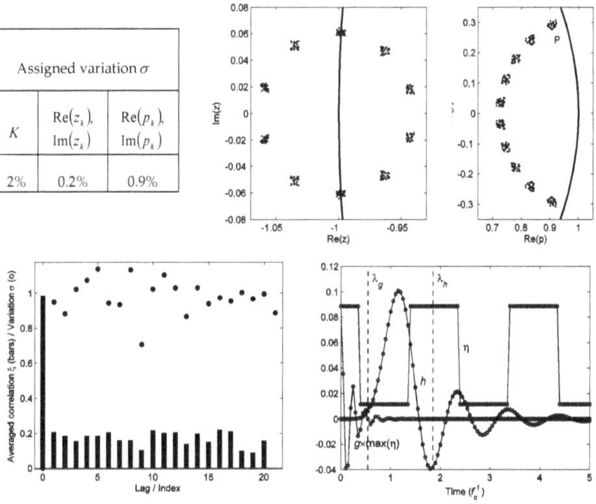

Figure 3. Top: Assigned variations (left) and resulting (z_k middle, p_k right) samples of the REF of the system model. Label P indicates the pole explored in Fig. 8. Bottom, left: Obtained variations σ_k (dots) and correlations ξ_k (bars) of parameters q (Eq. 33), with mixing (Eq. 11) $\Psi(U^TSUS^{-1})=0.22$ (adopted) and $\Psi(U^T)=0.39$. Bottom, right: Impulse responses $h(\langle q\rangle, t)$ and $g(t)$ and time-dependence $\eta(t)$ (Eq. 36) of noise intensity. The correlation lengths λ_h, λ_g were determined according to Eqs. 22-23, for $\beta=0.05$.

chosen. The impulse responses of the AR noise system and the system model, and the variation of the noise model are illustrated in Fig. 3, bottom right.

The 'true' result given by the response for the REFs for the different test signals is shown in Fig. 4. The propagated noise variation σ_N differs substantially from the input square wave η (top left) and is almost opposite in phase, due to the response time of about f_C^{-1}, see delay of μ_S (top, right and bottom). The signal distortion (μ_S) is strongly dependent on the input signal and decreases with increased regularity / differentiability. The propagated covariance σ_S has a more complex variation (top, right and bottom), as it is larger for the more regular Gaussian (bottom, right) than for the triangular pulse (bottom, left).

6.2. Deterministic sampling

The error of the scent and the standard deviation for the STD, SPX and BIN ensembles of the system model is displayed in Fig. 5, for all test signals. The low scent of the REF (left: thin, dotted) suggests the model is close to LP. Despite the relative errors are large they are quite small on an absolute scale. The SPX has the largest errors, for the scent as well as the variance. That is likely caused by its skewness being much larger than that of the REF. The BIN has the lowest errors and is thus the best approximate representation of the REF.

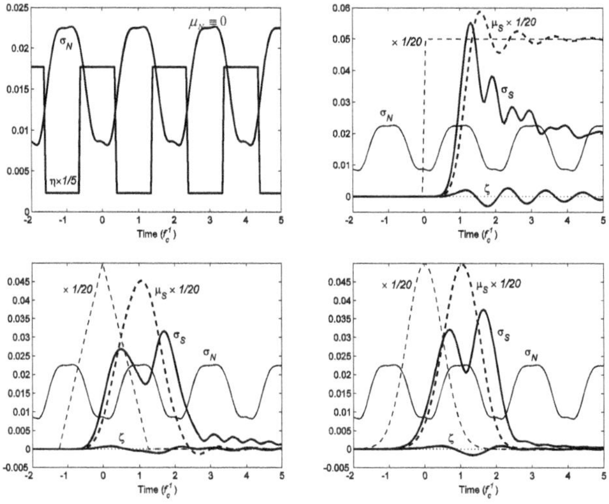

Figure 4. The mean μ_S (dashed), the standard deviations $\sigma_{S,N}$ (solid) and the scent ζ (thin, solid) for the REFs, for the different test signals (thin, dashed). The subscripts refer to the system (S) and noise (N) models. The variation of noise intensity is given by $\eta(t)$ (top, left).

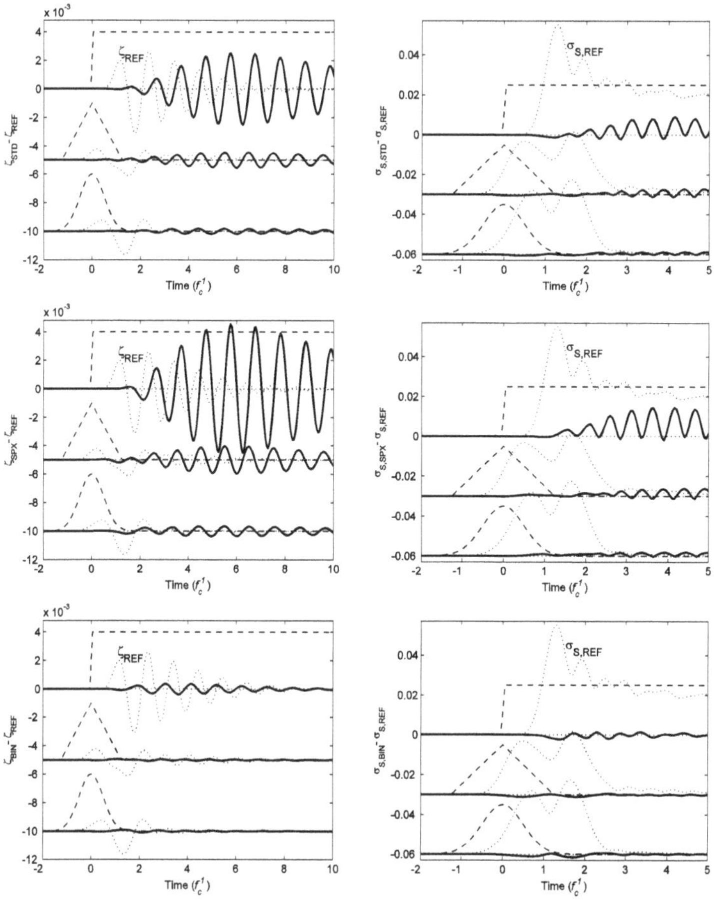

Figure 5. The errors of the scent $\zeta - \zeta_{REF}$ (left) and the standard deviation $\sigma_S - \sigma_{S,REF}$ (right) of the system model (solid) for the STD (top), SPX (middle) and BIN (bottom) and the three test signals (thin, dashed). The correct ζ_{REF} (left) and $\sigma_{S,REF}$ (right) are included for comparison (thin, dotted). The triangular and Gaussian signals are displaced for clarity.

The errors might appear large, considering *all* ensembles are 'correct', i.e. correctly represent (typically) available accurate information (mean and covariance of parameters). The errors reflect ambiguities caused by the ubiquitous lack of information in signal processing, rather than inadequacies of DS. RS can only produce better results by making further *assumptions*.

The result of applying the ESV and the CRS methods to reduce the SPX ensemble for propagating the noise is displayed in Fig. 6. By choosing sufficiently low thresholds α for elimination of singular values (ESV) and β for truncation of the correlation lengths (see Eqs. 20,22), the errors can be made arbitrarily low. As the reduction will decrease accordingly, there is a trade-off between accuracy and efficiency. For the chosen values, CRS is about twice as accurate and twice as efficient as ESV. In contrast to ESV, the number of samples for the CRS method is independent of the number of noise samples. The computational cost thus increases linearly with the length of the noise signal for CRS but quadratically (approximately) for ESV. For ESV to be most efficient, the model covariance needs to be strongly rank deficient. That is not as unlikely as it might appear, since the model usually is derived from a limited amount of experimental results.

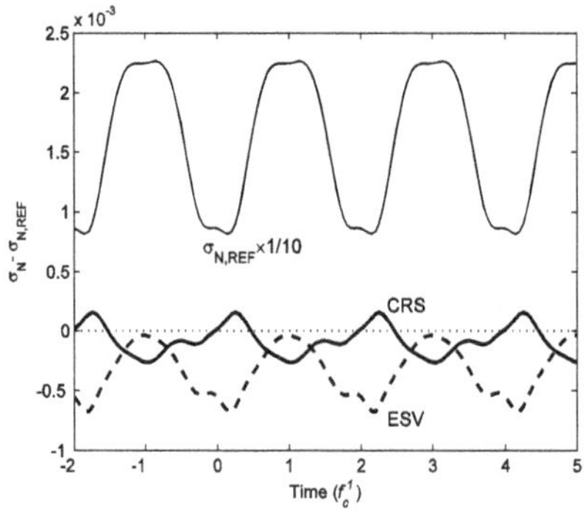

Figure 6. The error $\sigma_N - \sigma_{N,REF}$ of propagated noise, for the ESV (section 5.3) and CRS (section 5.4) ensemble reduction methods, and the correct $\sigma_{N,REF}$ (thin, $\times 1/10$). The thresholds were $\alpha = 0.1$ (Eq. 20) for ESV, and $\beta = 0.05$ (Eq. 22) for CRS. That resulted in $m = 142$ samples for ESV and $m = 75$ for CRS, compared to $m = 402$ of the original SPX.

The summation of the noise and the model covariance is illustrated in Fig. 7. The propagation of the covariance of the system model (q) is not LP. The quadratic summation rule (Eq. 27), or Gauss approximation formula [16], is therefore not applicable. Nevertheless, the low scent ζ (Fig. 5, left) suggests that both propagations are close to LP. The summation error (ε) is hence finite, but quite small. It differs qualitatively from both contributions, indicating that the summation is non-trivial.

Finally, the samples of one pole of the derived ensembles are compared to the reference samples of the REF in Fig. 8. The limit ($|z| = 1$) of stability is included to illustrate how close the samples are to be physically forbidden. The construction of the different ensembles is apparent, even though the transformation $T = U^T S U S^{-1}$ distorts the scatter plots (sections 3.2, 5.2), and tilts the principal axes (lines). The samples of the REF are almost evenly distributed. Only four samples of the STD, labelled p_1, p_2, p_3, p_4, deviate significantly from a dense central cluster, as described by the excitation matrix V_{STD} (Eq. 30). It also is evident that SPX originates from half the STD. A small translation required to achieve the correct mean is discernible, while the Gram-Schmidt orthogonalization renders a minor rotation and distortion. The BIN contains comparable variations in all samples and thus has no central cluster and its samples are repelled from the principal directions (lines). The statistical differences to the REF refer to the shape of the joint pdf. Choosing the best ensemble is thus equivalent of selecting the most appropriate pdf in RS. The BIN seems to resemble the REF scatter plot the most, as verified by its low errors in Fig. 5.

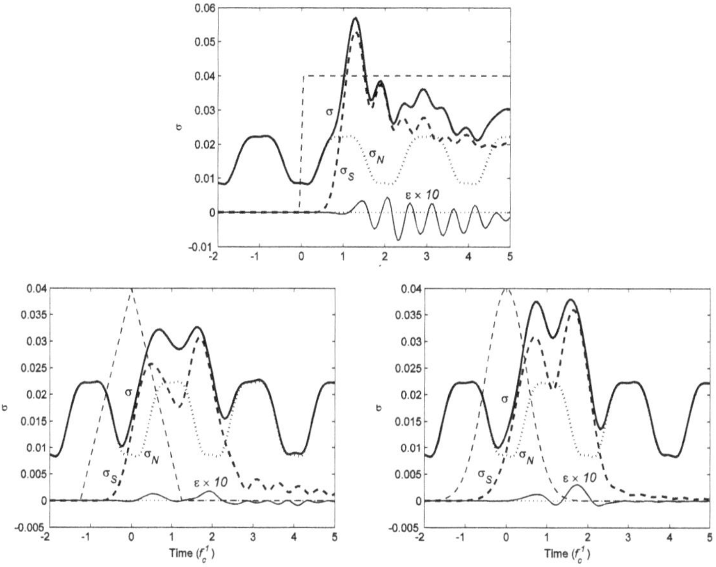

Figure 7. Summation of covariance: Total (solid), system (dashed) and noise (dotted), for the three test signals (thin, dashed), with the error (ε) of square summation (×10) (Eq. 27).

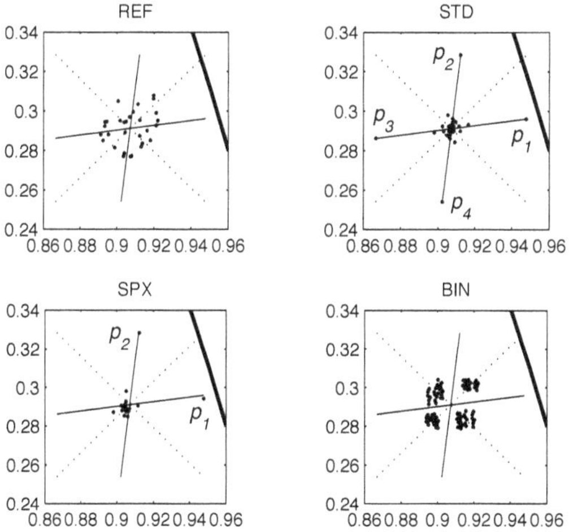

Figure 8. The different samples (dots) of the pole marked 'P' in Fig. 3, of the reference (REF), standard (STD), simplex (SPX) and binary (BIN) ensembles. The limit $|z|=1$ of stability (solid, thick) and lines connecting the primary variations p_1, p_2, p_3, p_4 of the STD as well as lines (dashed) to combined excitations of the BIN are included for reference.

7. Conclusions

Deterministic sampling remains controversial [27] while random sampling has qualified as a preferred state-of-the-art method for propagating uncertainty. Both result in finite *statistical* [17] ensembles, which are approximate finite representations of the primary statistical models. Their sampling strategies and convergence rates are dramatically different. While deterministic sampling humbly aims at representing the most relevant and best known statistical information, random sampling targets complete control of all features of the ensemble. Such detailed information is rarely known and must instead be more or less blindly assigned. The inevitable consequence is that critical computational resources are spent on propagating, at best, vaguely known details. The numerical power of modern computers is better spent on refinements of the signal processing model (longer time series, higher sampling rates, larger systems etc.). Refined methods of random sampling have therefore been proposed which either simplifies the model, or improve the sampling distributions. Compared to deterministic sampling though, their convergence rates remain low.

It is easy to confuse deterministic sampling with experimental design and optimization [28]. Even though any sample could be a possible outcome of an experiment, deterministic ensembles *represent* rather than *realize* (as random ensembles) statistical distributions. Instead of associating a joint distribution to the parameters of an uncertain model, it is possible to directly represent their statistics with a deterministic ensemble. That would eliminate the need of interpreting abstract distributions and result in complete reproducibility. The critical choice of ensemble would be assigned once and for all in the calibration experiment, with no further need of approximation.

The use of excitation matrices made it possible to construct universal generic ensembles. The efficiency of the minimal SPX ensemble is indeed high but so is also its third moment. While the STD maximizes the range of each parameter, the BIN minimizes it by varying all parameters in all samples. The STD is the simplest while the SPX is the most efficient ensemble. In the example, the BIN was most accurate. For non-parametric models with many parameters, reduction of samples may be required. Elimination of singular values (ESV) and correlated sampling (CRS) were two such techniques. The presented ensembles are not to be associated to random sampling as a method. They are nothing but a few examples of deterministic sampling, likely the best ensembles are yet to be discovered.

It is indeed challenging but also rewarding to find novel deterministic sampling strategies. Once the sampling rules are found, the application is just as simple as random sampling, but usually much more efficient. Deterministic sampling is one of very few methods capable of non-linear propagation of uncertainty through large signal processing models.

Author details

Jan Peter Hessling*

Measurement Technology, SP Technical Research Institute of Sweden, Borås, Sweden

References

[1] Kay S. Fundamentals of Statistical signal processing: Estimation Theory. New Jersey: Prentice Hall; 1993.

[2] Hessling JP. Propagation of dynamic measurement uncertainty. Meas. Sci. Technol. 2011; 22 (10) 105105 (13pp).

[3] Hessling JP. Integration of digital filters and measurements. In: Márquez FPG. (ed.) Digital Filters. Rijeka: InTech; 2011. p123-154. Available from http://www.intechopen.com/books/digital-filters/integration-of-digital-filters-and-measurements (accessed 4 July 2012).

[4] Pintelon R, Schoukens J. System Identification: A Frequency Domain Approach. Piscataway, New Jersey: IEEE Press; 2001.

[5] Witteveen JAS. Efficient and Robust Uncertainty Quantification for Computational Fluid Dynamics and Fluid-Structure Interaction. PhD thesis. Delft University of Technology; 2009.

[6] Hale PD, Dienstfrey A, Wang JCM, Williams DF, Lewandowski A, Keenan DA, Clement TS. Traceable Waveform Calibration With a Covariance-Based Uncertainty Analysis. IEEE Trans. Instrum. Meas. 2009; 58 (10) 3554-3568.

[7] Hessling JP. Metrology for non-stationary dynamic measurements. In: Sharma MK. (ed.) Advances in Measurement systems. Vukovar: InTech; 2010. p. 221-256. Available from http://www.intechopen.com/books/advances-in-measurement-systems/metrology-for-non-stationary-dynamic-measurements (accessed 4 July 2012).

[8] Metropolis N, Ulam S. The Monte Carlo Method. Journal of the American Statistical Association 1949; 44 (247) 335-341.

[9] Rubenstein RY, Kroese DP. Simulation and the Monte Carlo Method, 2nd Ed. New York: John Wiley & Sons Inc.; 2007.

[10] Hessling JP. A novel method of evaluating dynamic measurement uncertainty utilizing digital filters. Meas. Sci. Technol. 2009; 20 (5) 055106 (11pp).

[11] Råde L, Westergren B. Beta Mathematics Handbook, 2nd Ed. Lund, Sweden: Studentlitteratur; 1990.

[12] Hessling JP, Svensson T. Propagation of uncertainty by sampling on confidence boundaries, accepted for publication in *International Journal for Uncertainty Quantification*.

[13] Hessling JP. Deterministic sampling for propagating model covariance, submitted for publication.

[14] Lovett T. Polynomial Chaos Simulation of Analog and Mixed-Signal Systems: Theory, Modeling method, Application. Saarbrucken: Lambert Academic Publishing; 2010.

[15] Li H, Zhang D. Probabilistic collocation method for flow in porous media: Comparisons with other stochastic methods. Water Resources Research 2007; 43 W09409 (13 pp).

[16] ISO GUM. Guide to the Expression of Uncertainty in Measurement. Geneva: International Organisation for Standardisation; 1995.

[17] Metropolis N. The Beginning of the Monte Carlo Method. Los Alamos Science special issue 1987; 15 125-130.

[18] Helton J, Davis L. Latin hypercube sampling and the propagation of uncertainty in analyses of complex systems. Reliability Engineering and System Safety 2003; 81 23-69.

[19] Björk Å. Numerical methods for least squares problems. Philadelphia: Siam; 1996.

[20] Wikipedia: http://en.wikipedia.org/wiki/Latin_square (accessed 3 July 2012):

[21] Julier S, Uhlmann J, Durrant-Whyte HF. A new approach for filtering nonlinear systems. Proc IEEE American Control Conference June 21-23 1995; 1628-1632.

[22] Julier S, Uhlmann J. Unscented filtering and nonlinear estimation. Proceeding IEEE March 2004; 92 (3) 401-422.

[23] Simon D. Optimal State Estimation: Kalman, H∞ and non-linear approaches. New Jersey: Wiley; 2006.

[24] Matlab with Signal Processing Toolbox, The Mathworks, Inc.

[25] Julier S, Uhlmann J. The scaled unscented transformation, Proceedings of the IEEE American Control Conference 8-10 May 2002; 4555-4559.

[26] Hessling JP. Non-linear propagation and summation of covariance using deterministic sampling, in preparation.

[27] Gustafsson F, Hendeby G. Some Relations between Extended and Unscented Kalman Filters. IEEE Trans. sign. proc. 2012; 60 (2) 545-555.

[28] Fischer RA. Statistical Methods, Experimental design and Scientific Inference. New York: Oxford University Press; 1990.

Chapter 4

Direct Methods for Frequency Filter Performance Analysis

Alexey Mokeev

Additional information is available at the end of the chapter

1. Introduction

Analysis methods based on determining system performance specifications by step response, as well as indirect methods: pole-zero plot, magnitude response and integral analysis methods are applied in automatic control theory for performance estimation of linear systems [1,2,3]. However, in many cases the mentioned methods result in crude performance estimation of a linear system (filter) operation. Furthermore, direct methods of linear system performance specifications (settling time, accuracy, overshoot etc.) characterization require a huge amount of calculations being performed.

Specification or estimation of signal processing performance criteria are usual tasks in frequency filter analysis. In some cases it is considered to be enough to examine a filter behavior at average statistical parameters of a useful signal and its disturbance. In other cases, for instance for robust filters, it is rather more complicated – one needs to determine the limit of variables for signal processing performance specifications at any possible input signal parameters variation.

The author offers to use the filter analysis methods, developed by him on the basis of spectral representations of the Laplace transform, to solve efficiently the problem of signal processing performance specifications determination by frequency filters at different variations of input signal parameter [4,5,6]. The mentioned methods are based on using consistent mathematical models for input signals and filter impulse characteristics by means of a set of continuous/discrete semi-infinite or finite damped oscillatory components. Similar models can be applied for simple semi-infinite harmonic and aperiodic signals or filter impulse characteristics, compound signals of any form, including signals with composite envelopes, as well as pulse signals (radio and video pulses).

The application of signal/filter frequency and frequency-time representations, based on Laplace transform, allowed developing simple and effective direct methods for performance analysis of signal processing by analog and digital filters.

To simplify the task of analog and digital filter signal processing performance analysis the author offers two methods for performance express-analysis of signal processing by frequency filters using filter frequency responses based on Laplace transform: frequency and frequency-time analysis methods [7].

The frequency method from an indirect analysis method for signal processing, in fact, has transformed into a direct analysis method by means of Laplace transform spectral representations. This method is the most effective in cases, where only two main performance specifications: signal processing speed and accuracy – are required to be evaluated.

The frequency-time analysis method is being applied in cases, where there is a need to evaluate signal processing speed and accuracy, as well as the history of transient processes in a filter, for instance, to control oscillation of transient process in a filter. It is suggested to perform the analysis by using frequency responses based on filter transfer function, dependent on time, in sections of input signal complex frequencies.

In case of FIR filter an effective estimation of signal processing performance specifications can be carried out by using filter frequency response 3D analysis based on Laplace transform in sections of input signal complex frequencies, considering their change. To evaluate signal processing performance specifications for IIR filters one will need along with sections of 3D filter frequency responses to use sections of 3D signal spectrum on filter impulse response complex frequencies [5].

The issues about the application of the analog and digital filter analysis methods, developed by the author for signal processing performance analysis by frequency filters, are considered below.

2. IIR filters analysis

2.1. Signal processing performance analysis by analog IIR filters

Let us consider signal processing performance analysis by IIR filters at semi-infinite input signals on the basis of analysis methods based on the Laplace transform spectral representations.

Three methods of frequency filter analysis are suggested by the author for the time-and-frequency representations positions of signals and linear systems in coordinates of complex frequency [5,6]. Let us consider the first two methods for signal processing performance analysis by frequency filters.

The mathematical description for the generalized input signal and IIR filter at time and frequency domains for the first (item2) and the second (item 3) analysis methods, mathematical expressions for calculating forced and free components of a filter reaction by the first method (items 4,5), components of a filter reaction by the second method (item6) are given in the Table 1.

№ Name	Expression	Remark
1. Input signal	$x(t)=\mathrm{Re}(\dot{\boldsymbol{X}}^{\mathrm{T}}e^{\boldsymbol{p}t})$, $X(p)=\mathrm{Re}\left(\dot{\boldsymbol{X}}^{\mathrm{T}}\left[\frac{1}{p-p_n}\right]_N\right)$	$\dot{\boldsymbol{X}}=[\dot{X}_n]_N=[Xm_ne^{-j\varphi_n}]_N$, $\boldsymbol{p}=[p_n]_N=[-\beta_n+j\omega_n]_N$
2. Filter	$g(t)=\mathrm{Re}(\dot{\boldsymbol{G}}^{\mathrm{T}}e^{\boldsymbol{q}t})$, $K(p)=\mathrm{Re}\left(\dot{\boldsymbol{G}}^{\mathrm{T}}\left[\frac{1}{p-p_m}\right]_M\right)$	$\dot{\boldsymbol{G}}=[\dot{G}_m]_M=[k_me^{-j\phi_m}]_M$, $\boldsymbol{q}=[p_m]_M=[-a_m+jw_m]_M$
3. Time dependent transfer function	$K(p,t)=\mathrm{Re}\left(\dot{\boldsymbol{G}}^{\mathrm{T}}\left[\frac{1-e^{-(p-p_m)t}}{p-p_m}\right]_M\right)$	$K(p,t)=\int_0^t g(\tau)e^{-p\tau}d\tau$
4. Forced components	$y_1(t)=\mathrm{Re}(\dot{\boldsymbol{Y}}^{\mathrm{T}}e^{\boldsymbol{p}t})$	$\dot{\boldsymbol{Y}}=\mathrm{diag}(\dot{\boldsymbol{X}})K(\boldsymbol{p})$
5. Free components	$y_2(t)=\mathrm{Re}(\dot{\boldsymbol{V}}^{\mathrm{T}}e^{\boldsymbol{q}t})$	$\dot{\boldsymbol{V}}=\mathrm{diag}(\dot{\boldsymbol{G}})X(\boldsymbol{q})$
6. Filter reaction	$y(t)=\mathrm{Re}(\dot{\boldsymbol{Y}}(t)^{\mathrm{T}}e^{\boldsymbol{p}t})$	$\dot{\boldsymbol{Y}}(t)=\mathrm{diag}(\dot{\boldsymbol{X}})K(\boldsymbol{p},t)$

Table 1. IIR filters analysis

The operation of the real part extraction on the right side of the expression in the items 1,2,3 for $X(p)$, $K(p)$, $K(p, t)$ is solved in terms of the complex coefficients $\dot{G}_m$ and p_m with no relevance to the complex variable p.

The first method is a complex amplitude method generalization for definition of forced and free components for filter reaction at semi-infinite or finite input signals [6]. The advantages of this method are related to simple algebraic operations, which are used for determining the parameters of linear system reaction (filter, linear circuit) components to an input action described by a set of semi-infinite or finite damped oscillatory components. To analyze a filter it is needed to use simple algebraic operations and operate a set of complex amplitudes and frequencies of forced and free filter reaction components. In this case, there are simple relations between complex amplitudes of output signal forced components and complex amplitudes of an input signal (item 4 Table 1), between complex amplitudes of output signal free components and complex amplitudes of a filter impulse function (item5).

The time-and-frequency approach in the second analysis method applies to a filter transfer function, i.e. time dependent transfer function of the filter is used [6,8]. In that case, instead of two sets of filter reaction components only one of them may be used.

Analysis methods given in the Table 1 enable to reduce effectively the computational costs when performing a filter analysis by using simple algebraic operations to determine the forced and free components of a filter reaction to an input action as a set of damped oscillatory components. Therefore, the considered analysis methods for linear systems (filters) can be effectively applied for performance analysis of signal processing by frequency filters.

Let us consider a simple example of performance analysis of signal processing by a high-pass second-order filter relating to signal processing task in power system protection and automation devices [9,10]. The filter is used to extract a sinusoidal component of commercial frequency and eliminate disturbance as a free component of transient processes in a control object. In this case, a change of a useful signal initial phase is acceptable.

All the initial data and dependencies which are necessary for the analysis are represented in the Table 2. IIR filter parameters are specified, the mathematical description of an input signal with specified sizes of changing for useful signal and disturbance parameters affecting their spectrum is given in the Table 2 as well.

The impulse function of high-pass second-order filter contains a delta function of Dirac which is used for determining complex amplitudes of forced components when defining $K(p)$ by the impulse function (item 3 Table 2) and cannot be applied for determining complex amplitudes of filter reaction free components (item 5). To simplify the analysis the delta function can be represented as an extreme case of the exponential component $\alpha e^{-\alpha t}$ at $\alpha \rightarrow \infty$[6].

The analysis results should ensure the following performance criteria of signal processing by a filter:

1. a filter settling time should be less than 30 ms at 5% acceptable total error of signal processing at any value of disturbance parameters within the specified range,
2. an acceptable error at frequency deviation of useful signal from the nominal value of 50Hz within the range ±5 Hz should not be more than 5%,
3. an acceptable overshoot should not be more than 10%.

As it follows from the Table 1, simple algebraic operations are applied to determine complex amplitudes, as well as forced and free components of a filter output signal.

When using Mathematica, Mapple, Matlab, Mathcad and other state-of-art mathematical software for determining forced and free components of an output signal it is necessary to specify only complex amplitude vectors of an input signal and a filter impulse function, as well as complex frequency vectors of an input signal and a filter. In this case all the necessary calculations, related to a filter analysis, would be carried out automatically. If it is needed to determine complex amplitudes of a filter impulse function at specified transfer function the ready-made formulas may be used [6], which can be easily applied in the mathematical software mentioned above.

All the examples in the present chapter are given using the mathematical software Mathcad. Mathcad was chosen due to pragmatic considerations related to assuring the maximum visibility of the examples for filter analysis, as in Mathcad mathematical expressions are given in the form, closest to universally accepted mathematical notation [11,12].

An example of a filter computation using Mathcad at the specified filter parameters and the following input signal parameters: $X m_2 = X m_1 = 1$, $\omega_1 = 2\pi 50$rad/s, $\varphi_1 = 0$, $\beta_2 = 60$ s^{-1} is given on the Figure 1.

№ Name	Expression
1. Input signal	$x(t)=Xm_1\cos(\omega_1 t-\varphi_1)-Xm_2e^{-\beta_2 t}$, $\dot{\boldsymbol{X}}=\left[Xm_1e^{-j\varphi_1}\ \ Xm_2e^{-j\pi}\right]^T$, $\boldsymbol{p}=[j\omega_1\ \ -\beta_2]^T$, $\mu_2=Xm_2/Xm_1=0\div1,\ \omega_1=2\pi(45\div55),\ \varphi_1=0\div2\pi,\ \beta_2=2\div200$ $X(p)=Xm_1\frac{p\cos(\varphi_1)+\omega_1\sin(\varphi_1)}{p^2+\omega_1^2}-Xm_2\frac{1}{p+\beta_2}$
2. High-pass filter	$K(p)=\frac{k_0p^2}{p^2+2a_1p+w_2^2},\ g(t)=k_0\delta(t)+\mathrm{Re}\left(\dot{G}_1e^{p_1t}\right), k_0=1,206$, $\dot{\boldsymbol{G}}=\left[k_1e^{-j\Phi_1}\right]=[-424,\ 5e^{j0.342}]$, $\boldsymbol{q}=[-a_1+jw_1]=[-165,\ 9+j117,\ 1]$
3. Forced components of complex amplitudes	$\dot{\boldsymbol{Y}}=\mathrm{diag}(\dot{\boldsymbol{X}})K(\boldsymbol{p})=\begin{bmatrix}\dot{X}_1 & 0\\ 0 & \dot{X}_2\end{bmatrix}\begin{bmatrix}K(j\omega_1)\\ K(-\beta_2)\end{bmatrix}=\begin{bmatrix}\dot{X}_1K(j\omega_1)\\ \dot{X}_2K(-\beta_2)\end{bmatrix}=\begin{bmatrix}\dot{Y}_1\\ \dot{Y}_2\end{bmatrix}=$ $=Xm_1\begin{bmatrix}\frac{-1,044\omega_1^2e^{-j\varphi_1}}{100^2-\omega_1^2+j166,66\omega_1} & \frac{\mu_2\beta_2^2e^{-j\pi}}{100^2+\beta_2^2-166,66\beta_2}\end{bmatrix}^T$
4. Forced components	$y_1(t)=\mathrm{Re}(\dot{\boldsymbol{Y}}^T e^{\boldsymbol{p}t})=\mathrm{Re}\left(\begin{bmatrix}\dot{Y}_1\\ \dot{Y}_2\end{bmatrix}^T\begin{bmatrix}e^{j\omega_1t}\\ e^{-\beta_2t}\end{bmatrix}\right)=\mathrm{Re}\left(\dot{Y}_1e^{j\omega_1t}+\dot{Y}_2e^{-\beta_2t}\right)$, $y_1(t)=y_{11}(t)+y_{12}(t),\ y_{11}(t)=\mathrm{Re}\left(\dot{Y}_1e^{j\omega_1t}\right),\ y_{12}(t)=\mathrm{Re}\left(\dot{Y}_2e^{-\beta_2t}\right)$
5. Complex amplitudes of free components	$\dot{\boldsymbol{V}}=[\dot{V}_1]=[X(p_1)\dot{G}_1]=Xm_1\left[\frac{p_1\cos(\varphi_1)+\omega_1\sin(\varphi_1)}{p_1^2+\omega_1^2}-\mu_2\frac{1}{p_1+\beta_2}\right]$
6. Free components	$y_2(t)=\mathrm{Re}\left(\dot{V}_1e^{p_1t}\right)$
7. Error	$\varepsilon(t)=y_2(t)+y_{12}(t)$

Table 2. IIR filter analysis

For determining or estimating performance specifications of signal processing by the investigated IIR filter one need either to improve the software (Figure 1) or to reduce the amount of calculations by simplifying the analysis task. Let us consider the second option first.

The easiest operation is to define the error level in signal processing by a filter at frequency deviation of useful sinusoidal signal within the range ±5 Hz from the nominal value of 50 Hz. This error, as it is known, may be determine by an average amplitude-frequency response of a filter. In this case the value of a filter amplitude-frequency response in the areas of frequency $2\pi(45\div55)$ rad/s is between 0,95 and 1,038. Thus, the filter meets the signal processing performance requirement mentioned above.

A filter settling time can be defined by total damping of a free component τ_1 and a forced component of exponential disturbance τ_2 if the last component was not eliminated by the filter till the necessary level. Time τ_1 and τ_2 can be defined according to the Table 1 (item 6 and item 4).

A damping time of disturbance free component τ_1 and forced component τ_2 to the required level of 5% may be determined on the basis of the expressions given in the Table 2.

ORIGIN:= 1 $\quad j := \sqrt{-1}$

Input signal $\quad X := (1 \ \ -1)^T \quad p1 := (j \cdot 2\pi \cdot 50 \ \ -60)^T \quad N := \text{length}(X)$

IIR filter $\quad G := \left(-424.4 e^{j \cdot 0.342} \ \ 0\right)^T \quad \rho := (-165.9 + j \cdot 117.1 \ \ 0)^T \quad M := \text{length}(G)$

1 METOD $\quad X1(p) := \frac{1}{2} \sum_{i=1}^{N} \left[\frac{X_i}{p - p1_i} + \frac{(\overline{X})_i}{p - (\overline{p1})_i} \right] \quad K1(p) := \frac{1}{2} \sum_{i=1}^{M} \left[\frac{G_i}{p - \rho_i} + \frac{(\overline{G})_i}{p - (\overline{\rho})_i} \right] + 1.206$

Complex amplitudes: $\quad Y := \text{diag}(X) \cdot K1(p1) \quad Yf := \text{diag}(G) \cdot X1(\rho)$

$yp(t) := \text{Re}\left(Y^T e^{p1 \cdot t}\right) \quad Yp(t) := \text{Re}\left(\text{diag}(Y) e^{p1 \cdot t}\right) \quad yf(t) := \text{Re}\left(Yf^T e^{\rho \cdot t}\right) \quad y(t) := yp(t) + yf(t) \quad \varepsilon(t) := yf(t) + Yp(t)_2$

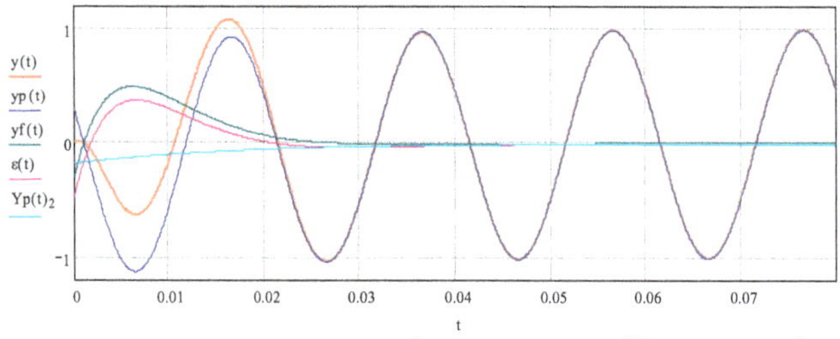

2 METOD $\quad K2(p,t) := \frac{1}{2} \sum_{i=1}^{M} \left[\frac{G_i}{p - \rho_i} \cdot \left[1 - e^{-(p - \rho_i)t}\right] + \frac{(\overline{G})_i}{p - (\overline{\rho})_i} \cdot \left[1 - e^{-\left[p - (\overline{\rho})_i\right]t}\right] \right] + 1.206$

$Y2(i,t) := (X_i) \cdot K2(p1_i, t) \quad y2(i,t) := \text{Re}\left(Y2(i,t) e^{p1_i \cdot t}\right) \quad YM2(i,t) := |Y2(i,t)| \cdot e^{\text{Re}(p1_i) \cdot t} \quad ys(t) := \sum_{i=1}^{N} y2(i,t)$

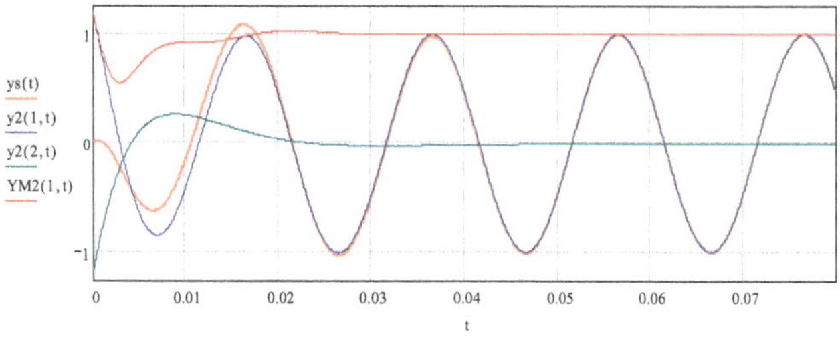

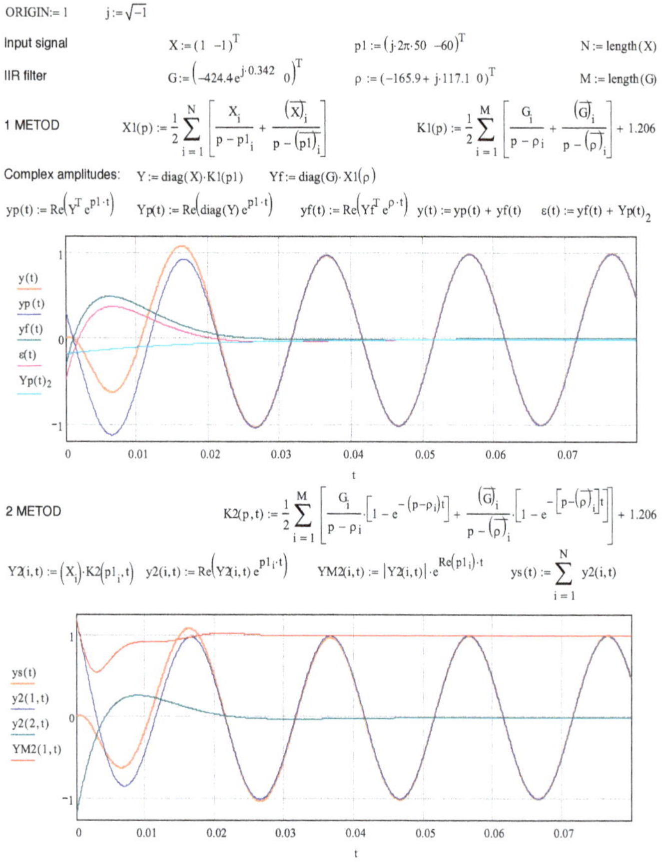

Figure 1. IIR filters analysis using Mathcad software

$$\tau_1 = -\frac{1}{\alpha_1} \ln\left(\frac{0{,}05}{k_1 |X(\rho_1)|} \right), \tau_2 = -\frac{1}{\beta_2} \ln\left(\frac{0{,}05}{\mu_2 |K(-\beta_2)|} \right). \tag{1}$$

Variables τ_1 and τ_2 depend not only on filter parameters which are constant, but also on signal parameters. Let us assume the worst option μ_2=1, φ_1=0, ω_1=2π50rad/s, taking into account the particularities of the controlled object [9,10]. Thus, let us consider the dependence $\tau_1$and τ_2 from β_2.

A presice settling time τ can be determined through the total error of signal processing $\varepsilon(t)$. In case of $t \geq \tau$the condition $|\varepsilon(t)| \leq \varepsilon_{\lim}$should be performed, when $\varepsilon_{\lim}$=0, 05.

Let us consider an estimation of total damping τ by the specified values of τ_1 and τ_2 5 %. In this case $\tau < \tau_1 + \tau_2$. If the filter is designed in a correct way, then $|y_{12}(t)| \leq \varepsilon_{\lim}$ at $t \geq \tau_1$, that is when disturbance is eliminated to the specified level by the moment of the end of transient process in a filter, then $\tau \approx \tau_1$. An estimation of a filter settling time can be performed with some conservative value on the basis of a sum of modules of free component envelopes of transient process in a filter and disturbance forced components [13]. The dependence τ from β_2 can be quite easily determined by an insignificant improvement of the program on the Mathcad example, represented on the Figure 1.

Dependencies τ, τ_1 and τ_2, depending on the value of exponential disturbance damping coefficient β_2 are shown on the Figure 2.

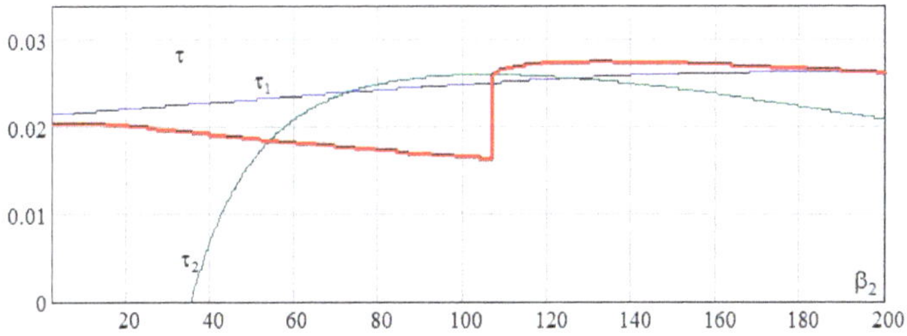

Figure 2. The dependence of a filter settling time from damping coefficient β_2

In case of β_2=2÷35 s^{-1} the initial level of a disturbance forced component is below the acceptable error, so $|\dot{Y}_2| \leq \varepsilon_{\lim}$, and τ_2=0. Then a filter settling time is mostly defined by damping transient process of its own in a filter, in other words by the value τ_1. Within the range β_2=74÷125 s^{-1} damping of a disturbance force component is longer than damping time of transient process of its own in a filter, that is $\tau_2 > \tau_1$. At $\beta_2 \leq 107$ s^{-1} a filter settling time τ is less than values τ_1 and τ_2, and at $\beta_2 > 107$ s^{-1} is longer than any of the values mentioned above. It is due to plus-minus signs of filter reaction components, defining an error of signal processing, as well as to values of the components complex frequencies.

An overshoot level in a filter can be determined by the program improvement on the Mathcad example, shown on the Figure 1.

The performance analysis results for signal processing of the investigated IIR filter are represented in the Table 3. Performance specifications determined by using one traditional method - by step response of a second-order low-pass filter are given in the Table 3 as well. In this case the description for the low-pass filter was obtained on the basis of the investigated second-order low-pass filter by applying a well known frequency transformation [14].

№ Name	Step response	Direct estimation
1. Settling time, s	0,0172	0,0275
2. Maximum error in the steady-state mode, %	0	5
3. Maximum overshoot level, %	1,17	11,81
4. Additional error at a frequency variation of an useful signal	-	5

Table 3. Signal processing performance for a second-order high-pass filter

As it follows from the Table 3, in the considered example there are some substantial differences in the performance specifications, gained by the traditional method and on the basis of their direct determination.

2.2. IIR filters analysis at dissemination of time-and-frequency approach to transfer function filter

As it follows from the Table 1 (item 3 and item 6), estimation of functioning performance for filters can be performed at dissemination of time-and-frequency approach to filter transfer function, in other words by using time dependent transfer function of a filter [5,6].

In that case, comparing to the first method where two groups of filter reaction components are using – forced and free components, only the first group of components is using. The information about the transient process is containing in the time dependent complex amplitudes of a filter reaction $\dot{Y}(t)$. In the case of $t \to \infty$ the mentioned set of complex amplitudes will be equal to complex amplitudes of forced components $\dot{Y}(t) \to \dot{Y}$.

The necessary dependences for determination of output signal components of a filter are represented in the Table 1, an example for high-pass filter analysis at the specified parameters of an useful signal is given on the bottom part of the fig.1.

The advantage of the considered method is connected to determination envelopes for every component of a filter output signal, based on which the total envelope of a filter output signal and a variation law of initial phase can be defined. This information can be effectively used for performance analysis of signal processing by frequency filters. For instance, when determining an overshoot level (oscillativity) of a transient process in a filter.

2.3. IIR filter express-analysis method

It follows from the Table 1, that quality indexes estimation for filter operation can be carried out on the basis of interim calculation results $-K(p)$ and $X(q)$, that means - based on spectral representations of signals and filter impulse functions in complex frequency coordinates. Another, not less effective approach, is related to the usage of the interim results of the second analysis method – the transfer function $K(p, t)$, which is dependent of time. Thus, the application of filter frequency characteristics and a signal spectrum in complex frequency coordinates increase significantly the effectiveness of using the frequency methods of performance analysis for frequency filter operation [7].

The express-analysis methods for filters, including performance analysis of signal processing, were developed based on investigation of 3D and 4D frequency responses [7]. It is enough to consider the sections $p = j\omega$ and $p = -\gamma$ of 3D frequency responses $K(\boldsymbol{p})$, as well as the section $p = -\alpha_1 + jw_1$ of a input signal spectrum according to the Laplace transform to estimate the settling time and accuracy of signal processing for the example given on the fig.1.

The express-analysis methods mentioned above can be effectively applied for FIR filters as well, the detailed explanation will be given further in the present chapter.

2.4. Digital IIR filters analysis

Under the definition of digital filters in the chapter discrete filters are ment. In many cases it is justified, for instance, in cases of using microcontrollers or digital signal processor with high digit capacity and especially for microprocessors with support for floating-point operations [15,16].

When using discrete filters their analysis has a lot of similarities with the analysis of analog filters-prototypes. There is a small difference only when it comes to transition from images to originals. The main expressions for determining components of a digital filter output signal when injecting on the filter input a signal as a set of discrete semi-infinite damped oscillatory components are given in the Table 4.

An example for digital filter analysis as a continuation of the example of the analog filter-prototype analysis (fig. 1) is represented on the fig. 3. The mathematical description for the digital filter was obtained by the method of invariant impulse responses at the discrete sampling step $T = 0,\ 0005$s.

№ Name	Expressions	Remark
1. Input signal	$x(k) = \mathrm{Re}(\dot{\boldsymbol{X}}^{\mathrm{T}} Z(\boldsymbol{P}, k))$, $X(z) = \mathrm{Re}\left(\dot{\boldsymbol{X}}^{\mathrm{T}}\left[\frac{z}{z - z_n}\right]_N\right)$	$\dot{\boldsymbol{X}} = \left[Xm_n e^{-j\varphi_n}\right]_N,\ \boldsymbol{p} = [-\beta_n + j\omega_n]_N$ $\boldsymbol{z} = e^{\boldsymbol{p}T},\ Z(p, k) = e^{pkT}$, T - discrete sampling step
2. Filter	$g(k) = \mathrm{Re}(\dot{\boldsymbol{G}}^{\mathrm{T}} Z(\boldsymbol{Q}, k))$, $K(z) = \mathrm{Re}\left(\dot{\boldsymbol{G}}^{\mathrm{T}}\left[\frac{z}{z - z_m}\right]_M\right)$	$\dot{\boldsymbol{G}} = \left[k_m e^{-j\phi_m}\right]_M$, $\boldsymbol{q} = [-a_m + jw_m]_M$, $\boldsymbol{z} = e^{\boldsymbol{q}T}$
3. Time dependent transfer function	$K(z, k) = \mathrm{Re}\left(\dot{\boldsymbol{G}}^{\mathrm{T}}\left[\frac{z\left(1 - e^{p_m kT} z^{-k}\right)}{z - z_m}\right]_M\right)$	$K(z, k) = \sum_{i=0}^{k} g(i) z^{-i}$
4. Forced components	$y_1(k) = \mathrm{Re}(\dot{\boldsymbol{Y}}^{\mathrm{T}} Z(\boldsymbol{p}, k))$	$\dot{\boldsymbol{Y}} = \mathrm{diag}(\dot{\boldsymbol{X}}) K(\boldsymbol{z})$,
5. Free components	$y_2(k) = \mathrm{Re}(\dot{\boldsymbol{V}}^{\mathrm{T}} Z(\boldsymbol{q}, k))$	$\dot{\boldsymbol{V}} = \mathrm{diag}(\dot{\boldsymbol{G}}) X(\boldsymbol{z})$
6. Filter reaction	$y(k) = \mathrm{Re}(\dot{\boldsymbol{Y}}(k)^{\mathrm{T}} Z(\boldsymbol{p}, k))$	$\dot{\boldsymbol{Y}}(k) = \mathrm{diag}(\dot{\boldsymbol{X}}) K(\boldsymbol{z}, k)$

Table 4. IIR digital filter analysis

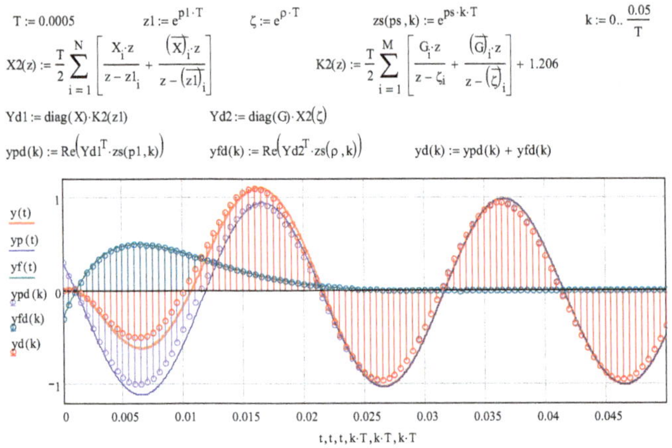

Figure 3. IIR digital filter analysis using Mathcad software

№ Name	Expressions	Remark
1. Input signal	$x(t)=\mathrm{Re}(\dot{\boldsymbol X}^{\mathrm T}e^{\boldsymbol P(\boldsymbol C t-\boldsymbol t)}-\dot{\boldsymbol X}'^{\mathrm T}e^{\boldsymbol P(\boldsymbol C t-\boldsymbol t')})$, $X'(p)=\mathrm{Re}\left(\dot{\boldsymbol X}^{\mathrm T}\left[\frac{1}{p-p_n}\right]_N\right)$, $X''(p)=\mathrm{Re}\left(\dot{\boldsymbol X}'^{\mathrm T}\left[\frac{1}{p-p_n}\right]_N\right)$, $X(p)=\mathrm{Re}\left(\dot{\boldsymbol X}^{\mathrm T}\left[\frac{e^{-pt_n}}{p-p_n}\right]_N+\dot{\boldsymbol X}'^{\mathrm T}\left[\frac{e^{-pt'_n}}{p-p_n}\right]_N\right)$	$\dot{\boldsymbol X}=[\dot X_n]_N=[Xm_n e^{-j\varphi_n}]_N$, $\dot{\boldsymbol X}'=\mathrm{diag}(\dot{\boldsymbol X})e^{\boldsymbol P(\boldsymbol t-\boldsymbol t')}$, $\boldsymbol p=[p_n]_N=[-\beta_n+j\omega_n]_N$, $\boldsymbol P=\mathrm{diag}(\boldsymbol p)$, $\boldsymbol t=[t_n]_N$, $\boldsymbol t'=[t'_n]_N$, $\boldsymbol C=[1]_N$
2. Filter	$g(t)=\mathrm{Re}(\dot{\boldsymbol G}^{\mathrm T}e^{\boldsymbol q t})$, $K(p)=\mathrm{Re}\left(\dot{\boldsymbol G}^{\mathrm T}\left[\frac{1}{p-p_m}\right]_M\right)$	$\dot{\boldsymbol G}=[\dot G_m]_M=[k_m e^{-j\phi_m}]_M$, $\boldsymbol q=[p_m]_M=[-\alpha_m+jw_m]_M$
3. Time dependent transfer function	$K(p,t)=\mathrm{Re}\left(\dot{\boldsymbol G}^{\mathrm T}\left[\frac{1-e^{-(p-p_m)t}}{p-p_m}\right]_M\right)$	$K(p,t)=\int_0^t g(\tau)e^{-p\tau}d\tau$
4. Forced components	$y_1(t)=\mathrm{Re}(\dot{\boldsymbol Y}^{\mathrm T}e^{\boldsymbol P(\boldsymbol Ct-\boldsymbol t)}-\dot{\boldsymbol Y}'^{\mathrm T}e^{\boldsymbol P(\boldsymbol Ct-\boldsymbol t')})$	$\dot{\boldsymbol Y}=\mathrm{diag}(\dot{\boldsymbol X}')K(\boldsymbol p)$, $\dot{\boldsymbol Y}'=\mathrm{diag}(\dot{\boldsymbol X}'')K(\boldsymbol p)$
5. Free components	$y_2(t)=\mathrm{Re}\left(\sum_n \dot{\boldsymbol V}^{\langle n\rangle\mathrm T}e^{\boldsymbol q(t-t_n)}-\sum_n\dot{\boldsymbol V}'^{\langle n\rangle\mathrm T}e^{\boldsymbol q(t-t'_n)}\right)$	$\dot{\boldsymbol V}=[\dot G_m X(p_m)_n]_{M,N}$, $\dot{\boldsymbol V}'=[\dot G_m X'(p_m)_n]_{M,N}$
6 Filter reaction	$y(t)=\mathrm{Re}(\dot{\boldsymbol Y}(t)^{\mathrm T}e^{\boldsymbol P(\boldsymbol Ct-\boldsymbol t)}-\dot{\boldsymbol Y}'(t)^{\mathrm T}e^{\boldsymbol P(\boldsymbol Ct-\boldsymbol t')})$	$\dot{\boldsymbol Y}(t)=\mathrm{diag}(\dot{\boldsymbol X}')K(\boldsymbol p,\boldsymbol Ct-\boldsymbol t)$, $\dot{\boldsymbol Y}'(t)=\mathrm{diag}(\dot{\boldsymbol X}'')K(\boldsymbol p,\boldsymbol Ct-\boldsymbol t')$

Table 5. IIR filter analysis at compound finite signals

2.5. IIR filters analysis at finite signals

Performance analysis of processing finite signals by IIR filters as a set of damped oscillatory components with finite duration may be performed on the basis of dependencies for IIR filters at semi-infinite signals [5,6].

All the needed expressions were obtained on the basis of the expressions from the Table 1 using a time shift and the principle of additivity.

Let us consider the IIR filter analysis at compound input signals as a set of sequentially adjacent finite signals [17]. The calculation of IIR filter reaction for this case is represented in the Table 5.

In this case every component of an input signal in a general way has a different shift and a different duration, where $\dot{V}^{\langle n \rangle}$ – n-th matrix column $\dot{V}$.

The expressions represented in the Table 5 can be significantly simplified, if the IIR filter analysis at a finite signal, for instance, at injection a finite signal on its input from N number of components with equal duration and the same time shift [6,17].

An example of filter calculation, analogous to the example on the fig.1, but using an input signal with finite duration is given on the fig.4.

The dependences given in the present section can be effectively used for not only analysis of passing through IIR filter one or another finite signal, but also for performance analysis of signal processing by filters.

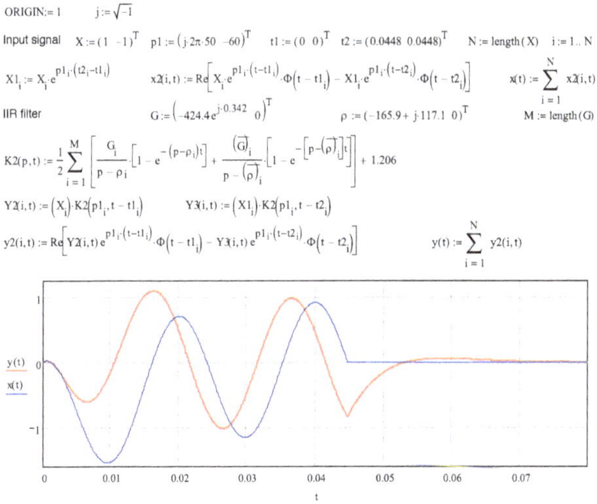

Figure 4. IIR filter analysis at a finite signal

For digital IIR filter analysis at injection finite signals on their inputs analogous mathematical operations are applied. All the necessary dependences may be obtained on the basis of the formulas from the Table 4 [13].

3. FIR filters performance analysis

3.1. Particularities of the analysis

Mathematical description for FIR filters can be obtained on the basis of the IIR filter description (Table 1) by using twice as many of filter impulse function components.

№ Name	Expressions	Remark
1. Input signal	$x(t)=\mathrm{Re}(\dot{\boldsymbol{X}}^{\mathrm{T}}e^{\boldsymbol{p}t})$, $X(p)=\mathrm{Re}\left(\dot{\boldsymbol{X}}^{\mathrm{T}}\left[\frac{1}{p-p_n}\right]_N\right)$	$\dot{\boldsymbol{X}}=[\dot{X}_n]_N=[Xm_n e^{-j\varphi_n}]_N$, $\boldsymbol{p}=[p_n]_N=[-\beta_n+j\omega_n]_N$
2. FIR filter	$g(t)=\mathrm{Re}(\dot{\boldsymbol{G}}^{\mathrm{T}}e^{\boldsymbol{q}t}-\dot{\boldsymbol{G}}'^{\mathrm{T}}e^{\boldsymbol{Q}(\boldsymbol{C}t-\boldsymbol{T})})$, $K_1(p)=\mathrm{Re}\left(\dot{\boldsymbol{G}}^{\mathrm{T}}\left[\frac{1}{p-p_m}\right]_M\right)$, $K_2(p)=\mathrm{Re}\left(\dot{\boldsymbol{G}}'^{\mathrm{T}}\left[\frac{1}{p-p_m}e^{-pT_m}\right]_M\right)$, $\boldsymbol{K_3(p)=\mathrm{Re}\left(\dot{G}'^{\mathrm{T}}\left[\frac{1}{p-p_m}\right]_M\right)}$	$\dot{\boldsymbol{G}}=[\dot{G}_m]_M=[k_m e^{-j\phi_m}]_M$, $\dot{\boldsymbol{G}}'=\mathrm{diag}(\dot{\boldsymbol{G}})e^{\boldsymbol{QT}}$, $\boldsymbol{q}=[p_m]_M=[-\alpha_m+jw_m]_M$, $\boldsymbol{Q}=\mathrm{diag}(\boldsymbol{q})$, $\boldsymbol{T}=[T_m]_M$, $\boldsymbol{C}=[1]_M$, $K(p)=K_1(p)-K_2(p)$
3. Forced components	$y_1(t)=\mathrm{Re}(\dot{\boldsymbol{Y}}^{\mathrm{T}}e^{\boldsymbol{p}t})$	$\dot{\boldsymbol{Y}}=\mathrm{diag}(\dot{\boldsymbol{X}})K(\boldsymbol{p})$
4. Free components	$y_2(t)=y_3(t)+y_4(t)$, $y_3(t)=\mathrm{Re}(\dot{\boldsymbol{V}}^{\mathrm{T}}e^{\boldsymbol{q}t}-\dot{\boldsymbol{V}}'^{\mathrm{T}}e^{\boldsymbol{Q}(\boldsymbol{C}t-\boldsymbol{T})})$, $y_4(t)=y_1(t)-\mathrm{Re}\left(\dot{\boldsymbol{U}}^{\mathrm{T}}e^{\boldsymbol{p}t}-\sum_m \dot{\boldsymbol{U}}'^{\mathrm{T}}e^{\boldsymbol{p}(t-T_m)}\right)$	$\dot{\boldsymbol{V}}=\mathrm{diag}(\dot{\boldsymbol{G}})X(\boldsymbol{q})$, $\dot{\boldsymbol{V}}'=\mathrm{diag}(\dot{\boldsymbol{G}}')X(\boldsymbol{q})$, $\dot{\boldsymbol{U}}=\mathrm{diag}(\dot{\boldsymbol{X}})K_1(\boldsymbol{p})$, $\dot{\boldsymbol{U}}'=[\dot{X}_n K_3(p_n)_m]_{N,M}$
5. Filter reaction	$y(t)=y_1(t)+y_2(t)$	
6. Time dependent transfer function	$K_1(p,t)=\mathrm{Re}\left(\dot{\boldsymbol{G}}^{\mathrm{T}}\left[\frac{1-e^{-(p-p_m)t}}{p-p_m}\right]_M\right)$, $K_2(p,t)=\mathrm{Re}\left(\dot{\boldsymbol{G}}'^{\mathrm{T}}\left[\frac{1-e^{-(p-p_m)(t-T_m)}}{p-p_m}\right]_M\right)$, $K(p,t)=K_1(p,t)-K_2(p,t)e^{-pT_1}$	
7. Filter reaction	$y(t)=\mathrm{Re}(\dot{\boldsymbol{Y}}(t)^{\mathrm{T}}e^{\boldsymbol{p}t}-\dot{\boldsymbol{Y}}'(t)^{\mathrm{T}}e^{\boldsymbol{P}(\boldsymbol{C}t-\boldsymbol{T})})$	$\dot{\boldsymbol{Y}}(t)=\mathrm{diag}(\dot{\boldsymbol{X}})K_1(\boldsymbol{p},t)$, $\dot{\boldsymbol{Y}}'(t)=\mathrm{diag}(\dot{\boldsymbol{X}})K_2(\boldsymbol{p},t)$

Table 6. FIR filters analysis

The additional components have the same set of complex frequencies and differ by time shift and values of complex amplitudes in a way to ensure the finitude of a filter impulse characteristic [5,6]. According to this approach to mathematical description, IIR filters are special cases of FIR filters.

The input-output dependences for FIR filters can be obtained on the basis of analogous dependences of FIR filters by using time shift and principle of additivity operations [6].

Comparing to IIR filters, FIR filters have finite duration of transient processes of their own, which are defined by a filter length. This to a certain extent simplifies performance analysis for signal processing of this type of filters, especially when using the suggested express-analysis methods for signal processing performance by FIR filters.

3.2. Analog FIR filters

Basic expressions for FIR filter analysis at injection on the filter input a set of semi-infinite damped oscillatory components are given in the Table 6.

Due to characteristics of FIR filters, among forced and free components in an filter output signal there is the third group of components $y_4(t)$, which is conventionally referred to free components in the Table 6 [6].

№ Name	Expressions	Remark
1. Input signal	$x(t)=\mathrm{Re}\left(\dot{\boldsymbol{X}}^{\mathrm{T}}e^{\boldsymbol{P}(\boldsymbol{C}t-\boldsymbol{t})}-\dot{\boldsymbol{X}}^{'\mathrm{T}}e^{\boldsymbol{P}(\boldsymbol{C}t-\boldsymbol{t}')}\right)$	$\dot{\boldsymbol{X}}=[\dot{X}_n]_N$, $\boldsymbol{p}=[p_n]_N$, $\dot{\boldsymbol{X}}'=\mathrm{diag}(\dot{\boldsymbol{X}})e^{\boldsymbol{P}(\boldsymbol{t}-\boldsymbol{t}')}$, $\boldsymbol{P}=\mathrm{diag}(\boldsymbol{p}), \boldsymbol{C}=[1]_N$, $\boldsymbol{t}=[t_n]_N$, $\boldsymbol{t}'=[t_n']_N$
2. FIR filter: impulse function, time dependent transfer function	$g(t)=\mathrm{Re}\left(\dot{\boldsymbol{G}}^{\mathrm{T}}e^{\boldsymbol{q}t}-\dot{\boldsymbol{G}}^{'\mathrm{T}}e^{\boldsymbol{q}(t-T_1)}\right)$, $K_1(p,\ t)=\mathrm{Re}\left(\dot{\boldsymbol{G}}^{\mathrm{T}}\left[\dfrac{1-e^{-(p-p_m)t}}{p-p_m}\right]_M\right)$, $K_2(p,\ t)=\mathrm{Re}\left(\dot{\boldsymbol{G}}^{'\mathrm{T}}\left[\dfrac{1-e^{-(p-p_m)(t-T_1)}}{p-p_m}\right]_M\right)$, $K(p,\ t)=K_1(p,\ t)-K_2(p,\ t)e^{-pT_1}$	$\dot{\boldsymbol{G}}=[\dot{G}_m]_M$, $\boldsymbol{q}=[p_m]_M$, $\dot{\boldsymbol{G}}'=\mathrm{diag}(\dot{\boldsymbol{G}})e^{\boldsymbol{QT}}$, $\boldsymbol{Q}=\mathrm{diag}(\boldsymbol{q}), \boldsymbol{T}=[T_m]_M$
3. Filter reaction	$y(t)=\mathrm{Re}\left(\dot{\boldsymbol{Y}}_1(t)^{\mathrm{T}}e^{\boldsymbol{P}(\boldsymbol{C}t-\boldsymbol{t})}-\dot{\boldsymbol{Y}}_2(t)^{\mathrm{T}}e^{\boldsymbol{P}(\boldsymbol{C}t-\boldsymbol{t}')}-\right.$ $\left.-\dot{\boldsymbol{Y}}_3(t)^{\mathrm{T}}e^{\boldsymbol{P}(\boldsymbol{C}(t-T_1)-\boldsymbol{t})}+\dot{\boldsymbol{Y}}_3(t)^{\mathrm{T}}e^{\boldsymbol{P}(\boldsymbol{C}(t-T_1)-\boldsymbol{t}')}\right)$	$\dot{\boldsymbol{Y}}_1(t)=\mathrm{diag}(\dot{\boldsymbol{X}})K_1(\boldsymbol{p},\ \boldsymbol{C}t-\boldsymbol{t})$, $\dot{\boldsymbol{Y}}(t)=\mathrm{diag}(\dot{\boldsymbol{X}})K_1(\boldsymbol{p},\ \boldsymbol{C}t-\boldsymbol{t}')$, $\dot{\boldsymbol{Y}}_3(t)=\mathrm{diag}(\dot{\boldsymbol{X}})K_2(\boldsymbol{p},\ \boldsymbol{C}t-\boldsymbol{t})$, $\dot{\boldsymbol{Y}}_4(t)=\mathrm{diag}(\dot{\boldsymbol{X}})K_2(\boldsymbol{p},\ \boldsymbol{C}t-\boldsymbol{t}')$

Table 7. FIR filter analysis at input signal as a set of finite components

Basic expressions for FIR filter analysis at signal injection on a filter input as a set of damped oscillatory components with finite duration are represented in the Table 7. The calculation of a filter reaction is given in the Table 7 using only the second analysis method for the case when duration of all the components of a filter impulse function is equal T_1. The filter analysis based on the first method is represented in details in the author's papers [6,18].

Let us consider an analysis example of FIR filters which are used in one of the most perspective intelligent electronic devices (IED) - Phasor Measurement Units (PMU) [19].

A brief description of a basic algorithm for PMU signal processing on the example of an analog system-prototype is given in the item 1 of the Table 8 [20].

№	Name	Diagram/expressions
1.	Block scheme of algorithm	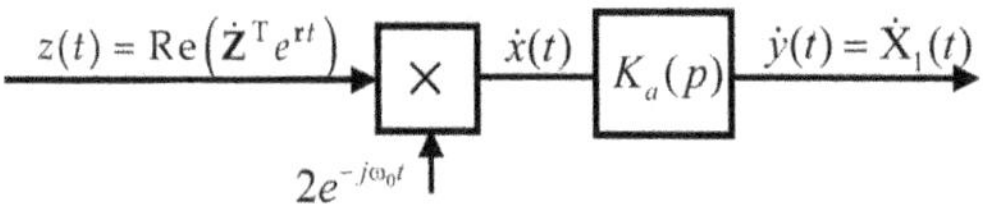
2.	Signal description $z(t)$	Semi-infinite signal $z(t)=\mathrm{Re}(\dot{\mathbf{Z}}^{\mathrm{T}}e^{\mathbf{r}t})$ or $z(t)=0,5(\dot{\mathbf{Z}}^{\mathrm{T}}e^{\mathbf{r}t}+\bar{\mathbf{Z}}^{\mathrm{T}}e^{\bar{\mathbf{r}}t})$, $\dot{\mathbf{Z}}=[Z_0\ \ \dot{Z}_1\ \ \dot{Z}_2\ \ \dots\ \ \dot{Z}_{K-1}]^{\mathrm{T}}$, $\mathbf{r}=[-\beta_0\ \ j\omega_1\ \ j2\omega_1\ \ \dots\ \ j(K-1)\omega_1]^{\mathrm{T}}$, $\bar{\mathbf{Z}},\bar{\mathbf{r}}$ – are complex – adjoint vectors Signal as a set of finite components $z(t)=0,5(\dot{\mathbf{Z}}^{\mathrm{T}}e^{\mathbf{R}(\mathbf{C}t-\mathbf{t})}-\dot{\mathbf{Z}}'^{\mathrm{T}}e^{\mathbf{R}(\mathbf{C}t-\mathbf{t}')}+\bar{\mathbf{Z}}^{\mathrm{T}}e^{\bar{\mathbf{R}}(\mathbf{C}t-\mathbf{t})}-\bar{\mathbf{Z}}'^{\mathrm{T}}e^{\bar{\mathbf{R}}(\mathbf{C}t-\mathbf{t}')})$ $\mathbf{R}=\mathrm{diag}(\mathbf{r})$
3.	Input signal of a filter	$\dot{x}(t)=2e^{-j\omega_0 t}z(t)$, At semi-infinite input signal device $\dot{x}(t)=\dot{\mathbf{Z}}^{\mathrm{T}}e^{(\mathbf{r}-\mathbf{C}j\omega_0 t)t}+\bar{\mathbf{Z}}^{\mathrm{T}}e^{(\bar{\mathbf{r}}-\mathbf{C}j\omega_0 t)t}$
4.	Algorithm	$\dot{X}_1(t)=\int_{t-T_1}^{t}z(\tau)e^{-j\omega_0\tau}g(t-\tau)d\tau=\int_{t-T_1}^{t}\dot{x}(\tau)g(t-\tau)d\tau$
5.	Average FIR filter	$\dot{\mathbf{G}}=[80.48e^{j4.273}\ \ 37.93e^{j0.5887}]^{\mathrm{T}}$, $\dot{\mathbf{G}}'=\mathrm{diag}(\dot{\mathbf{G}})e^{\mathbf{q}T_1}$ $\mathbf{q}=[-22.99+j62.30\ \ -23.26+j186.9]^{\mathrm{T}}$, $\mathbf{T}=[T_1\ \ T_1]^{\mathrm{T}}$, $T_1=0.051c$, $g(t)=\mathrm{Re}(\dot{\mathbf{G}}^{\mathrm{T}}e^{\mathbf{q}t}-\dot{\mathbf{G}}'^{\mathrm{T}}e^{\mathbf{q}(t-T_1)})$, $K(p)=\mathrm{Re}\left(\dot{\mathbf{G}}^{\mathrm{T}}\left[\frac{1}{p-\rho_m}\right]_M-\dot{\mathbf{G}}'^{\mathrm{T}}\left[\frac{1}{p-\rho_m}e^{-pT_m}\right]_M\right)$

Table 8. IED algorithm

An input signal of intelligent electronic devices is represented by a set of complex amplitudes $\dot{\mathbf{Z}}$ and frequencies $\mathbf{r}$, as well as time parameters when using a signal model as a set of finite components. Let us constrain the signal models to one model only as an exponential component, useful sinusoidal component of commercial frequency ω_1 (nominal value is ω_0=2π50rad/sec) and higher harmonics. More complicated signal models are considered in the papers [6,20].

An average FIR filter $K_a(p)$ should isolate the constant ($\dot{y}(t)=\dot{X}_1$ at $\omega_1=\omega_0$) or low-frequency component ($\dot{y}(t)=\dot{X}_1(t)$ at $\omega_1\neq\omega_0$). The filters should suppress higher harmonics and a damped oscillatory component with the complex frequency $p=-\beta_0+j\omega_0$.

The considered filters should have low sensitivity to a change of damping coefficient β_0 in the range from 10÷200 sec^{-1} and the frequency $\omega_1=2\pi(50\pm5)$ rad/sec. The acceptable static error of signal processing should not be more than 0,5%, and the acceptable dynamic error at $t\geq T_1$ should not be higher than 3%.

FIR filter analysis is performed at input signal of a device as a set of semi-infinite or finite damped oscillatory components according to the algebraic expressions from the Table 7 and the Table 8.

An example of FIR filter analysis using Mathcad at compound input signals as a set of sequentially adjacent finite signals is given on the fig.6.

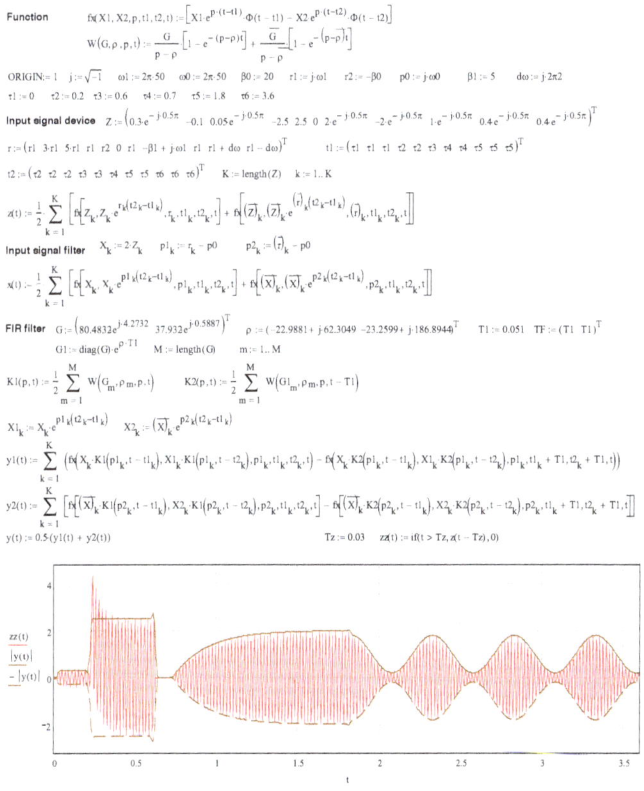

Figure 5. FIR filter analysis using Mathcad software

Each of five sets of finite components represented comply with a particular power system regime. The first set is related to a normal regime of a power system and is represented by sections of sinusoidal component of industrial frequency and higher harmonics. The second set of finite signals corresponds to an accidental regime and contains of finite sinusoidal and exponential component, the third set shows no-current regime, the fourth set is connected to a change of PMU input signals envelope due to automation operation, the fifth set represents a swings regime.

A filter processes a complex signal $\dot{x}(t)$, formed after multiplication of PMU input signal $z(t)$ by a reference signal $e^{-j\omega_0 t}$ to shift the signal spectrum to the left for the formation of orthogonal components of complex amplitude of sinusoidal signal with industrial frequency ω_1.

A plot for PMU input signal $z(t)$and a plot for a module of filter output signal $|\dot{y}(t)|$ taking into account filter group delay time are represented on the fig.6. As it follows from the fig.1, the plot $|\dot{y}(t)|$ is close to the envelope of sinusoidal component of PMU input signal.

It appears from the plot that the investigated filter has a feature which is connected to the absence of overshoot (oscillation) of transient process in the filter in its traditional definition, in other words at a stepwise growth of a signal. and presence of overshoot at a stepwise reduction of signal.

In fact, due to particular characteristics of impulse function, from the traditional point of view, an overshoot is absent in the considered FIR filter, as at the end of transient process in the filter signal behavior is close to aperiodic process. In this case oscillation is noticeable on the initial stage of transient process.

Analysis of FIR filter and signal processing algorithm in total is carried out only at fixed values of input signal parametres in the example. With some further ordinary improvement, analog to the example on the fig.1, it is possible to determine performance specifications for FIR filter signal processing at any variation of useful signal and disturbance parameters. However, it is much easier to use the specific express-analysis methods.

3.3. Express-analysis methods for signal processing performance

Let us consider an example of express-analysis of signal processing performance of a FIR filter, which mathematical description is given in the Table 7. In addition, let us consider performance analysis for signal processing of a filter with the impulse function $g_2(t)=g_1(T_1-t)$ (Fig.7).

To check the adherence to the mentioned conditions, it is enough to consider amplitude-frequency response of the filters in complex frequency coordinates.

The amplitude-frequency response for the first and second filters is given in the Fig.8. The multiplier $e^{-\beta T_1}$ at $\beta \neq 0$ allows for attenuation of forced damped components by the moment of the transient process ending in the filter. The speed of the FIR filter is determining by its length T_1 in the case of noise damped components are suppressed to the level of the acceptable dynamic error during the time, that is equal to the filter length.

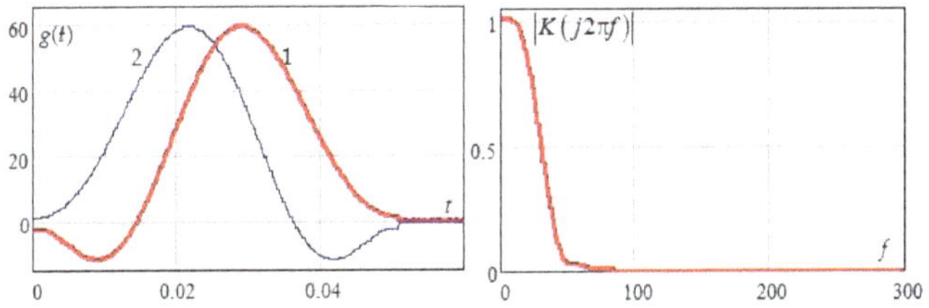

Figure 6. Impulse characteristics of the filters

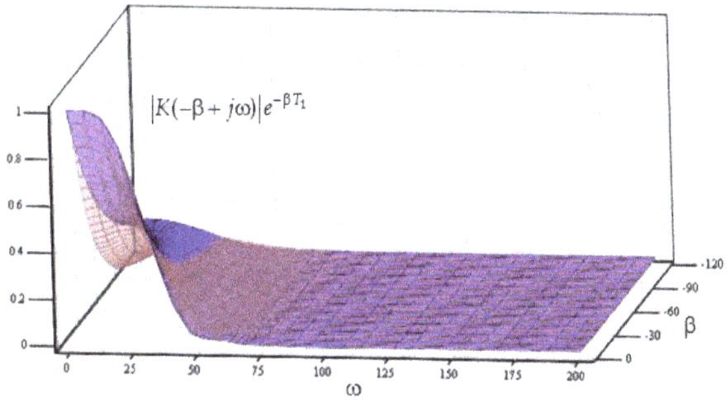

Figure 7. Amplitude-frequency response of the filter

To check the compliance with the requirements for filters to the specified quality indexes of signal processing for the specified input signal, it would be enough to consider two sections of 3D amplitude-frequency response: in the sections $p = j2\pi f$ (Fig.6) and $p = -\beta + j\omega_0$ (Fig.8). As it follows from the Fig.6, the examined filters have the same amplitude-frequency response in the section $p = j2\pi f$ and ensure the required quality of useful signal processing and suppressing higher harmonics. Absolutely different situation appears in the case of eliminating by filters a damped oscillatory component with frequency $p_0 = -\beta_0 + j\omega_0$.

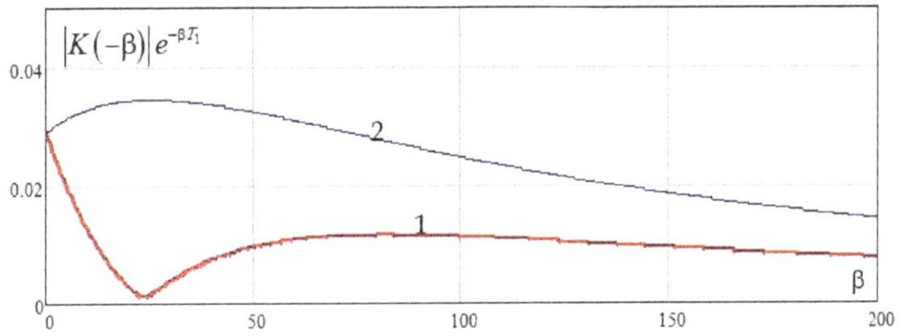

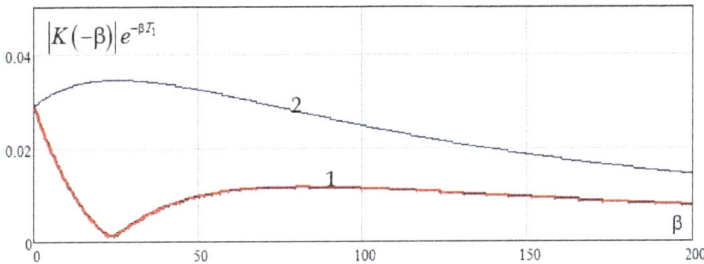

Figure 8. Frequency response of the filters in the section $p = -\beta + j\omega_0$

The filters 1 and 2 in the section of 3D frequency response $p = -\beta + j\omega_0$ (Fig.7) have different characteristics, and the second filter does not ensure compliance with the specified requirements to suppress the component of an input signal with the complex frequency $p_0 = -\beta_0 + j\omega_0$. It results in ambiguity of using the traditional frequency response of filters FIR filters with asymmetric form of the impulse response for analysis aperiodic signals.

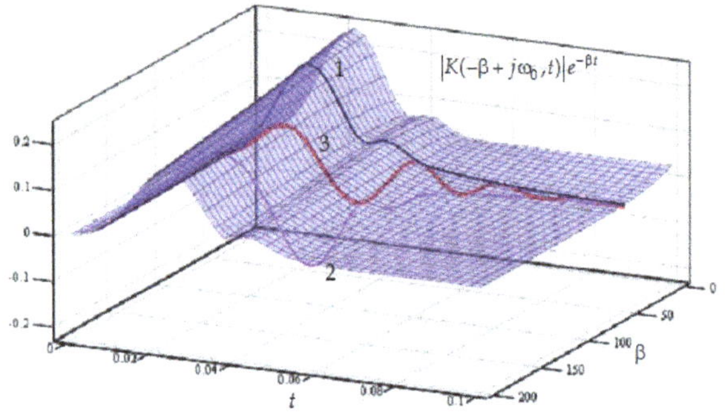

Figure 9. Amplitude-frequency response of the filter

In the case of a need along with estimating the speed and accuracy of signal processing to appraise the history of transient processes in a filter, for instance, to control oscillation of transient process in a filter, it is eligible to conduct the analysis using amplitude-frequency response based on the filter transfer function, which depends on time.

It is necessary for analysis to depress a damped oscillatory component with the complex frequency $p_0 = -\beta_0 + j\omega_0$ to fix the imaginary part, due to ω_0=const. The plot, proportional to the product $|K_2(-\beta + j\omega_0, t)| e^{-\beta t}$, is given in the Fig.10. In the case of complex frequency $p = -\beta + j\omega_0$ the plot will be equal to the envelope (curves 1 and 2) of the reaction of the second filter (curve 3) to the examined input impact at $\beta_0 = 50$ sec^{-1}.

The graph presented describes graphically the fact, that the second filter does not ensure the suppressing of an input signal component with the complex frequency $p_0 = -\beta_0 + j\omega_0$ by the moment of free components completion in the filter at the most specified values of β_0.

3.3. Digital FIR filters

As for IIR filters (Table 1 and 4), input-output dependencies for digital (discrete) FIR filter may be obtained by discretization of expressions for analog FIR filter and transition from the Laplace transform to Z transform [13].

In case of using the mathematical software Matlab, Mathcad etc. for digital FIR filter analysis it is enough to get the data about complex amplitudes and input signal frequencies $\dot{X}$ ($\dot{X}'$ at finite signals), p (or $z=e^{pT}$) and a filter $\dot{G}$, $\dot{G}'$, q (or $z=e^{qT}$), as well as a set of parameters, which define the duration of filter impulse function components and the beginning and duration of input signal components.

In some cases only mathematical description of analog filter-prototypeis specified. In case of FIR filters, different methods of transition from analog filter-prototype description to digital filter description, for instance, method of differential equation discretization, method of invariant impulseresponses, bilinear transformation are applied [14]. The mentioned methods expand also for the case of FIR filters in the author's paper [13].

The transfer function of analog filter-prototype with finite impulse response according to the Table 6 can be described in the following way

$$K(p) = \mathrm{Re}\left(\dot{\mathbf{G}}^{\mathrm{T}} \left[\frac{1}{p - \rho_m} \right]_M - \dot{\mathbf{G}}'^{\mathrm{T}} \left[\frac{1}{p - \rho_m} e^{-pT_m} \right]_M \right) \tag{2}$$

The expression for a digital filter transfer function

$$K(z) = \mathrm{Re}\left(\dot{\mathbf{G}}^{\mathrm{T}} \left[\frac{k_m (z + a)}{z - z_m} \right]_M - \dot{\mathbf{G}}'^{\mathrm{T}} \left[\frac{k_m (z + a)}{z - z_m} z^{-N_m} \right]_M \right) \tag{3}$$

where k_m, a - coefficients; z_m- m-th pole of system function, $N_m T$ - duration of the m-th impulse function component, T - discrete sampling step.

All the mentioned constants, except N_m, depend on the transition method being applied. The values of the given coefficients for the three transition methods mentioned above are represented in the Table 9

№	Method	Meaning
1.	Method of differential equation discretization	$k_m = T$, $a = 0$, $z_{1m} = 1/(1 - \rho_{1m})$, $\rho_{1_m} = \ln(z_m)/T$
2.	Method of invariant impulse responses	$k_m = T$, $a = 0$, $z_{2m} = e^{\rho_m T}$, $\rho_{2_m} = \rho_m$
3.	Bilinear transformation	$k_m = T/(2 - \rho_m T)$, $a = 1$, $z_{3m} = (2 + \rho_m T)/(2 - \rho_m T)$, $\rho_{3_m} = \ln(z_{3m})/T$

Table 9. Transition methods

The additional indexes of the constants in the Table 9, corresponding to order numbers of transition methods, are given for the constants, which do not coincide with the analog filter-prototype parameters.

4. Performance express-analysis at modulated signal

4.1. Mathematical description of signals

The signals as a set of semi-infinite or finite damped oscillatory components were considered above. Compound signals of different forms, including compound periodical and quasi-periodic signals, nonstationary signals and signals with compound envelopes can be synthesized on the basis of the collection of components mentioned above [5,6]. The mentioned models also make it possible to describe the majority of impulse signals, which are widely applicable in radio engineering (radio pulse and video pulse).

The analysis methods considered above can be applied for that kind of signals.

However, the more general case is obviously more interesting, when signals with compound dependencies of envelopes and signal total phase are applied. Semi-infinite or finite signals with compound envelopes, the most frequently used signals in radio engineering, can be described by the following model

$$x(t) = \mathrm{Re}\left(\dot{X}_1(t)e^{p_1(t)t}\right) \tag{4}$$

or in general case it would be

$$x(t) = \mathrm{Re}\left(\dot{\mathbf{X}}(t)^{\mathrm{T}} e^{\mathbf{P}(t)(\mathrm{C}t-\mathrm{t})}\right) \tag{5}$$

For the considered signal models, for instance, of the signal (1),their finite duration can be specified by the finite duration of a complex amplitude $\dot{X}_1(t)$.

Using the model (1), signals with amplitude, phase and frequency modulation which are commonly used in radio engineering signals can be described. Among with the similar models input signals in sound processing systems, automation devices of power systems at electromechanical transients process can be described.

It is known, that using the signal (1), in case when amplitude, frequency or initial phase are time functions, it is impossible to define the law of amplitude variation and a filter initial phase by values of amplitude-frequency response and phase-frequency response on an input signal frequency [21,22]. Thus, the analysis methods considered above cannot be applied directly.

There are exceptional cases when at some variation laws $\dot{X}_1(t)$ and $p_1(t)$ signals of the type (2) can be transformed into signals, described by a set of semi-infinite or finite damped oscillatory components [5,6].

However, signals (1) and (2) can be decomposed into components of a set of damped oscillatory components using the Prony's method and its modifications [23]. Leaving behind the issue concerning decomposition into "real" components by using the mentioned methods it is important to note that even general signal approximation (1) and (2) by Prony's method for calculations enables to apply the considered filter analysis methods.

Another option for solution of the filter analysis at signal type (1) is connected to modification of the suggested analysis methods. Modification of the second analysis method is represented below.

4.2. Filter analysis

Let us consider IIR filter analysis at input signals (1), as well as for special cases, when amplitude $\dot{X}(t)=X_m(t)e^{-j\varphi}$ and phase $\dot{X}(t)=X_m e^{-j\varphi(t)}$ are corresponding to a modulation signal.

To develop the necessary dependences one can use expressions, obtained for IIR filter at injection on its input a set of finite damped oscillatory components. Let us decompose the signal into time steps, during which signal time dependent parameters are mostly constant. In this case time step Δt can be even and uneven.

In case of an even step of signal decomposition one will obtain the following expression to determine a filter ouput signal

$$y(t)=\mathrm{Re}\left(\sum_{n=0}^{N-1}\dot{X}_{n1}e^{p_{n1}n\Delta t}\left(K\left(p_{n1},t-n\Delta t\right)e^{p_{n1}(t-n\Delta t)}-e^{p_{n1}\Delta t}K\left(p_{n1},t-(n+1)\Delta t\right)e^{p_{n1}(t-(n+1)\Delta t)}\right)\right) \tag{6}$$

where $\dot{X}_{n1}=\dot{X}_1(n\Delta t)$, $p_{n1}=p_1(n\Delta t)$, $N=t_k/\Delta t$, t_k- duration of a finite signal (beginning of the signal coincides with zero reading on time).

As before, only algebraic operations with complex amplitudes, frequencies and values of time dependent transfer function on complex frequency of an input signal are applied for the filter analysis at the input signals type (1).

The same approach can be used for digital filters as well – one should replace continuous time t to discrete time kT, and instead of transfer function $K(p,\ t)$ of an analog filter one should use a transfer function $K(z,\ k)$in the expression. In this case, if one assumes Δt to be equal to discrete sampling step T, it enables to take into account the errors of analog-to-digital converter and finite digit capacity microprocessor in filter analysis.

The filter analysis on the basis of the expression (3) is approximate and can be considered as numerical method. To determine explicit dependencies one need to perform passage to the limit $\Delta t \to 0$.

The required dependency of input-output can be defined also by using the convolution integral by substitution of the expression for the input signal (1). Performing discretization with the passage to the limit, the following input-output dependency for an analog filter-prototype can be obtained

$$y(t) = \mathrm{Re}\left(\int_0^t \dot{X}_1(t-\tau)e^{p_1(\tau)t}K'(p_1(\tau),\tau)d\tau\right) \tag{7}$$

where

$$K'(p_1(t),t) = \frac{dK(p_1(t),t)}{d\tau} \tag{8}$$

In case of amplitude, phase modulation or their combination, the expression for input-output would significantly simplify

$$y(t) = \mathrm{Re}\left(\left(\int_0^t \dot{X}_1(t-\tau)K'(p_1,\tau)d\tau\right)e^{p_1 t}\right) \tag{9}$$

For IIR filter the following expression for a derivative of time dependent transfer function takes place

$$K'(p_1,t) = \frac{dK(p_1,t)}{d\tau} = \mathrm{Re}\left(\left[\dot{G}_m e^{-(p_1-p_m)t}\right]_M\right) \tag{10}$$

In case of $\dot{X}_1(t) = \dot{X}_1$ the input-output dependency coincides with input-output dependency for IIR filter obtained before (item 6 Table 1)

$$y(t) = \mathrm{Re}\left(\dot{X}_1 K(p_1,t)e^{p_1 t}\right) \tag{11}$$

It follows from the comparison of the input-output dependencies (4) and (5), that in the second case complex amplitude of IIR filter output signal on the input injection as a damped oscillatory component is determined by multiplication of a signal complex amplitude by the value of time dependent transfer function on the input signal frequency $\dot{Y}(t) = \dot{X}_1 K(p_1, t)$, in the first case more complicated dependency takes place $\dot{Y}(t) = \int_0^t \dot{X}_1(t-\tau)K'(p_1, \tau)d\tau$.

The dependencies enable to perform filter analysis, including performance analysis of signal processing, at input signals with a composite form. Solving problems of this kind is relevamt not only for radio engineering and communication systems, but also for other industries.

For the example of considered in the sections 3.2 and 3.3 analysis of frequency filters, which are used in PMU, according to a new version of IEEE C37.118.2 standard it is necessary to perform testing for the mentioned devices not only at stepwise change of amplitude and initial phase of sinusoidal component of input signal with commercial frequency, but also at PMU input signals at electromechanical transient process in a power system. In the mentioned regimes of power system operation in the controlled currents and voltage envelopes and total phases are time functions.

For FIR filter analysis at input actions (1) the input - output expression can be obtained on the basis of the dependences given in the Table 6. Such-like dependences for IIR filters and FIR filters can be obtained based on the dependences given before for the first analysis method.

5. The application of the analysis methods

Let us consider possible areas of application for the suggested performance analysis methods for signal processing by frequency filters, which are used in intelligent electronic devices of power systems, in automation devices, in radio engineering and communication systems, as well as in other fields of engineering where digital signal processing is commonly applied.

The prospectives of power system development in nearest future are related to technology Smart Grid implementation and the application of automatic control and regulation systems of a new generation. Power system control improvement involves a wide application of fast action IED based on synchronized measurement of current and voltage phasors of a fundamental harmonic on the basis of IEEE C37.118-2011 and IEC 61850-90-5 standards.

Up to date IED should ensure performance signal processing in conditions of an intense electromagnetic and electromechanical transients process. Mathematical model of an input signal IED in a normal and an accidental regimes of power systems in some cases can be represented by a set of semi-infinite or finite damped oscillatory components, in other cases – by analogous models, in which complex amplitudes and frequencies of mentioned components are time functions.

Most of IED power systems should ensure performance of signal processing at any possible combination of input signal parameters. The suggested analysis methods enable to solve effectively problems of determination of performance specifications for frequency filters, used in IED power systems. The examples for the performance analysis of signal processing by frequency filters and algorithms of signal processing for IED of power systems based on the phasor measurement technology are considered in the sections 3 and 4 of the present chapter. The example for IIR filter analysis for general devices of relay protection and automation is given in the section 2.

The analysis methods for analog IIR filters considered in the chapter can be applied also for the linear circuit analysis. It is important to note that for absolute majority of microprocessor control systems and measuring systems the information sources have analogue nature. In this case for analysis of controlled objects equivalent circuits based on linear electric circuits are used. The illustrative examples: equivalent circuit of power systems, power plants, electric grids and power-supply systems. Application of the suggested analysis methods enables a consistent approach for the regime analysis of the controlled object operation, analysis of analog filters-prototype and digital filters [5,6].

The suggested analysis methods may be applied for performance analysis of signal processing by frequency filters in up-to-date measuring devices, automation devices, radio devices, in communications systems, sound processing systems and other devices, where digital signal processing is commonly used.

Advantages and particularities of the suggested analysis methods are related to uniform analysis methods for analog filter-prototypes (IIR filters and FIR filters), as well as for digital filters.

The majority of pulse signals (radio and video pulses), which are commonly used in radio engineering, can be described by a set of finite damped oscillatory components. For performance analysis of pulse signal processing by IIR filters and FIR filters the analysis methods considered in sections 2 and 3 of the chapter can be applied.

The author suggests to use the methods, considered in the section 4 of the present chapter for performance analysis of modulated signal processing by filters.

The synthesis methods mentioned above can be also effectively applied for typical signal filtering problems, including problems of a useful signal extraction against the white noise.

In this case, the white noise realizations can be described as a set of time-shifted fast damping exponents of different digits. The initial values and appearance time of the mentioned exponential components are random variables, which variation law ensures the white noise to have specified spectral characteristics.

6. Performance analysis of signal processing as a step of filter synthesis

Guaranteeing the necessary quality of signal processing is utterly important at frequency filter synthesis. The application of the considered above approaches for the signal type (1) and the filter impulse characteristics (2) enables to reject the traditional approach, related to formulation the requirements to a filter amplitude-frequency response in different fields (band pass, stop band, transition region). In the case of filter synthesis, it is enough to lay down the requirements to the filter frequency characteristics on the basis of Laplace transform on complex frequencies of an input signal with the allowance for their change. Thus, according to the approach described above, the signal processing performance analysis is one of the steps of filter synthesis.

This approach enables to formalize the task of filter synthesis and to gain optimal solutions in combination with methods of multicriterion optimization with limitations[5]. Using methods of multicriterion optimization, by limitations values of filter frequency responses are ment and in some cases values of input signal spectrum based on the Laplace transform spectral representations on complex frequencies of impulse function of a filter and input signal [5,20].

Synthesized in that way filters ensure the specified performance specifications for signal processing at any possible variation of input signal parameters.

Among with frequency filter synthesis this approach gives an opportunity to perform time window synthesis for hort-time Fourier transform, as well as synthesis of father and mother wavelets for cases when an input signal can be represented by a set of semi-infinite or finite damped oscillatory components [5,20].

7. Conclusion

The effective methods of performance analysis for signal processing based on the Laplace transform spectral representations and the uniformity of mathematical models of input signals as well as the filter impulse functions as a sets of continuous/discrete semi-infinite or finite damped oscillatory components were developed for a direct determination of IIR filters and FIR filters performance specification. input signals as well as the filter impulse functions as a set of continuous/discrete semi-infinite or finite damped oscillatory components were developed. Simple semi-infinite harmonic, aperiodic signals, compound signals and impulse characteristics of any form can be synthesized on the basis of components set mentioned above, including signal composite envelopes, as well as pulse signals (radio and video pulses). The uniformity of the mathematical description of the signals and filters enables, on one hand, allows to employ both a consistent and a compact form of their characterization in the configuration of a set of complex amplitudes, complex frequencies and time parameters. On the other hand, it simplifies significantly performance analysis for signal processing by analog or digital filters at any possible useful signal and disturbance parameters variation by reducing significantly the amount of calculations.

The analysis methods can be used in case of mathematical models as well - where complex amplitudes and/or complex frequencies are time functions.

To simplify the task of the analysis, two methods are suggested for performance express-analysis of signal processing by frequency filters using filter frequency responses based on Laplace transform: frequency and frequency-time analysis methods.

The application of the suggested methods for performance analysis of signal processing as one of the steps of filters synthesis enables to automatize filter design for filters with low sensitivity to a signal parameter change within the specified range.

Author details

Alexey Mokeev

Northern (Arctic) Federal University, Russia

References

[1] Voronov, A. A. (1985). Basic principles of automatic control theory: Special linear and nonlinear systems. *Moscow: Mir Publishers*, 319.

[2] Nise, N. S. (2004). Control System Engineering. *NJ: John Wiley & Sons.*, 969.

[3] Ogata, K. (1997). Modern Control Engineering. *NJ: Prentice Hall.*

[4] Mokeev, A. V. (2007). Spectral expansion in coordinates of complex frequency application to analysis and synthesis filters. *Tampere: TICSP Report 37.*, 159-167.

[5] Mokeev, A. V. (2011). Application of spectral representations in coordinates of complex frequency for the digital filter analysis and synthesis. *Márquez F.P.G, editor. Digital Filters. Rijeka: InTech*, 27-52.

[6] Mokeev, A. V. (2008). Signal processing in intellectual electronic devices of electric power systems. 3, *The signal and system spectral expansion in coordinates of complex frequency. Arkhangelsk: ASTU*, 196.

[7] Mokeev, A. V. (2011). Quality analysis of signal processing using digital filters. *International Siberian IEEE Conference. Krasnoyarsk.*, 106-109.

[8] Kharkevich, A. A. (1960). Spectra and Analysis. *New York: Consultants Bureau.*, 222.

[9] Vanin, V. K., & Pavlov, G. M. (1991). Relay Protection of Computer Components. *Énergoatomizdat, Moscow.*

[10] Prévé, C. (2006). Protection of Electrical Networks. *AREVA: Mâcon*, 512.

[11] Maxfield, B. (2006). Engineering with MathCad: Using MathCad to Create and Organize Your Engineering Calculations. *Butterworth-Heinemann*, 494.

[12] Pritchard, P. (2011). Mathcad: A Tool for Engineering Problem Solving. *McGraw-Hill*, 203.

[13] Mokeev, A. V. (2008). Signal processing in intellectual electronic devices of electric power systems. 4, *The mathematical description of digital systems. Arkhangelsk: ASTU*, 201.

[14] Smith, S. W. (2002). Digital Signal Processing: A Practical Guide for Engineers and Scientists. *Newnes*, 672.

[15] Ifeachor, E. C., & Jervis, B. W. (2002). Digital Signal Processing: A Practical Approach. *2nd edition. Pearson Education.*, 933.

[16] Lyons, R. G. (2004). Understanding Digital Signal Processing. *Prentice Hall PTR.*, 665.

[17] Mokeev, A. V. (2009). Analysis of digital filters used for preprocessing of signals of relay protection. *Electromehanica.*, 4, 37-42.

[18] Mokeev, A. V. (2009). Frequency filters analysis on the basis of features of signal spectral representations in complex frequency coordinates. *Scientific and Technical Bulletin of SPbSPU.*, 2, 61-68.

[19] Hector, J., Altuve, Ferrer. H. J. A., Edmund, O., & Schweitzer, E. O. (2010). Modern Solutions for Protection, Control, and Monitoring of Electric Power Systems. *SEL.*, 400.

[20] Mokeev, A. V. (2011). Signal processing algorithms for intelligent electronic devices using phasor measurement technology. *In 2011 Proc. Int. Actual Trends in Development of Power System Protection and Automation (CIGRE-2011).*

[21] Fink, L. M. (1984). Signals, Interferences and Errors. *Moscow: Radio and Svyaz*, 256.

[22] Gonorovsky, I. S. (1981). Radio circuits and signals. *Moscow: Mir Publishers.*, 639.

[23] Marple, S. L. (1987). Digital Spectral Analysis with Application. *NJ: Prentice Hall.*

Chapter 5

Frequency Transformation for Linear State-Space Systems and Its Application to High-Performance Analog/Digital Filters

Shunsuke Koshita, Masahide Abe and
Masayuki Kawamata

Additional information is available at the end of the chapter

1. Introduction

Frequency transformation is one of the well-known techniques for design of analog and digital filters [1, 2]. This technique is based on variable substitution in a transfer function and allows us to easily convert a given prototype low-pass filter into any kind of frequency selective filter such as low-pass filters of different cutoff frequencies, high-pass filters, band-pass filters, and band-stop filters. It is also well-known that the transformed filters retain some properties of the prototype filter such as the stability and the shape of the magnitude response. For example, if a prototype filter is stable and has the Butterworth magnitude response, any filter given by the frequency transformation is also stable and of the Butterworth characteristic. Due to this useful fact, the frequency transformation is suitable not only to the filter design but also to the real-time tuning of cutoff frequencies, which can be applied to design of variable filters [3] and to adaptive notch filtering [4, 5]. Hence the frequency transformation plays important roles in many modern applications of signal processing from both the theoretical and practical points of view.

The purpose of this chapter is to provide further insights into the theory of frequency transformation from the viewpoint of *internal* properties of filters. In many textbooks on digital signal processing, the frequency transformation is discussed in terms of only the input-output properties, i.e. properties on the transfer function. In other words, few results have been reported about the relationship between the frequency transformation and the internal properties. As is well-known, the internal properties of filters are closely related to the problem of how we should construct a filter structure of a given transfer function, and this problem must be carefully considered in order to obtain analog filters of high dynamic range and low sensitivity [6–12] or digital filters of high accuracy with respect to finite wordlength effects [13–25]. Hence it is worthwhile to investigate the frequency transformation from the viewpoint of the internal properties, and to extend the results to some practical applications.

In order to discuss the frequency transformation from the viewpoint of the internal properties of filters, we make use of the state-space representation. The state-space representation is one of the well-known internal descriptions of linear systems and, in addition, it provides a powerful tool for synthesis of analog/digital filter structures with the aforementioned high-performance. The results from our discussion are twofold. First, we reveal many useful properties of frequency transformation in terms of the state-space representation. The properties to be presented here are closely related to the following three elements of linear state-space systems: the controllability Gramian, the observability Gramian, and the second-order modes. These three elements are known to be very important in characterization of internal properties of analog/digital filters and synthesis of high-performance filter structures. Second, we apply this result to the technique of design and synthesis of analog and digital filters with high performance structures. To be more specific, we present simple and unified frameworks for design and synthesis of analog/digital filters that simultaneously realize the change of frequency characteristics and attain the aforementioned high-performance. Furthermore, we extend this result to variable filters with high-performance structures.

The chapter is organized as follows. Section 2 reviews the fundamentals of the state-space representation of linear systems, including analog filters and digital filters. Section 3 introduces the classical theory of frequency transformation. Sections 4 and 5 are the main theme of this chapter. In Section 4 we discuss the frequency transformation by using the state-space representation and reveal insightful relationships between the frequency transformation and the internal properties of filters. In Section 5 we extend this theory and present new useful methods for design and synthesis of high-performance analog/digital filters.

2. State-space representation, Gramians and second-order modes

In this section we introduce state-space representation of linear systems. In addition, we introduce the aforementioned three elements on the internal properties—controllability Gramian, observability Gramian, and second-order modes—and we address how these elements are applied to synthesis of high-performance filter structures. We will present these topics for digital filters and analog filters, respectively.

2.1. State-space representation of digital filters

Consider the following state-space equations for an N-th order stable single-input/single-output linear discrete-time system:

$$\begin{aligned} \boldsymbol{x}(n+1) &= \boldsymbol{A}\boldsymbol{x}(n) + \boldsymbol{b}u(n) \\ y(n) &= \boldsymbol{c}\boldsymbol{x}(n) + du(n) \end{aligned} \tag{1}$$

where $u(n)$, $y(n)$ and $\boldsymbol{x}(n) \in \Re^{N\times 1}$ denote the scalar input, the scalar output and the state vector, respectively, and $\boldsymbol{A} \in \Re^{N\times N}, \boldsymbol{b} \in \Re^{N\times 1}, \boldsymbol{c} \in \Re^{1\times N}$ and $d \in \Re^{1\times 1}$ are constant coefficients. Throughout this chapter we assume that the system is stable, controllable and observable. If this state-space system represents a digital filter, each entry of $\boldsymbol{x}(n)$

corresponds to each output of delay elements of the filter. Taking the z-transform of (1), we have

$$\begin{aligned} z\boldsymbol{X}(z) &= \boldsymbol{A}\boldsymbol{X}(z) + \boldsymbol{b}U(z) \\ Y(z) &= \boldsymbol{c}\boldsymbol{X}(z) + dU(z) \end{aligned} \tag{2}$$

from which the transfer function $H(z)$ is described in terms of $(\boldsymbol{A}, \boldsymbol{b}, \boldsymbol{c}, d)$ as

$$H(z) = d + \boldsymbol{c}(z\boldsymbol{I}_N - \boldsymbol{A})^{-1}\boldsymbol{b} \tag{3}$$

where $\boldsymbol{I}_N$ denotes the $N \times N$ identity matrix.

It is well-known that the transfer function $H(z)$ is invariant under nonsingular transformation matrices $\boldsymbol{T} \in \Re^{N \times N}$ of the state: if $\boldsymbol{x}(n)$ is transformed into $\overline{\boldsymbol{x}}(n) = \boldsymbol{T}^{-1}\boldsymbol{x}(n)$, then the state-space system $(\boldsymbol{A}, \boldsymbol{b}, \boldsymbol{c}, d)$ is also transformed into the following set $(\overline{\boldsymbol{A}}, \overline{\boldsymbol{b}}, \overline{\boldsymbol{c}}, \overline{d})$:

$$(\overline{\boldsymbol{A}}, \overline{\boldsymbol{b}}, \overline{\boldsymbol{c}}, \overline{d}) = (\boldsymbol{T}^{-1}\boldsymbol{A}\boldsymbol{T}, \boldsymbol{T}^{-1}\boldsymbol{b}, \boldsymbol{c}\boldsymbol{T}, d). \tag{4}$$

It is easy to show that the transfer function of this new set is the same as that of $(\boldsymbol{A}, \boldsymbol{b}, \boldsymbol{c}, d)$. Therefore, many structures exist for a digital filter with a given transfer function $H(z)$. This nonsingular transformation is called similarity transformation.

We next introduce the controllability Gramian, the observability Gramian, and the second-order modes. For the system $(\boldsymbol{A}, \boldsymbol{b}, \boldsymbol{c}, d)$, the solutions $\boldsymbol{K}$ and $\boldsymbol{W}$ to the following Lyapunov equations are called the controllability Gramian and the observability Gramian, respectively:

$$\begin{aligned} \boldsymbol{K} &= \boldsymbol{A}\boldsymbol{K}\boldsymbol{A}^T + \boldsymbol{b}\boldsymbol{b}^T \\ \boldsymbol{W} &= \boldsymbol{A}^T\boldsymbol{W}\boldsymbol{A} + \boldsymbol{c}^T\boldsymbol{c}. \end{aligned} \tag{5}$$

The Gramians $\boldsymbol{K}$ and $\boldsymbol{W}$ are symmetric and positive definite, i.e. $\boldsymbol{K} = \boldsymbol{K}^T > 0$ and $\boldsymbol{W} = \boldsymbol{W}^T > 0$, because the system $(\boldsymbol{A}, \boldsymbol{b}, \boldsymbol{c}, d)$ is assumed to be stable, controllable and observable. Then, the eigenvalues of the matrix product $\boldsymbol{K}\boldsymbol{W}$ are all positive. We denote these eigenvalues as $\theta_1^2, \theta_2^2, \cdots, \theta_N^2$ and assume that $\theta_1^2 \geq \theta_2^2 \geq \cdots \geq \theta_N^2$. Their positive square roots $\theta_1 \geq \theta_2 \geq \cdots \geq \theta_N$ are called the second-order modes of the system. In the literature on control system theory, the second-order modes are also called Hankel singular values because $\theta_1, \theta_2, \cdots, \theta_N$ are equal to the nonzero singular values of the Hankel operator of $H(z)$.

The two Gramians and the similarity transformation $\overline{\boldsymbol{x}}(n) = \boldsymbol{T}^{-1}\boldsymbol{x}(n)$ are simply related as follows: the controllability/observability Gramians $(\overline{\boldsymbol{K}}, \overline{\boldsymbol{W}})$ of the system in (4) are given by

$$(\overline{\boldsymbol{K}}, \overline{\boldsymbol{W}}) = (\boldsymbol{T}^{-1}\boldsymbol{K}\boldsymbol{T}^{-T}, \boldsymbol{T}^T\boldsymbol{W}\boldsymbol{T}). \tag{6}$$

On the other hand, the second-order modes are invariant under similarity transformation because of the following relationship

$$\overline{KW} = T^{-1}(KW)T. \tag{7}$$

Hence it follows that the Gramians depend on realizations of the system, while the second-order modes depend only on the transfer function.

In the literature on synthesis of filter structures [13–25], it is shown that the two Gramians and the second-order modes play central roles in analysis and optimization of filter performance such as the roundoff noise and the coefficient sensitivity. In other words, given the transfer function of a digital filter, we can formulate some cost functions with respect to the aforementioned filter performance in terms of the two Gramians $(\boldsymbol{K}, \boldsymbol{W})$, and a filter structure of high performance can be obtained by constructing the two Gramians appropriately in such a manner that they optimize or sub-optimize the corresponding cost functions.

An example of high-performance digital filter structures is the balanced form [15, 16, 18, 23, 25]. This form consists of the two Gramians given by

$$\boldsymbol{K} = \boldsymbol{W} = \boldsymbol{\Theta} \tag{8}$$

where Θ is the diagonal matrix consisting of the second-order modes, i.e.

$$\boldsymbol{\Theta} = \mathrm{diag}(\theta_1, \theta_2, \cdots, \theta_N). \tag{9}$$

Another example is the minimum roundoff noise structure [13, 14, 16, 17], which consists of the two Gramians that satisfy the following relationships

$$\begin{aligned} \boldsymbol{W} &= \left(\frac{1}{N}\sum_{i=1}^{N}\theta_i\right)^2 \boldsymbol{K} \\ K_{ii} &= 1 \end{aligned} \tag{10}$$

where K_{ii} denotes the i-th diagonal entry of $\boldsymbol{K}$.

Finally, we address the significance of the second-order modes from two practical aspects. First, it is known in the literature that the second-order modes describe the optimal values of the aforementioned cost functions. Therefore, it follows that the optimal performance is determined by the second-order modes of a given transfer function. Another important feature of the second-order modes can be seen in the field of the balanced model reduction [26–28], where it is shown that the second-order modes provide the upper bound of the approximation error between the reduced-order system and the original system.

2.2. State-space representation of analog filters

An N-th order linear continuous-time system (including analog filter) can be described by the following state-space representation

$$\begin{aligned}\frac{dx(t)}{dt} &= Ax(t) + bu(t) \\ y(t) &= cx(t) + du(t)\end{aligned} \tag{11}$$

where $u(t)$, $y(t)$ and $x(t) \in \Re^{N\times 1}$ are the scalar input, the scalar output and the state vector of the system, respectively, and $A \in \Re^{N\times N}, b \in \Re^{N\times 1}, c \in \Re^{1\times N}$ and $d \in \Re^{1\times 1}$ are constant coefficients. The system (A, b, c, d) is assumed to be stable, controllable and observable. If this system represents a continuous-time analog filter that comprises N integrators, the state vector corresponds to the output signals of these integrators.

Taking the Laplace transform of (11) leads to

$$\begin{aligned}sX(s) &= AX(s) + bU(s) \\ Y(s) &= cX(s) + dU(s),\end{aligned} \tag{12}$$

which results in the following transfer function

$$H(s) = d + c(sI_N - A)^{-1}b. \tag{13}$$

As similar to the discrete-time case, the transfer function is invariant under similarity transformation: if $x(t)$ is transformed by a nonsingular matrix $T \in \Re^{N\times N}$ into $T^{-1}x(t)$, then the new state-space system $(T^{-1}AT, T^{-1}b, cT, d)$ is an equivalent realization to (A, b, c, d) of the transfer function $H(s)$. Therefore, many circuit topologies exist for an analog filter with a given transfer function $H(s)$.

The controllability Gramian K and the observability Gramian W of a continuous-time state-space system are respectively obtained as the solutions to the following Lyapunov equations:

$$\begin{aligned}AK + KA^T + bb^T &= 0_{N\times N} \\ A^TW + WA + c^Tc &= 0_{N\times N}\end{aligned} \tag{14}$$

where $0_{N\times N}$ denotes the $N \times N$ zero matrix. By the assumption of the stability, controllability and observability of (A, b, c, d), the Gramians K and W are shown to be symmetric and positive definite. Then, as in the discrete-time case, the second-order modes $\theta_1, \theta_2, \cdots, \theta_N$ are obtained as the positive square roots of the eigenvalues of KW.

The relationship of similarity transformations to the Gramians and the second-order modes in the continuous-time case is the same as that in the discrete-time case. The new Gramians $(\overline{K}, \overline{W})$ of the transformed continuous-time system given by a similarity transformation T are shown to be $(T^{-1}KT^{-T}, T^TWT)$, and thus the Gramians depend on realizations of the system. On the other hand, the second-order modes are invariant because $\overline{KW} = T^{-1}(KW)T$ holds.

As in the discrete-time case, the Gramians and the second-order modes of continuous-time systems play important roles in synthesis of filter structures of high performance [6–12]. A high-performance structure can be obtained by optimizing or sub-optimizing a prescribed cost function in terms of the controllability and observability Gramians. Such a cost function can be seen as a measure of the dynamic range and the sensitivity of an analog filter. In addition, the optimal values of such cost functions are determined by the second-order modes.

3. Frequency transformation

3.1. Frequency transformation of digital filters

Frequency transformation of digital filters can be seen in the work of Oppenheim [29] and Constantinides [2]. The work of Oppenheim is applied to finite impulse response (FIR) transfer functions, whereas the work of Constantinides is applied to infinite impulse response (IIR) transfer functions. In this chapter, the frequency transformation of digital filters is restricted to the work of Constantinides.

Now let $H(z)$ be the transfer function of a given N-th order digital low-pass filter. The frequency transformation in the discrete-time case is defined as

$$H(F(z)) = H(z)|_{z^{-1} \leftarrow 1/F(z)} \tag{15}$$

which results in a new composite transfer function $H(F(z))$. The function $1/F(z)$ for this transformation is defined as an M-th order stable all-pass function of the form

$$\begin{aligned} \frac{1}{F(z)} &= \pm z^{-M} \frac{G(z^{-1})}{G(z)} \\ G(z) &= 1 + \sum_{k=1}^{M} g_k z^{-k}. \end{aligned} \tag{16}$$

The well-known typical frequency transformations make use of the following four types of all-pass functions

$$\begin{aligned} \frac{1}{F_{\text{LP}}(z)} &= \frac{z^{-1} - \xi}{1 - \xi z^{-1}} \\ \frac{1}{F_{\text{HP}}(z)} &= -\frac{z^{-1} + \xi}{1 + \xi z^{-1}} \\ \frac{1}{F_{\text{BP}}(z)} &= -\frac{z^{-2} - \frac{2\xi\eta}{\eta+1} z^{-1} + \frac{\eta-1}{\eta+1}}{1 - \frac{2\xi\eta}{\eta+1} z^{-1} + \frac{\eta-1}{\eta+1} z^{-2}} \\ \frac{1}{F_{\text{BS}}(z)} &= \frac{z^{-2} - \frac{2\xi\eta}{1+\eta} z^{-1} + \frac{1-\eta}{1+\eta}}{1 - \frac{2\xi\eta}{1+\eta} z^{-1} + \frac{1-\eta}{1+\eta} z^{-2}} \end{aligned} \tag{17}$$

which respectively correspond to the low-pass-low-pass (LP-LP), low-pass-high-pass (LP-HP), low-pass-band-pass (LP-BP) and low-pass-band-stop (LP-BS) transformations. The parameters ξ and η determine the cutoff frequencies of the transformed filters. On the block diagram of a digital filter, the frequency transformation means that each delay element z^{-1} in $H(z)$ is replaced[1] with an all-pass filter $1/F(z)$.

3.2. Frequency transformation of analog filters

Let $H(s)$ be the transfer function of a given N-th order analog low-pass filter. The frequency transformation of analog filters is defined as the following variable substitution [1]

$$H(F(s)) = H(s)|_{s^{-1} \leftarrow 1/F(s)}. \tag{18}$$

Hence the frequency transformation yields a new composite transfer function $H(F(s))$ from the prototype transfer function $H(s)$. In general, the cutoff frequency of the prototype low-pass filter is set to be 1 rad/s. From a circuit point of view, the substitution $s^{-1} \leftarrow 1/F(s)$ means that each integrator $1/s$ in the prototype filter $H(s)$ is replaced with another system with the transfer function $1/F(s)$.

The transformation function $1/F(s)$ is defined as the following Foster reactance function [1]

$$\frac{1}{F(s)} = \frac{z(s)}{p(s)} = G\frac{(s^2+\omega_{z1}^2)(s^2+\omega_{z2}^2)(s^2+\omega_{z3}^2)\cdots}{s(s^2+\omega_{p1}^2)(s^2+\omega_{p2}^2)(s^2+\omega_{p3}^2)\cdots} \tag{19}$$

where $G > 0$ and $0 \le \omega_{z1} < \omega_{p1} < \omega_{z2} < \omega_{p2} < \omega_{z3} < \omega_{p3} < \cdots$. The Foster reactance functions are determined in such a manner that the degree of difference of $p(s)$ and $z(s)$ is 1, i.e. $|\deg p(s) - \deg z(s)| = 1$. In the case of the well-known typical LP-LP, LP-HP, LP-BP and LP-BS transformations, the reactance functions are respectively given by

$$\begin{aligned}
\frac{1}{F_{LP}(s)} &= \frac{G}{s} \\
\frac{1}{F_{HP}(s)} &= Gs \\
\frac{1}{F_{BP}(s)} &= \frac{Gs}{s^2+\omega_{p1}^2} \\
\frac{1}{F_{BS}(s)} &= \frac{G(s^2+\omega_{z1}^2)}{s}.
\end{aligned} \tag{20}$$

The parameters G, ω_p and ω_s determine the cutoff frequencies of the transformed filters.

---

[1] To be precise, replacing z^{-1} with another transfer function often yields a delay-free loop. In this case, some extra processing such as reformulation of the coefficients of the transformed filter is required after this replacement.

It is important to note that the Foster reactance functions are classified into two categories—strictly proper reactance functions and improper reactance functions[2]. In the typical frequency transformations of (20), $1/F_{\text{LP}}(s)$ and $1/F_{\text{BP}}(s)$ correspond to strictly proper reactance functions, whereas $1/F_{\text{HP}}(s)$ and $1/F_{\text{BS}}(s)$ are improper reactance functions.

4. State-space analysis of frequency transformation

In this section, we discuss the frequency transformation from the viewpoint of the internal properties. In other words, we show many interesting results of the frequency transformation in terms of the state-space representation.

This research has its roots in the work of Mullis and Roberts [30], where they presented a simple state-space formulation of frequency transformation for digital filters and they proved an important property of the second-order modes—they are invariant under frequency transformation. In addition, they provided practical impacts of these results on the design and synthesis of high-performance digital filters.

In this chapter we start with introducing this work, and then we further extend this result and present other theoretical results on the relationship between the frequency transformation and the state-space representation of discrete-time systems. In addition, we also present similar results for continuous-time systems.

4.1. State-space formulation of frequency transformation for digital filters and invariance of second-order modes

Mullis and Roberts [30] first presented an explicit state-space representation of frequency transformation as follows. Let $(\boldsymbol{A}, \boldsymbol{b}, \boldsymbol{c}, d)$ be a state-space representation of a given prototype filter $H(z)$. Then, the transfer function $H(F(z))$ that is given by the frequency transformation (15) with an M-th order all-pass function $1/F(z)$ can be explicitly described by

$$H(F(z)) = \mathcal{D} + \mathcal{C}(z\boldsymbol{I}_{MN} - \mathcal{A})^{-1}\mathcal{B} \tag{21}$$

with the following coefficients

$$\begin{aligned}
\mathcal{A} &= \boldsymbol{I}_N \otimes \boldsymbol{\alpha} + [\boldsymbol{A}(\boldsymbol{I}_N - \delta\boldsymbol{A})^{-1}] \otimes (\boldsymbol{\beta}\boldsymbol{\gamma}) \\
\mathcal{B} &= [(\boldsymbol{I}_N - \delta\boldsymbol{A})^{-1}\boldsymbol{b}] \otimes \boldsymbol{\beta} \\
\mathcal{C} &= [\boldsymbol{c}(\boldsymbol{I}_N - \delta\boldsymbol{A})^{-1}] \otimes \boldsymbol{\gamma} \\
\mathcal{D} &= d + \delta\boldsymbol{c}(\boldsymbol{I}_N - \delta\boldsymbol{A})^{-1}\boldsymbol{b}
\end{aligned} \tag{22}$$

where $(\boldsymbol{\alpha}, \boldsymbol{\beta}, \boldsymbol{\gamma}, \delta)$ is an arbitrary state-space representation of $1/F(z)$, and $\otimes$ stands for the Kronecker product for matrices.

[2] A rational function $G(s) = N(s)/D(s)$ is called strictly proper if $\deg N(s) < \deg D(s)$. On the other hand, $G(s)$ is called improper if $\deg N(s) > \deg D(s)$. Since the Foster reactance functions given by (19) always satisfy $|\deg p(s) - \deg z(s)| = 1$, there does not exist any reactance function such that $\deg p(s) = \deg z(s)$.

The significance of the description given by (22) lies in the fact that, by using this description, we can easily carry out the frequency transformation on a state-space structure as well as a transfer function. Also, note that this description does not include any delay-free loop.

In addition to the above state-space formulation, Mullis and Roberts also described the Gramians and the second-order modes of the transformed system $(\mathcal{A}, \mathcal{B}, \mathcal{C}, \mathcal{D})$. The two Gramians, which are respectively denoted by $\mathcal{K}$ and $\mathcal{W}$, are given as follows:

$$\begin{aligned} \mathcal{K} &= K \otimes Q \\ \mathcal{W} &= W \otimes Q^{-1} \end{aligned} \tag{23}$$

where Q is the controllability Gramian of the all-pass system $(\alpha, \beta, \gamma, \delta)$. From this relationship we easily see

$$\mathcal{K}\mathcal{W} = (KW) \otimes I_M \tag{24}$$

which means that the matrix product $\mathcal{K}\mathcal{W}$ have the same eigenvalues as KW with multiplicity M. This shows that the second-order modes of transformed filters are the same as those of a given prototype filter. Hence the second-order modes of digital filters are invariant under frequency transformation.

The practical benefit of this invariance property is discussed as follows. As stated in Section 2, the second-order modes determine the optimal values of cost functions with respect to finite wordlength effects. In [30], using the fact that the minimum roundoff noise is characterized by the second-order modes, it was proved that the minimum attainable value of the roundoff noise of digital filters is independent of the filter characteristics that are controlled by the frequency transformation. A similar conclusion can be drawn for the balanced model reduction: the upper bound of the approximation error due to the balanced model reduction is invariant under frequency transformation.

Furthermore, in the case of the LP-LP transformation, the work of [30] also presents the specific state-space-based frequency transformation that can preserve the optimal realizations. This specific transformation is given by

$$\begin{aligned} \mathcal{A} &= (\xi I_N + A)(I_N + \xi A)^{-1} \\ \mathcal{B} &= \sqrt{1-\xi^2}(I_N + \xi A)^{-1} b \\ \mathcal{C} &= \sqrt{1-\xi^2} c (I_N + \xi A)^{-1} \\ \mathcal{D} &= d - \xi c (I_N + \xi A)^{-1} b. \end{aligned} \tag{25}$$

By setting the prototype state-space filter (A, b, c, d) to be the optimal realization and applying (25), we can obtain arbitrary low-pass filters that have the same optimal realization as the prototype filter.

In the rest of this section, we will provide our results that are derived by further extending these results.

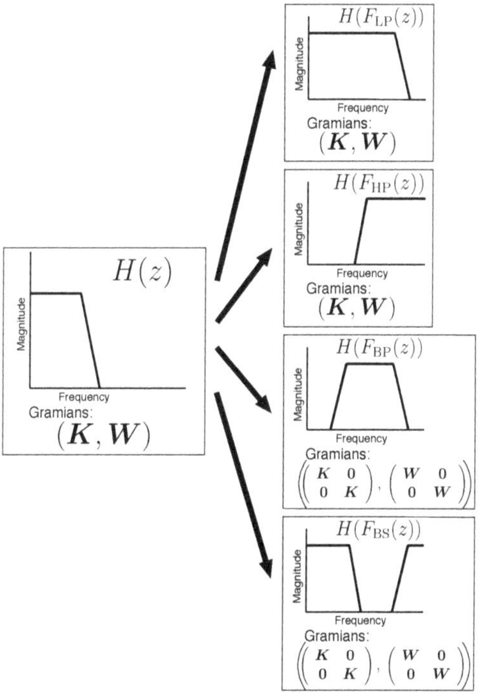

Figure 1. Gramian-preserving frequency transformation.

4.2. Gramian-preserving frequency transformation for digital filters

Here we pay special attention to the controllability and observability Gramians, and we provide a new state-space formulation of frequency transformation that can keep these Gramians invariant. This new state-space-based frequency transformation is called the Gramian-preserving frequency transformation [31] and includes the formulation of (25) as a special case.

Before showing the mathematical formulation of the Gramian-preserving frequency transformation, we first discuss how the Gramian-preserving frequency transformation is related to design and synthesis of digital filters. Simple examples for design/synthesis of low-pass, high-pass, band-pass and band-stop filters are given in Fig. 1. Here, suppose that we are given a prototype low-pass filter with the transfer function $H(z)$, as shown at the left of this figure. Also, let the controllability/observability Gramians of this prototype filter be $\boldsymbol{K}$ and $\boldsymbol{W}$, respectively. Then, by applying the Gramian-preserving frequency transformation to this prototype filter, we can convert this filter into other arbitrary low-pass, high-pass, band-pass and band-stop filters that consist of the *same* controllability/observability Gramians as those of the prototype filter[3]. Now, recalling that high-performance structures can be obtained by appropriate choice of the Gramians, we notice that the Gramian-preserving frequency transformation is a very powerful technique for simultaneous design and synthesis of high-performance digital fitlers. That is, if we prepare the structure of a given prototype low-pass filter as a high-performance one such as the balanced form and the minimum roundoff noise form, the Gramian-preserving frequency transformation enables us to obtain other types of filters with the same high-performance structure. This fact is also true for analog filters, as will be shown later in the next subsection.

We now present the mathematical formulation of the Gramian-preserving frequency transformation. Given a prototype state-space digital filter $(\boldsymbol{A}, \boldsymbol{b}, \boldsymbol{c}, d)$ with the transfer function $H(z)$ and an M-th order all-pass function $1/F(z)$, the following description provides the Gramian-preserving frequency transformation to produce the composite transfer function $H(F(z))$:

$$\begin{aligned}
\widetilde{\mathcal{A}} &= \widetilde{\boldsymbol{\alpha}} \otimes \boldsymbol{I}_N + (\widetilde{\boldsymbol{\beta}}\widetilde{\boldsymbol{\gamma}}) \otimes [\boldsymbol{A}(\boldsymbol{I}_N - \widetilde{\delta}\boldsymbol{A})^{-1}] \\
\widetilde{\mathcal{B}} &= \widetilde{\boldsymbol{\beta}} \otimes [(\boldsymbol{I}_N - \widetilde{\delta}\boldsymbol{A})^{-1}\boldsymbol{b}] \\
\widetilde{\mathcal{C}} &= \widetilde{\boldsymbol{\gamma}} \otimes [\boldsymbol{c}(\boldsymbol{I}_N - \widetilde{\delta}\boldsymbol{A})^{-1}] \\
\widetilde{\mathcal{D}} &= d + \widetilde{\delta}\boldsymbol{c}(\boldsymbol{I}_N - \widetilde{\delta}\boldsymbol{A})^{-1}\boldsymbol{b}
\end{aligned} \tag{26}$$

where the set $(\widetilde{\boldsymbol{\alpha}}, \widetilde{\boldsymbol{\beta}}, \widetilde{\boldsymbol{\gamma}}, \widetilde{\delta})$ is a state-space representation of $1/F(z)$ with the controllability/observability Gramians equal to the identity matrix, i.e.

$$\widetilde{\boldsymbol{\alpha}}\widetilde{\boldsymbol{\alpha}}^T + \widetilde{\boldsymbol{\beta}}\widetilde{\boldsymbol{\beta}}^T = \widetilde{\boldsymbol{\alpha}}^T\widetilde{\boldsymbol{\alpha}} + \widetilde{\boldsymbol{\gamma}}^T\widetilde{\boldsymbol{\gamma}} = \boldsymbol{I}_M. \tag{27}$$

This relationship means that the set $(\widetilde{\boldsymbol{\alpha}}, \widetilde{\boldsymbol{\beta}}, \widetilde{\boldsymbol{\gamma}}, \widetilde{\delta})$ is a balanced form. It should be noted that such a set always exists if $1/F(z)$ is stable.

Now we turn our attention to the mathematical formulation of the Gramians of $(\widetilde{\mathcal{A}}, \widetilde{\mathcal{B}}, \widetilde{\mathcal{C}}, \widetilde{\mathcal{D}})$, which are respectively denoted by $\widetilde{\mathcal{K}}$ and $\widetilde{\mathcal{W}}$. They are given in terms of the Gramians of the prototype filter as follows:

$$\begin{aligned}
\widetilde{\mathcal{K}} &= \boldsymbol{I}_M \otimes \boldsymbol{K} \\
\widetilde{\mathcal{W}} &= \boldsymbol{I}_M \otimes \boldsymbol{W}
\end{aligned} \tag{28}$$

which means that $\widetilde{\mathcal{K}}$ and $\widetilde{\mathcal{W}}$ become block diagonal matrices with M diagonal blocks all equal to $\boldsymbol{K}$ and $\boldsymbol{W}$. Therefore, as stated earlier, $\widetilde{\mathcal{K}}$ and $\widetilde{\mathcal{W}}$ respectively become the same as $\boldsymbol{K}$ and $\boldsymbol{W}$ with multiplicity M. Hence (26) preserves the Gramians under frequency transformation.

---

[3] In the case of LP-BP and LP-BS transformations, the transformed filters have the same Gramians with multiplicity 2 as those of the prototype filter. This is because the all-pass functions $1/F_{BP}(z)$ and $1/F_{BS}(z)$ are second-order functions and the order of $H(F_{BP}(z))$ and $H(F_{BS}(z))$ become twice as high as that of $H(z)$.

We next discuss the Gramian-preserving frequency transformation from a realization point of view. From (27), we first see that realization of the Gramian-preserving frequency transformation requires us to construct the structure of the all-pass filter $1/F(z)$ appropriately such that its state-space representation becomes a balanced form. Although formulation of the balanced form is known to be non-unique for a given transfer function, we presented a useful technique [31]: given an all-pass transfer function $1/F(z)$, its normalized lattice structure becomes a balanced form, which enables us to realize the Gramian-preserving frequency transformation. This is derived from the fact that $1/F(z)$ is all-pass. Now, recall that the frequency transformation of digital filters means that each delay element in a prototype filter is replaced with an all-pass filter (and delay-free loops, if any, are eliminated after this replacement)[4]. In view of this, we can conclude that the Gramian-preserving frequency transformation is interpreted as the replacement of each delay element in the prototype filter with the all-pass filter that has the normalized lattice structure. Figure 2 illustrates this scheme. Given a state-space prototype filter as in Fig. 2(a), we carry out the aforementioned replacement and we obtain the transformed state-space filter as in Fig. 2(b). The all-pass filter that is included in this structure consists of M lattice sections $\Phi_1, \cdots, \Phi_M$, and each section Φ_i is given as in Fig. 2(c). The variable ξ_i for $1 \leq i \leq M$ denotes the i-th lattice coefficient for $1/F(z)$, and $\widehat{\xi}_i = \sqrt{1-\xi_i^2}$.

Finally, we provide the mathematical formulation of the Gramian-preserving frequency transformation based on the normalized lattice structure. The normalized lattice structure of $1/F(z)$ can be given by the following state-space representation:

$$
\begin{aligned}
\widetilde{\alpha} &= \begin{pmatrix}
-\xi_1 & -\widehat{\xi}_1\xi_2 & -\widehat{\xi}_1\widehat{\xi}_2\xi_3 & \cdots & -\widehat{\xi}_1\widehat{\xi}_2\widehat{\xi}_3\cdots\widehat{\xi}_{M-3}\xi_{M-2} & -\widehat{\xi}_1\widehat{\xi}_2\widehat{\xi}_3\cdots\widehat{\xi}_{M-2}\xi_{M-1} & -\widehat{\xi}_1\widehat{\xi}_2\widehat{\xi}_3\cdots\widehat{\xi}_{M-1}\xi_M \\
\widehat{\xi}_1 & -\xi_1\xi_2 & -\xi_1\widehat{\xi}_2\xi_3 & \cdots & -\xi_1\widehat{\xi}_2\widehat{\xi}_3\cdots\widehat{\xi}_{M-3}\xi_{M-2} & -\xi_1\widehat{\xi}_2\widehat{\xi}_3\cdots\widehat{\xi}_{M-2}\xi_{M-1} & -\xi_1\widehat{\xi}_2\widehat{\xi}_3\cdots\widehat{\xi}_{M-1}\xi_M \\
0 & \widehat{\xi}_2 & -\xi_2\xi_3 & \cdots & -\xi_2\widehat{\xi}_3\widehat{\xi}_4\cdots\widehat{\xi}_{M-3}\xi_{M-2} & -\xi_2\widehat{\xi}_3\widehat{\xi}_4\cdots\widehat{\xi}_{M-2}\xi_{M-1} & -\xi_2\widehat{\xi}_3\widehat{\xi}_4\cdots\widehat{\xi}_{M-1}\xi_M \\
\vdots & \vdots & \vdots & \ddots & \vdots & \vdots & \vdots \\
0 & 0 & 0 & \cdots & 0 & \widehat{\xi}_{M-1} & -\xi_{M-1}\xi_M
\end{pmatrix} \\
\widetilde{\beta} &= \begin{pmatrix}
\widehat{\xi}_1\widehat{\xi}_2\widehat{\xi}_3\cdots\widehat{\xi}_{M-1}\widehat{\xi}_M \\
\xi_1\widehat{\xi}_2\widehat{\xi}_3\cdots\widehat{\xi}_{M-1}\widehat{\xi}_M \\
\xi_2\widehat{\xi}_3\cdots\widehat{\xi}_{M-1}\widehat{\xi}_M \\
\vdots \\
\xi_{M-2}\widehat{\xi}_{M-1}\widehat{\xi}_M \\
\xi_{M-1}\widehat{\xi}_M
\end{pmatrix} \\
\widetilde{\gamma} &= \begin{pmatrix} 0 & 0 & 0 & \cdots & 0 & \pm\widehat{\xi}_M \end{pmatrix} \\
\widetilde{\delta} &= \pm\xi_M.
\end{aligned} \tag{29}
$$

Therefore, substitution of (29) into (26) carries out the Gramian-preserving frequency transformation. Note that the state-space representation $(\widetilde{\mathcal{A}}, \widetilde{\mathcal{B}}, \widetilde{\mathcal{C}}, \widetilde{\mathcal{D}})$ given in this way becomes sparse due to many zero entries in $\widetilde{\alpha}$ and $\widetilde{\gamma}$. To be precise, the set $(\widetilde{\mathcal{A}}, \widetilde{\mathcal{B}}, \widetilde{\mathcal{C}}, \widetilde{\mathcal{D}})$

[4] Note that the mathematical formulation of the Gramian-preserving frequency transformation (26) is derived after elimination of delay-free loops. Therefore, (26) does not have the problem of delay-free loops. See [30] for the details.

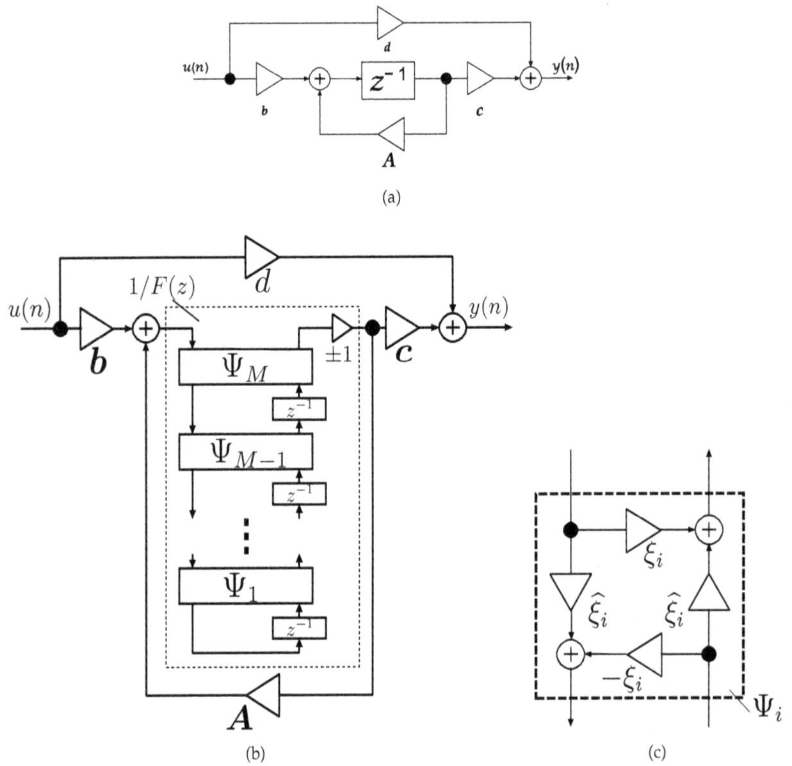

Figure 2. Gramian-preserving frequency transformation: (a) prototype state-space filter, (b) transformed state-space filter, and (c) a normalized lattice section Ψ_i.

has in total $(M-1)N(MN-M/2)$ zero entries. Hence this state-space filter is very suitable to implementation.

4.3. Results for analog filters

In the case of analog filters, little had been reported about the state-space analysis of frequency transformation. On the other hand, our work [32–34] has derived many results that are similar to the discrete-time case. Here we will introduce these results.

We first present a state-space formulation of frequency transformation for analog filters. One thing to be noted here is that, as stated in Section 3.2, the frequency transformation functions (i.e. Foster reactance functions) are classified into strictly proper functions and improper functions. In this chapter we focus on the case of strictly proper reactance functions, which include the LP-LP and the LP-BP transformations.

Now consider a state-space representation (A, b, c, d) of a given prototype low-pass filter with the transfer function $H(s)$. Also, let $(\mathcal{A}, \mathcal{B}, \mathcal{C}, \mathcal{D})$ be a state-space representation of $H(F(s))$, where $1/F(s)$ denotes a strictly proper Foster reactance function. Then, $(\mathcal{A}, \mathcal{B}, \mathcal{C}, \mathcal{D})$ can be given in terms of (A, b, c, d) as follows:

$$\begin{aligned} \mathcal{A} &= I_N \otimes \alpha + A \otimes (\beta\gamma) \\ \mathcal{B} &= b \otimes \beta \\ \mathcal{C} &= c \otimes \gamma \\ \mathcal{D} &= d \end{aligned} \tag{30}$$

where the set (α, β, γ) shown here is an arbitrary state-space representation of $1/F(s)$, i.e.

$$1/F(s) = \gamma(sI_M - \alpha)^{-1}\beta \tag{31}$$

and M is the order of $1/F(s)$, i.e. $M = \deg p(s)$ in (19). Note that the d-term in a state-space representation of $1/F(s)$ becomes zero because the reactance function is strictly proper. Therefore, the state-space-based frequency transformation given here is simpler than the discrete-time case (22).

Next we discuss the second-order modes of analog filters under frequency transformation. Let (K, W) and $(\mathcal{K}, \mathcal{W})$ be the controllability/observability Gramians of (A, b, c, d) and $(\mathcal{A}, \mathcal{B}, \mathcal{C}, \mathcal{D})$, respectively. Using (30), we can prove the following property:

$$\begin{aligned} \mathcal{K} &= K \otimes P^{-1} \\ \mathcal{W} &= W \otimes P \end{aligned} \tag{32}$$

where P is the positive definite matrix that satisfies the following relationship called the lossless positive-real lemma:

$$\begin{aligned} \alpha^T P + P\alpha &= 0_{M\times M} \\ P\beta &= \gamma^T. \end{aligned} \tag{33}$$

From (32) we easily see

$$\mathcal{K}\mathcal{W} = (KW) \otimes I_M \tag{34}$$

which proves that the second-order modes of analog filters are invariant under frequency transformation.

We now present the Gramian-preserving frequency transformation for analog filters. Let $(\widetilde{\mathcal{A}}, \widetilde{\mathcal{B}}, \widetilde{\mathcal{C}}, \widetilde{\mathcal{D}})$ be the state-space filter that is given by this transformation. Then, $(\widetilde{\mathcal{A}}, \widetilde{\mathcal{B}}, \widetilde{\mathcal{C}}, \widetilde{\mathcal{D}})$ is formulated as

$$\begin{aligned}\widetilde{\mathcal{A}} &= \widetilde{\alpha} \otimes I_N + (\widetilde{\beta}\widetilde{\gamma}) \otimes A \\ \widetilde{\mathcal{B}} &= \widetilde{\beta} \otimes b \\ \widetilde{\mathcal{C}} &= \widetilde{\gamma} \otimes c \\ \widetilde{\mathcal{D}} &= d\end{aligned} \tag{35}$$

where $(\widetilde{\alpha}, \widetilde{\beta}, \widetilde{\gamma})$ is a state-space representation of $1/F(s)$ that satisfies $P = I_M$ in (33), i.e.

$$\begin{aligned}\widetilde{\alpha}^T + \widetilde{\alpha} &= 0_{M\times M} \\ \widetilde{\beta} &= \widetilde{\gamma}^T.\end{aligned} \tag{36}$$

For $(\widetilde{\mathcal{A}}, \widetilde{\mathcal{B}}, \widetilde{\mathcal{C}}, \widetilde{\mathcal{D}})$ described as above, the controllability/observability Gramians $(\widetilde{\mathcal{K}}, \widetilde{\mathcal{W}})$ are found to be

$$\begin{aligned}\widetilde{\mathcal{K}} &= I_M \otimes K \\ \widetilde{\mathcal{W}} &= I_M \otimes W.\end{aligned} \tag{37}$$

Needless to say, this relationship is the same as in the discrete-time case (28). Hence the Gramians of a prototype state-space filter are preserved under this transformation.

As in the discrete-time case, formulation of $(\widetilde{\alpha}, \widetilde{\beta}, \widetilde{\gamma})$ is known to be non-unique. In [34], we presented a closed-form representation of $(\widetilde{\alpha}, \widetilde{\beta}, \widetilde{\gamma})$ that will be very suitable to circuit implementation. In order to derive this representation, we first rewrite the Foster reactance function (19) as the following partial fraction

$$\frac{1}{F(s)} = \sum_{i=1}^{L} \frac{G_i s}{s^2 + \omega_{pi}^2} + \frac{G_0}{s} \tag{38}$$

where $G_1, \cdots G_L$ and G_0 are all real and nonnegative, and $L = \lfloor M/2 \rfloor$, i.e. L is the largest integer less than or equal to $M/2$. Note that $G_0 = 0$ holds if M is even. Also, note that the first term on the right-hand side of (38) vanishes if $M = 1$. Now we can formulate the desired state-space representation of $1/F(s)$ by using the parameters of (38). The formulation depends on the value of M, i.e. the order of $1/F(s)$. For even M, we give the desired state-space representation, which is denoted by $(\widetilde{\alpha}_{\text{even}}, \widetilde{\beta}_{\text{even}}, \widetilde{\gamma}_{\text{even}})$, as follows:

$$\begin{aligned}\widetilde{\alpha}_{\text{even}} &= \text{block diag}\left(\Omega_{p1}, \Omega_{p2}, \cdots \Omega_{pL}\right) \\ \widetilde{\beta}_{\text{even}} &= \left(\widetilde{\psi}_1^T\ \widetilde{\psi}_2^T\ \cdots\ \widetilde{\psi}_L^T\right)^T \\ \widetilde{\gamma}_{\text{even}} &= \widetilde{\beta}_{\text{even}}^T\end{aligned} \tag{39}$$

where $\boldsymbol{\Omega}_{\mathrm{p}i} \in \Re^{2\times 2}$ and $\widetilde{\boldsymbol{\psi}}_i \in \Re^{2\times 1}$ for $M = 1, 2, \cdots, L$ are respectively given by

$$\boldsymbol{\Omega}_{\mathrm{p}i} = \begin{pmatrix} 0 & \omega_{\mathrm{p}i} \\ -\omega_{\mathrm{p}i} & 0 \end{pmatrix}$$
$$\widetilde{\boldsymbol{\psi}}_i = \begin{pmatrix} \sqrt{G_i} \\ 0 \end{pmatrix}. \tag{40}$$

If M is odd, we give the desired state-space representation $(\widetilde{\boldsymbol{\alpha}}_{\mathrm{odd}}, \widetilde{\boldsymbol{\beta}}_{\mathrm{odd}}, \widetilde{\boldsymbol{\gamma}}_{\mathrm{odd}})$ as

$$\widetilde{\boldsymbol{\alpha}}_{\mathrm{odd}} = \begin{pmatrix} \widetilde{\boldsymbol{\alpha}}_{\mathrm{even}} & \mathbf{0}_{2L\times 1} \\ \mathbf{0}_{1\times 2L} & \mathbf{0}_{1\times 1} \end{pmatrix}$$
$$\widetilde{\boldsymbol{\beta}}_{\mathrm{odd}} = \begin{pmatrix} \widetilde{\boldsymbol{\beta}}_{\mathrm{even}}^T & \sqrt{G_0} \end{pmatrix}^T$$
$$\widetilde{\boldsymbol{\gamma}}_{\mathrm{odd}} = \widetilde{\boldsymbol{\beta}}_{\mathrm{odd}}^T. \tag{41}$$

Note that the above expression reduces to $(\widetilde{\boldsymbol{\alpha}}_{\mathrm{odd}}, \widetilde{\boldsymbol{\beta}}_{\mathrm{odd}}, \widetilde{\boldsymbol{\gamma}}_{\mathrm{odd}}) = (0, \sqrt{G_0}, \sqrt{G_0})$ if $M = 1$. By direct calculation it is easy to prove that the state-space representations (39) and (41) satisfy the transfer function $1/F(s)$ given by (38) for even M and odd M, respectively, and that they also satisfy $\boldsymbol{P} = \boldsymbol{I}_M$ in the lossless positive-real lemma, i.e.

$$\widetilde{\boldsymbol{\alpha}}_{\mathrm{even}}^T + \widetilde{\boldsymbol{\alpha}}_{\mathrm{even}} = \mathbf{0}_{M\times M}$$
$$\widetilde{\boldsymbol{\beta}}_{\mathrm{even}} = \widetilde{\boldsymbol{\gamma}}_{\mathrm{even}}^T$$
$$\widetilde{\boldsymbol{\alpha}}_{\mathrm{odd}}^T + \widetilde{\boldsymbol{\alpha}}_{\mathrm{odd}} = \mathbf{0}_{M\times M}$$
$$\widetilde{\boldsymbol{\beta}}_{\mathrm{odd}} = \widetilde{\boldsymbol{\gamma}}_{\mathrm{odd}}^T. \tag{42}$$

This result shows that (39) and (41) offer the closed-form expression for the Gramian-preserving frequency transformation.

Finally, we discuss the physical interpretation of the Gramian-preserving frequency transformation, which will bring further insight into the circuit theory. As in the discrete-time case, we first discuss the Gramian-preserving frequency transformation in terms of the block diagram. As illustrated in Fig. 3, the Gramian-preserving frequency transformation for analog filters is derived from the model of Fig. 3(b), which is given by replacing the integrators in the prototype filter of Fig. 3(a) with an appropriate state-space representation $(\widetilde{\boldsymbol{\alpha}}, \widetilde{\boldsymbol{\beta}}, \widetilde{\boldsymbol{\gamma}})$ of the Foster reactance function $1/F(s)$. Here, we have to consider how the circuit topology of the set $(\widetilde{\boldsymbol{\alpha}}, \widetilde{\boldsymbol{\beta}}, \widetilde{\boldsymbol{\gamma}})$ is constructed. In order to answer this, consider again the partial fraction of strictly proper Foster reactance functions $1/F(s)$ as in (38). This expression is well-known as the LC driving-point impedance functions corresponding to the first Foster canonical form [1], which is realized by the series connection of a capacitor of capacitance $1/G_0$ and L parallel combinations of an inductor of inductance $G_i/\omega_{\mathrm{p}i}^2$ and a capacitor of capacitance $1/G_i$.

Figure 4(a) shows the circuit representation of $1/F(s)$, where $1/F(s)$ is related to V and I as $1/F(s) = V(s)/I(s)$. This circuit is easily expressed in state-space form as

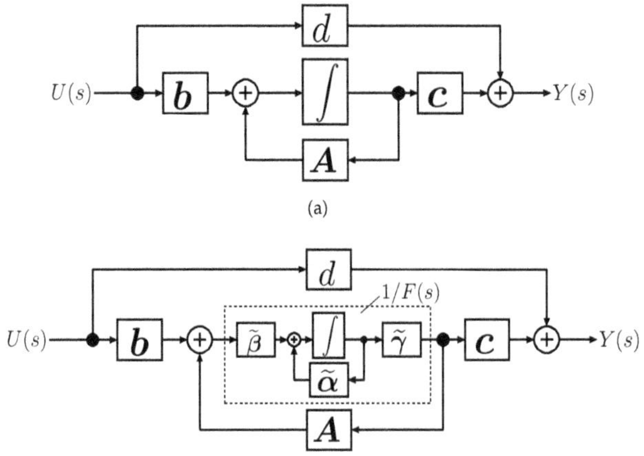

Figure 3. Gramian-preserving frequency transformation for analog filters: (a) prototype state-space filter, and (b) transformed state-space filter.

$$\frac{1}{F(s)} = \sum_{i=1}^{L} \gamma_i (s\boldsymbol{I}_2 - \boldsymbol{\alpha}_i)^{-1} \boldsymbol{\beta}_i + \gamma_0 (s\boldsymbol{I}_1 - \alpha_0)^{-1} \beta_0 \tag{43}$$

where the subsystems $(\boldsymbol{\alpha}_i, \boldsymbol{\beta}_i, \boldsymbol{\gamma}_i)$ for $1 \leq i \leq L$ and $(\alpha_0, \beta_0, \gamma_0)$ are found to be

$$\begin{aligned}
\boldsymbol{\alpha}_i &= \begin{pmatrix} 0 & G_i \\ -\frac{\omega_{pi}^2}{G_i} & 0 \end{pmatrix} \\
\boldsymbol{\beta}_i &= \begin{pmatrix} G_i \\ 0 \end{pmatrix} \\
\boldsymbol{\gamma}_i &= \begin{pmatrix} 1 \; 0 \end{pmatrix} \\
\alpha_0 &= 0 \\
\beta_0 &= G_0 \\
\gamma_0 &= 1
\end{aligned} \tag{44}$$

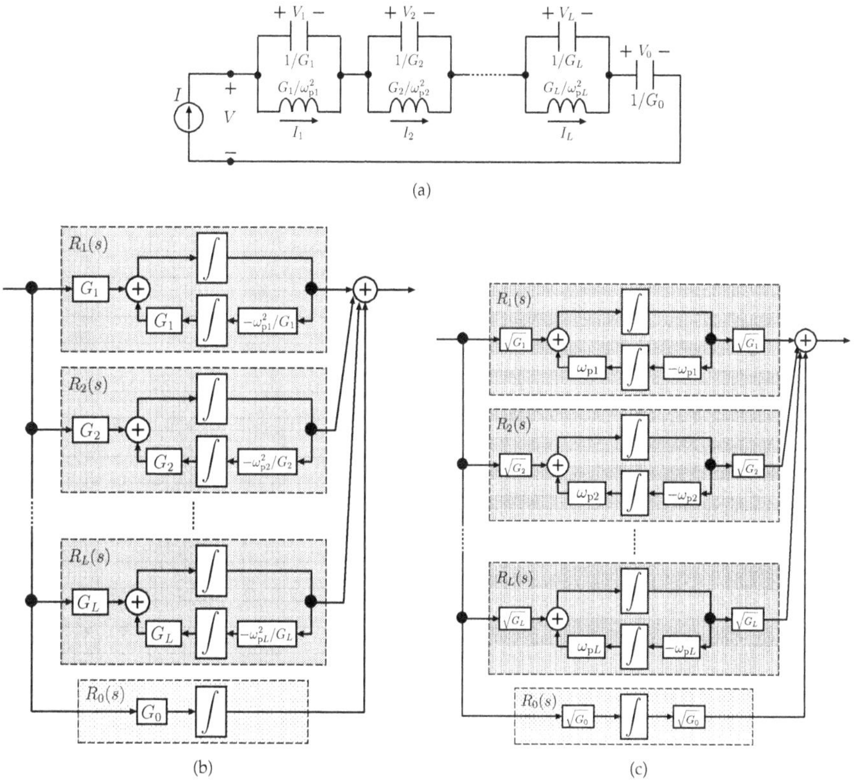

Figure 4. Construction of desired state-space model of $1/F(s)$ for Gramian-preserving frequency transformation: (a) *LC* circuit representation of $1/F(s)$, (b) state-space model of the *LC* circuit, and (c) desired state-space model.

with their state vectors $\boldsymbol{X}_i(s)$ and $\boldsymbol{X}_0(s)$ defined as

$$\begin{aligned} \boldsymbol{X}_i(s) &= \begin{pmatrix} V_i(s) & -I_i(s) \end{pmatrix}^T \\ \boldsymbol{X}_0(s) &= V_0. \end{aligned} \tag{45}$$

Figure 4(b) shows the state-space model of $1/F(s)$ described as above. Substituting (44) into (33), we obtain the solutions $\boldsymbol{P}_i$ and $\boldsymbol{P}_0$ to the lossless positive-lemma for $(\boldsymbol{\alpha}_i, \boldsymbol{\beta}_i, \gamma_i)$ and

$(\boldsymbol{\alpha}_0, \boldsymbol{\beta}_0, \boldsymbol{\gamma}_0)$ as follows:

$$\begin{aligned} \boldsymbol{P}_i &= \mathrm{diag}(1/G_i,\ G_i/\omega_{pi}^2) \\ \boldsymbol{P}_0 &= 1/G_0. \end{aligned} \tag{46}$$

From (46) we see that the state-space model of Fig. 4(b) does not satisfy $\boldsymbol{P} = \boldsymbol{I}_M$ in (33). Hence it is necessary to modify the structure of this model such that $\boldsymbol{P}_i = \boldsymbol{I}_2$ and $\boldsymbol{P}_0 = \boldsymbol{I}_1$ hold. To this end, we consider the following nonsingular matrices

$$\begin{aligned} \boldsymbol{T}_i &= \mathrm{diag}(\sqrt{G_i},\ \omega_{pi}/\sqrt{G_i}), \quad 1 \le i \le L \\ \boldsymbol{T}_0 &= \sqrt{G_0}. \end{aligned} \tag{47}$$

Note that these matrices satisfy $\boldsymbol{T}_i\boldsymbol{T}_i^T = \boldsymbol{P}_i^{-1}$ and $\boldsymbol{T}_0\boldsymbol{T}_0^T = \boldsymbol{P}_0^{-1}$. Using these matrices, we apply the similarity transformation to (44), which results in the new structure $(\boldsymbol{\alpha}'_i, \boldsymbol{\beta}'_i, \boldsymbol{\gamma}'_i)$ and $(\boldsymbol{\alpha}'_0, \boldsymbol{\beta}'_0, \boldsymbol{\gamma}'_0)$ as

$$\begin{aligned} \boldsymbol{\alpha}'_i &= \boldsymbol{T}_i^{-1}\boldsymbol{\alpha}_i\boldsymbol{T}_i = \begin{pmatrix} 0 & \omega_{pi} \\ -\omega_{pi} & 0 \end{pmatrix} \\ \boldsymbol{\beta}'_i &= \boldsymbol{T}_i^{-1}\boldsymbol{\beta}_i = \begin{pmatrix} \sqrt{G_i} \\ 0 \end{pmatrix} \\ \boldsymbol{\gamma}'_i &= \boldsymbol{\gamma}_i\boldsymbol{T}_i = \begin{pmatrix} \sqrt{G_i} & 0 \end{pmatrix} \\ \boldsymbol{\alpha}'_0 &= \boldsymbol{T}_0^{-1}\boldsymbol{\alpha}_0\boldsymbol{T}_0 = 0 \\ \boldsymbol{\beta}'_0 &= \boldsymbol{T}_0^{-1}\boldsymbol{\beta}_0 = \sqrt{G_0} \\ \boldsymbol{\gamma}'_0 &= \boldsymbol{\gamma}_0\boldsymbol{T}_0 = \sqrt{G_0} \end{aligned} \tag{48}$$

and its corresponding model is given by Fig. 4(c). Then, it immediately follows that this modified structure satisfies $\boldsymbol{P} = \boldsymbol{I}_M$ in (33) and coincides with the desired state-space representations (39) and (41).

The above discussion shows that the desired structure of Fig. 4(c) is obtained by applying the similarity transformation based on (47) to the first Foster canonical form for LC impedance networks. Here, it turns out that the nonsingular matrices $\boldsymbol{T}_i$'s and $\boldsymbol{T}_0$ serve as the scaling matrices that convert the matrices $\boldsymbol{P}_i$'s and $\boldsymbol{P}_0$ into the identity matrices. Therefore, we conclude that our proposed Gramian-preserving frequency transformation is derived from a state-space system of which integrators are replaced with $1/F(s)$, where the structure of $1/F(s)$ is constructed as the scaled version of the first Foster canonical form for LC impedance networks. It is interesting to note that this construction of Fig. 4(c) is similar to the realization of orthonormal ladder filters [7]: the orthonormal ladder filters are obtained by applying the

L_2 scaling to the structure of singly-terminated *LC* ladder networks, whereas the structures of Fig. 4(c) is obtained by applying another type of scaling, which makes use of the solutions to the lossless positive-real lemma, to the Foster canonical form for *LC* networks.

Before concluding this section, it should be noted again that the above results apply to the case of strictly proper reactance functions that include the LP-LP and the LP-BP transformations. For details of the improper reactance functions such as the LP-HP and the LP-BS transformations, see [32–34].

5. Application to design and synthesis of high-performance filters

This section applies the results of the previous section to design and synthesis of high-performance analog and digital filters. Emphasis is on the tunable filters, and we present a simple method to obtain state-space-based tunable filters with high-performance structures.

5.1. High-performance digital filters

Here we apply the Gramian-preserving frequency transformation to design and synthesis of a variable band-pass filter of high-performance structure [35]. The variable band-pass filter to be presented here is assumed to have the fixed bandwidth and the tunable center-frequency. Such a band-pass filter requires the simplified LP-BP transformation with the following all-pass function:

$$\frac{1}{F_{\mathrm{BP}}(z)} = -z^{-1}\frac{z^{-1} - \xi_{\mathrm{BP}}}{1 - \xi_{\mathrm{BP}} z^{-1}} \tag{49}$$

where $\xi_{\mathrm{BP}} = \cos\omega_{\mathrm{BP}}$ and ω_{BP} is the desired center-frequency of the passband in the variable band-pass filter. The desired state-space representation of (49) in order to carry out the Gramian-preserving frequency transformation (i.e. the state-space representation of (49) with the normalized lattice structure) is found to be

$$\begin{aligned}
\widetilde{\alpha} &= \begin{pmatrix} \xi_{\mathrm{BP}} & 0 \\ \sqrt{1-\xi_{\mathrm{BP}}^2} & 0 \end{pmatrix} \\
\widetilde{\beta} &= \begin{pmatrix} \sqrt{1-\xi_{\mathrm{BP}}^2} \\ -\xi_{\mathrm{BP}} \end{pmatrix} \\
\widetilde{\gamma} &= \begin{pmatrix} 0 & -1 \end{pmatrix} \\
\widetilde{\delta} &= 0.
\end{aligned} \tag{50}$$

Substituting (50) into (26), we obtain the state-space representation of the variable band-pass filter as

$$\widetilde{\mathcal{A}} = \begin{pmatrix} \xi_{\mathrm{BP}} I_N & -\sqrt{1-\xi_{\mathrm{BP}}^2} A \\ \sqrt{1-\xi_{\mathrm{BP}}^2} I_N & \xi_{\mathrm{BP}} A \end{pmatrix}$$
$$\widetilde{\mathcal{B}} = \begin{pmatrix} \sqrt{1-\xi_{\mathrm{BP}}^2} b \\ -\xi_{\mathrm{BP}} b \end{pmatrix}$$
$$\widetilde{\mathcal{C}} = \begin{pmatrix} 0_{1\times N} & -c \end{pmatrix}$$
$$\mathcal{D} = d \tag{51}$$

and we can easily control the center-frequency of this filter by changing the value of ξ_{BP} in (51).

Now we present a design/synthesis example. The prototype filter used here is the fourth-order elliptic low-pass filter with the following transfer function:

$$H(z) = \frac{0.0101 - 0.0362z^{-1} + 0.0524z^{-2} - 0.0362z^{-3} + 0.0101z^{-4}}{1 - 3.7895z^{-1} + 5.4142z^{-2} - 3.4553z^{-3} + 0.8310z^{-4}}. \tag{52}$$

The peak-to-peak ripple, the minimum stopband attenuation and the passband-edge frequency of this filter are 0.5 dB, 40 dB and 0.05π rad, respectively. We choose the state-space representation (A, b, c, d) of this prototype filter as follows:

$$A = \begin{pmatrix} 0.9838 & -0.1007 & -0.0165 & -0.0171 \\ 0.1007 & 0.9582 & -0.1029 & -0.0273 \\ -0.0165 & 0.1029 & 0.9336 & -0.1015 \\ 0.0171 & -0.0273 & 0.1015 & 0.9139 \end{pmatrix}$$
$$b = \begin{pmatrix} 0.1490 & -0.1953 & 0.1669 & -0.0995 \end{pmatrix}^T$$
$$c = \begin{pmatrix} 0.1490 & 0.1953 & 0.1669 & 0.0995 \end{pmatrix}$$
$$d = 0.0101. \tag{53}$$

The controllability/observability Gramians of this realization are calculated as

$$K = W = \mathrm{diag}(0.8850, 0.6124, 0.2761, 0.0817), \tag{54}$$

which shows that this realization is the balanced realization.

Applying (51) to (53) yields the eighth-order variable band-pass filter. It can be easily checked that, for any ξ_{BP}, the Gramians of this band-pass filter become the same as (54) with multiplicity 2, i.e.

$$K = W = \mathrm{diag}(0.8850, 0.6124, 0.2761, 0.0817, 0.8850, 0.6124, 0.2761, 0.0817). \tag{55}$$

Therefore, the variable band-pass filter keeps the balanced form regardless of the location of the center-frequency.

Figures 5(a), (b), (c) and (d) show the magnitude responses of our proposed variable filter for $\xi_{BP} = -0.8, -0.4, 0.5$ and 0.9, respectively. For comparison purpose, the magnitude responses in the case of the cascaded direct form are also shown here, and all the coefficients of these two variable filters are quantized to 10 fractional bits. From Figs. 5(a), (b), (c) and (d) we know that our proposed variable filter shows very good agreement with the ideal magnitude responses for all ξ_{BP}. This result confirms that, our proposed variable filter exhibits high accuracy for all tunable characteristics by constructing the state-space representation of the prototype filter appropriately with respect to the Gramians. On the other hand, the magnitude responses of the cascaded direct form are degraded in all cases and the degradation is extremely large for $\xi_{BP} = 0.9$. As is well-known, direct form digital filters are very sensitive to quantization effects. In addition, since variable digital filters with direct form do not take into account the controllability/observability Gramians, the performance of the direct form with respect to quantization effects highly depends on the frequency characteristics. These facts show the utility of our proposed method.

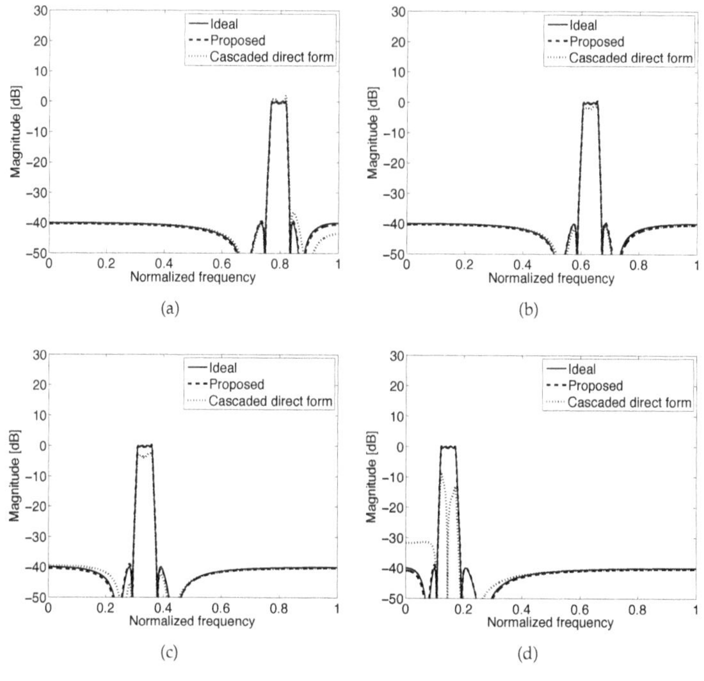

Figure 5. Magnitude responses of the eighth-order variable band-pass digital filters: (a) Responses for $\xi_{BP} = -0.8$. (b) Responses for $\xi_{BP} = -0.4$. (c) Responses for $\xi_{BP} = 0.5$. (d) Responses for $\xi_{BP} = 0.9$.

5.2. High-performance analog filters

Here we will design and synthesize a variable analog band-pass filter by using the Gramian-preserving frequency transformation. In the LP-BP transformation, we use the second-order Foster reactance function $1/F_{\mathrm{BP}}(s)$ as in (20). Therefore we apply (39) to (35), which results in the following state-space formulation of the desired variable analog band-pass filter:

$$\begin{aligned}\widetilde{\mathcal{A}} &= \begin{pmatrix} GA & \omega_{\mathrm{p}1}\boldsymbol{I}_N \\ -\omega_{\mathrm{p}1}\boldsymbol{I}_N & \boldsymbol{0}_{N\times N}\end{pmatrix}\\ \widetilde{\mathcal{B}} &= \begin{pmatrix}\sqrt{G}\boldsymbol{b} \\ \boldsymbol{0}_{N\times 1}\end{pmatrix}\\ \widetilde{\mathcal{C}} &= \begin{pmatrix}\sqrt{G}\boldsymbol{c} \; \boldsymbol{0}_{1\times N}\end{pmatrix}\\ \widetilde{\mathcal{D}} &= d.\end{aligned} \tag{56}$$

As a design/synthesis example, here we use the following prototype low-pass filter

$$H(s) = \frac{1}{s^3 + 2s^2 + 2s + 1}. \tag{57}$$

This transfer function is the third-order Butterworth low-pass filter with a cutoff frequency of 1 rad/s. We give the state-space representation of this prototype filter as the following orthonormal ladder structure [7]:

$$\begin{aligned}\boldsymbol{A} &= \begin{pmatrix} 0 & a_1 & 0 \\ -a_1 & 0 & a_2 \\ 0 & -a_2 & -a_3\end{pmatrix}\\ \boldsymbol{b} &= \begin{pmatrix}0\\0\\b_3\end{pmatrix}\\ \boldsymbol{c} &= \begin{pmatrix}c_1 \; 0 \; 0\end{pmatrix}\\ d &= 0\end{aligned} \tag{58}$$

with

$$(a_1, a_2, a_3, b_3, c_1) = (0.7071, 1.2247, 2.0000, 0.7979, 1.4472). \tag{59}$$

From (58) and (59), the controllability/observability Gramians of this filter are found to be

$$\begin{aligned}\boldsymbol{K} &= \boldsymbol{I}_3\\ \boldsymbol{W} &= \begin{pmatrix}16.4493 & 9.3052 & 3.7988\\ 9.3052 & 9.8696 & 5.3723\\ 3.7988 & 5.3723 & 3.2899\end{pmatrix}.\end{aligned} \tag{60}$$

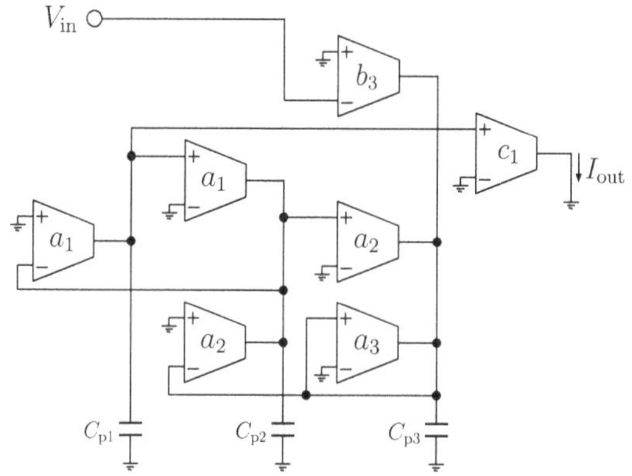

Figure 6. Prototype filter based on transconductance-capacitor integrators.

As seen above, the controllability Gramian of the orthonormal ladder structure becomes the identity matrix. This property brings the high-performance with respect to the dynamic range and the sensitivity. Figure 6 illustrates the block diagram of this filter structure based on transconductance-capacitor integrators, where the normalized capacitance distribution is given by

$$(C_{\text{p1}}, C_{\text{p2}}, C_{\text{p3}}) = C_{\text{p}}(0.3091, 0.3957, 0.2952) \tag{61}$$

and C_{p} is the unit-less value of the total capacitance when expressed in F. The specification of (61) is determined according to the following rule [10]:

$$\begin{aligned} C_{pi} &= \frac{\sqrt{\eta_i w_{ii} k_{ii}}}{\sum_j \sqrt{\eta_j w_{jj} k_{jj}}} \\ \eta_i &= \sum_j |a_{ij}|. \end{aligned} \tag{62}$$

As is seen from (58) and Fig. 6, the structure of this prototype filter is very sparse and suitable for circuit implementation. This is another benefit of the orthonormal ladder structure.

Applying (56) to this prototype filter, we finally obtain the state-space representation of the variable band-pass filter, and its corresponding circuit realization is given by Fig. 7. It can be easily shown that the controllability/observability Gramians $(\widetilde{\mathcal{K}}, \widetilde{\mathcal{W}})$ of this band-pass filter become

$$\begin{aligned} \widetilde{\mathcal{K}} &= \text{block diag}(\boldsymbol{K}, \boldsymbol{K}) = \boldsymbol{I}_6 \\ \widetilde{\mathcal{W}} &= \text{block diag}(\boldsymbol{W}, \boldsymbol{W}) \end{aligned} \tag{63}$$

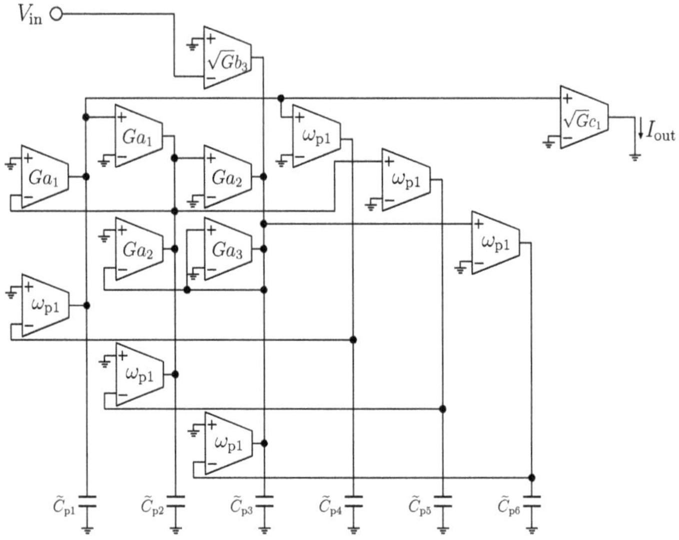

Figure 7. Band-pass filter given by Gramian-preserving frequency transformation.

for arbitrary values of G and $\omega_{\rm p1}$. It follows from this result that the Gramian-preserving frequency transformation easily produces the band-pass filter with the orthonormal ladder structure for arbitrary center frequency and bandwidth. Therefore, by controlling the parameters of G and $\omega_{\rm p1}$, we can realize tunable band-pass filters with the orthonormal ladder structure.

The high-performance of this band-pass filter can be demonstrated by not only calculation of the Gramians, but also numerical evaluation of the dynamic range. For details, see [34] and the references therein.

6. Conclusion

In this chapter we have introduced insightful and useful results on the classical frequency transformation of analog filters and digital filters. While most of the known results on the frequency transformation are described in terms of the transfer functions, the results given in this chapter are based on the state-space representation, which have revealed many useful properties with respect to the performance of filters that is dominated by the internal properties as well as the input-output relationship. In particular, the Gramian-preserving frequency transformation is very attractive to design and synthesis of high-performance filters. Using this new frequency transformation, we have presented variable analog/digital filters that retain high-performance regardless of the change of the frequency characteristics.

In addition to the aforementioned work, some other results on the frequency transformation have been reported in the literature. One of them is the state-space formulation of 2-D frequency transformation [36], which presents an explicit state-space-based frequency transformation for 2-D digital filters. Also, Yan et al. [37, 38] extended this work to formulations of more general 2-D frequency transformation. Moreover, in [39] we have revealed the invariance property of the second-order modes of 2-D separable denominator digital filters under frequency transformation. Proof of this invariance property in the case of 2-D non-separable denominator digital filters is still an open problem. Derivation of the Gramian-preserving frequency transformation in the 2-D case is also an open problem.

Another interesting topic is the transformations based on "lossy" functions. In both the cases of analog frequency transformation and digital frequency transformation, the required transformation functions have the lossless property. On the other hand, it is theoretically possible to use lossy functions for transformation. Motivated by this, in [33, 40] we presented the state-space analysis of lossy transformations and revealed that the second-order modes are decreased under such transformations. Development of a practical application of this property is a future work.

Author details

Shunsuke Koshita*, Masahide Abe and Masayuki Kawamata

* Address all correspondence to: kosita@mk.ecei.tohoku.ac.jp

Department of Electronic Engineering, Graduate School of Engineering, Tohoku University, Sendai, Japan

References

[1] W.-K. Chen, Ed., *The Circuits and Filters Handbook.* CRC Press, 1995.

[2] A. G. Constantinides, "Spectral transformations for digital filters," *Proc. IEE*, vol. 117, no. 8, pp. 1585–1590, Aug. 1970.

[3] G. Stoyanov and M. Kawamata, "Variable digital filters," *RISP Journal of Signal Processing*, vol. 1, no. 4, pp. 275–289, July 1997.

[4] J. A. Chambers and A. G. Constantinides, "Frequency tracking using constrained adaptive notch filters synthesised from allpass sections," *Proc. IEE (part F)*, vol. 137, no. 6, pp. 475–481, Dec. 1990.

[5] V. DeBrunner and S. Torres, "Multiple fully adaptive notch filter design based on allpass sections," *IEEE Signal Processing Lett.*, vol. 48, no. 2, pp. 550–552, Feb. 2000.

[6] W. M. Snelgrove and A. S. Sedra, "Synthesis and analysis of state-space active filters using intermediate transfer functions," *IEEE Trans. Circuits Syst.*, vol. CAS-33, no. 3, pp. 287–301, Mar. 1986.

[7] D. A. Johns, W. M. Snelgrove, and A. S. Sedra, "Orthonormal ladder filters," *IEEE Trans. Circuits Syst.*, vol. CAS-36, no. 3, pp. 337–343, Mar. 1989.

[8] G. Groenewold, "The design of high dynamic range continuous-time integratable bandpass filters," *IEEE Trans. Circuits Syst.*, vol. CAS-38, no. 8, pp. 838–852, Aug. 1991.

[9] J. Harrison and N. Weste, "Energy storage and gramians of ladder filter realisations," in *Proc. IEEE Int. Symp. Circuits and Systems*, May 2001, pp. I–29–I–32.

[10] D. P. W. M. Rocha, "Optimal design of analogue low-power systems, a strongly directional hearing-aid adapter," Ph.D. dissertation, Delft University of Technology, Delft, The Netherlands, Apr. 2003.

[11] S. A. P. Haddad, S. Bagga, and W. A. Serdijn, "Log-domain wavelet bases," *IEEE Trans. Circuits Syst. I*, vol. 52, no. 10, pp. 2023–2032, Oct. 2005.

[12] A. N. Akansu, W. A. Serdijn, and I. W. Selesnick, "Emerging applications of wavelets: A review," *Physical Communication*, vol. 3, no. 1, pp. 1–18, Mar. 2010.

[13] C. T. Mullis and R. A. Roberts, "Synthesis of minimum roundoff noise fixed point digital filters," *IEEE Trans. Circuits Syst.*, vol. CAS-23, no. 9, pp. 551–562, Sept. 1976.

[14] S. Y. Hwang, "Minimum uncorrelated unit noise in state-space digital filtering," *IEEE Trans. Acoust., Speech, Signal Processing*, vol. ASSP-25, no. 4, pp. 273–281, Aug. 1977.

[15] V. Tavşanoğlu and L. Thiele, "Optimal design of state-space digital filters by simultaneous minimization of sensitivity and roundoff noise," *IEEE Trans. Circuits Syst.*, vol. 31, no. 10, pp. 884–888, Oct. 1984.

[16] M. Kawamata and T. Higuchi, "A unified approach to the optimal synthesis of fixed-point state-space digital filters," *IEEE Trans. Acoust., Speech, Signal Processing*, vol. ASSP-33, no. 4, pp. 911–920, Aug. 1985.

[17] L. Thiele, "On the sensitivity of linear state-space systems," *IEEE Trans. Circuits Syst.*, vol. 33, no. 5, pp. 502–510, May 1986.

[18] M. Iwatsuki, M. Kawamata, and T. Higuchi, "Statistical sensitivity and minimum sensitivity structures with fewer coefficients in discrete time linear systems," *IEEE Trans. Circuits Syst.*, vol. CAS-37, no. 1, pp. 72–80, Jan. 1990.

[19] G. Li, B. D. O. Anderson, M. Gevers, and J. E. Perkins, "Optimal FWL design of state-space digital systems with weighted sensitivity minimization and sparseness consideration," *IEEE Trans. Circuits Syst. I*, vol. 39, no. 5, pp. 365–377, May 1992.

[20] W.-Y. Yan and J. B. Moore, "On L^2-sensitivity minimization of linear state-space systems," *IEEE Trans. Circuits Syst. I*, vol. 39, no. 8, pp. 641–648, Aug. 1992.

[21] T. Hinamoto, S. Yokoyama, T. Inoue, W. Zeng, and W.-S. Lu, "Analysis and minimization of L_2-sensitivity for linear systems and two-dimensional state-space filters using general controllability and observability Gramians," *IEEE Trans. Circuits Syst. I*, vol. 49, no. 9, pp. 1279–1289, Sept. 2002.

[22] T. Hinamoto, K. Iwata, and W.-S. Lu, "L_2-sensitivity minimization of one- and two-dimensional state-space digital filters subject to L_2-scaling constraints," *IEEE Trans. Signal Processing*, vol. 54, no. 5, pp. 1804–1812, May 2006.

[23] S. Yamaki, M. Abe, and M. Kawamata, "A closed form solution to L_2-sensitivity minimization of second-order state-space digital filters," *IEICE Trans. Fundamentals*, vol. E91-A, no. 5, pp. 1268–1273, May 2008.

[24] ——, "A closed form solution to L_2-sensitivity minimization of second-order state-space digital filters subject to L_2-scaling constraints," *IEICE Trans. Fundamentals*, vol. E91-A, no. 7, pp. 1697–1705, July 2008.

[25] ——, "Derivation of the class of digital filters with all second-order modes equal," *IEEE Trans. Signal Processing*, vol. 59, no. 11, pp. 5236–5242, Nov. 2011.

[26] B. C. Moore, "Principal component analysis in linear systems: Controllability, observability, and model reduction," *IEEE Trans. Automat. Contr.*, vol. AC-26, no. 1, pp. 17–32, Feb. 1981.

[27] L. Pernebo and L. M. Silverman, "Model reduction via balanced state space representations," *IEEE Trans. Automat. Contr.*, vol. AC-27, no. 2, pp. 382–387, Apr. 1982.

[28] A. C. Antoulas, *Approximation of large-scale dynamical systems.* SIAM Book series "Advances in Design and Control", 2005, vol. DC-06.

[29] A. V. Oppenheim, W. F. G. Mecklenbrauker, and R. M. Mersereau, "Variable cutoff linear phase digital filters," *IEEE Trans. Circuits Syst.*, vol. CAS-23, no. 4, pp. 199–203, Apr. 1976.

[30] C. T. Mullis and R. A. Roberts, "Roundoff noise in digital filters: Frequency transformations and invariants," *IEEE Trans. Acoust., Speech, Signal Processing*, vol. ASSP-24, no. 6, pp. 538–550, Dec. 1976.

[31] S. Koshita, S. Tanaka, M. Abe, and M. Kawamata, "Gramian-preserving frequency transformation for linear discrete-time state-space systems," *IEICE Trans. Fundamentals*, vol. E91-A, no. 10, pp. 3014–3021, Oct. 2008.

[32] M. Kawamata, Y. Mizukami, and S. Koshita, "Invariance of second-order modes of linear continuous-time systems under typical frequency transformations," *IEICE Trans. Fundamentals*, vol. E90-A, no. 7, pp. 1481–1486, July 2007.

[33] S. Koshita, Y. Mizukami, T. Konno, M. Abe, and M. Kawamata, "Analysis of second-order modes of linear continuous-time systems under positive-real transformations," *IEICE Trans. Fundamentals*, vol. E91-A, no. 2, pp. 575–583, Feb. 2008.

[34] S. Koshita, M. Abe, and M. Kawamata, "Gramian-preserving frequency transformation and its application to analog filter design," *IEEE Trans. Circuits Syst. I*, vol. 58, no. 3, pp. 493–506, Mar. 2011.

[35] S. Koshita, K. Miyoshi, M. Abe, and M. Kawamata, "Realization of variable band-pass/band-stop IIR digital filters using Gramian-preserving frequency transformation," in *Proc. IEEE Int. Symp. on Circuits Syst.*, May 2010, pp. 2698–2701.

[36] S. Koshita and M. Kawamata, "State-space formulation of frequency transformation for 2-D digital filters," *IEEE Signal Processing Lett.*, vol. 11, no. 10, pp. 784–787, Oct. 2004.

[37] S. Yan, L. Xu, and Y. Anazawa, "A two-stage approach to the establishment of state-space formulation of 2-D frequency transformation," *IEEE Signal Processing Lett.*, vol. 14, no. 12, pp. 960–963, Dec. 2007.

[38] S. Yan, N. Shiratori, and L. Xu, "Simple state-space formulations of 2-D frequency transformation and double bilinear transformation," *Multidimensional Systems and Signal Processing*, vol. 21, no. 1, pp. 3–23, Mar. 2010.

[39] S. Koshita and M. Kawamata, "Invariance of second-order modes under frequency transformation in 2-D separable denominator digital filters," *Multidimensional Systems and Signal Processing*, vol. 16, no. 3, pp. 305–333, July 2005.

[40] S. Koshita, M. Abe, and M. Kawamata, "Analysis of second-order modes of linear discrete-time systems under bounded-real transformations," *IEICE Trans. Fundamentals*, vol. E90-A, no. 11, pp. 2510–2515, Nov. 2007.

Chapter 6

Digital Filter Implementation of Orthogonal Moments

Barmak Honarvar Shakibaei Asli and
Raveendran Paramesran

Additional information is available at the end of the chapter

1. Introduction

Accuracy in detection, robustness in performance and the ability to perform accurately and robustly are all integral considerations in image related research. However, since applications of such research often have a real-time constraint, speed of computation and its corollary - time-saving- have also become increasingly important in the research agenda.

The importance of geometric moment (GM) invariants first received research attention when Hu [1] introduced them in his study. As they capture global features have been widely used in applications such as object classification, image and shape analysis and edge detection [2-8]. However, since GMs are non-orthogonal, a set of continuous orthogonal moments including Zernike and Legendre moments was introduced by Teague [9]. Their desirable qualities of orthogonality, speed of computation and robustness have found a wider range of applications such as character recognition, two-dimensional (2D) direction-of-arrival estimation, and trademark segmentation and retrieval [10-13]. However, these continuous moments pose certain problems in computation: they require coordinate transformation and suitable approximation of the continuous moment integrals. The transformation and approximation they require create additional opportunities for the occurrence of error in the computation of the feature descriptors. This recognition of the error potential in continuous orthogonal moments led to a new set of discrete orthogonal moments such as Tchebichef, Krawtchouk and Hahn moments as these do not require coordinate transformation and the discretization step in the moment computation [14-18].

The computation of both continuous and discrete orthogonal moments can be realized directly or via the GMs. The complexity in deriving higher order GMs using a direct calculation approach raises a significant challenge for real-time applications. A number of studies address this issue using different approaches. Hatamian [19] proposed a novel approach that uses 2D all-pole digital filters with separable impulse responses to implement the first 16 GMs. Wong and Siu [20] moved the delay element in the basic filter structure as proposed in [19] from the feed-forward to the feedback path. Their method considered moment computation of up to the third order. Using a similar structure of cascaded digital filters, Kotoulas and Andreadis showed orthogonal moments can be generated from their outputs [21]. To reduce the chip area size, they introduced an overflow counter in each of the basic filter structures [22]. However, the plurality of large bit-width adders in the digital filter structure presents a major challenge to chip synthesis, placement and the routing process. Taking a different tack, Al-Rawi [23] generalized the relationship between the moments and the digital filter outputs using a recurrence formula.

Two facts demand our attention at this point. One, in all the aforementioned literature, the basic concept involved in the computation of moments using the digital filter structure has remained unchanged over the past three decades [19]-[20] even though this digital filter structure was formulated when only low orders of moments were used. Two, many recent works have involved increasingly higher moment orders. They include, among others, invariant image watermarking (30 orders) [24], moving object reconstruction (55 orders) [25], and hand shape verification (60 orders) [26].

Two models have been proposed to deal with higher order moments. The usage of cascaded all-pole digital filters in generating higher order GMs for the formulation of orthogonal moments has been successfully explored by Kotoulas and Andreadis. In their work [21]-[22], the 40th and 70th high order Zernike and Tchebichef moments, respectively were obtained from the digital filter outputs and their transform coefficients. Their digital filter was based on the feedforward model, and for a $N \times N$ image, the digital filter output values for row filtering are sampled at $N+2,\ N+3,\ N+4,\ \ldots,\ N+p+2$, where p is the maximum order. The algorithm they propose, however, is undermined by a computational problem. At these time instances, the digital filter output values are much larger when compared to the earlier time instances. This is because the digital filter operates as an accumulator: their output values increase as the number of digital filters directly related to the order increases. The sample of digital filter output values obtained from the row filtering for each row are then used as inputs to the respective digital filters arranged in the column filtering. This further increases the size of the final digital filter outputs. Additionally, since the digital filter used is an approximation except for the first two orders, coefficients are then multiplied from second order onwards to make them exact. The coefficients, both positive and negative, are determined from the impulse response of the digital filter. The model proposed by Wong and Siu [20], though the digital filter outputs for all orders are sampled at their respective N, also suffers from a similar problem. The current approaches, it is clear, thus suffer from an increase in computational complexity arising from the large increase in the digital filter output values as the order increases [33]-[34].

The method of computation we propose in this work attempts to reduce computational complexity – and save time through a reduction in the number of additions – by addressing the problems arising from: 1) The increase in digital filter output values as the order of moment's increases and 2) the consequent use of positive and negative coefficient multipliers to make them exact. To circumvent the problems arising from these two facts, we are proposing the use of only positive coefficient multipliers which then makes it possible to use lower digital filter outputs as the order of the moment increases in generation of GMs. As will be discussed in greater detail in Section 3, the proposed method is based on the formulation and understanding of the impulse response of the digital filters, the unit step function to be used and their relationship with GMs. This formulation makes it possible for the digital filter outputs to be evaluated at earlier instances at $N,\ N-1,\ N-2,\ \ldots,\ N-p,$ where the lowest digital filter output value, sampled at $N-p$ is for the highest order. Meanwhile, this set of output values starts to decrease after $p/2$ moment orders.

Recently, another types of discrete orthogonal moments such as Krawtchouk, dual Hahn, Racah, Meixner and Hahn moments have been introduced in image analysis community [16]–[17]. It was shown that they have better image representation capability than the continuous orthogonal moments.

One main difficulty concerning the use of moments as feature descriptors is their high computational complexity. To solve this problem, a number of fast algorithms have been reported in the literature [14]–[19]. Most of them concentrated on the fast computation of geometric moments and continuous orthogonal moments. This work examines various aspects; both theory and applications of image moment implementation using digital filter structures. Since these aspects can be discussed rather independently, we devote each chapter to the discussion of one particular aspect of moment structures. The following is a summary of contents of the sections.

- **Section 2: Orthogonal Moments.** Numerous types of orthogonal polynomial, both in 1D and 2D, have been described in traditional mathematical literature. In this chapter we present a survey of orthogonal moments that are of importance in image analysis. The literature on orthogonal moments is very broad, namely in the area of practical applications, and our survey has no claim on completeness. We divide orthogonal polynomials and orthogonal moments into two basic groups. The polynomials *orthogonal on a rectangle* originate from 1D orthogonal polynomials whose 2D versions were created as products of 1D polynomials in x and y. The main advantage of the moments orthogonal on a rectangle is that they preserve the orthogonality even on the sampled image. They can be made scale-invariant but creating rotation invariants from them is very complicated. The polynomials *orthogonal on a disk* are intrinsically 2D functions. They are constructed as products of a radial factor (usually a 1D orthogonal polynomial) and angular factor which is usually a kind of harmonic function. When implementing these moments, an image must be mapped into a disk of orthogonality which creates certain resampling problems. On the other hand, moments orthogonal on a disk can easily be used for construction of rotation invariants because they change under rotation in a simple way.
- **Section 3: A New Formulation of Geometric Moments from Lower Output Values of Digital Filters.** In this chapter we propose a new method to accelerate geometric moment's computation using digital filters based on the lower output values. It is shown in this chapter a brief reviews of the digital filter methods employed in [19] and [20]. First, a description of the proposed method that includes the theoretical formulation of the relationship between

the digital filter outputs and GMs is provided. Second, we discuss the computational complexity with both an artificial and two real images, and the computational time. Finally, this study with some suggestions for future research is concluded.

- **Section 4: A Reduced 2D digital Filter Structure for Fast Computation of Geometric Moments.** It is shown in this chapter how to design a reduced 2D digital filter grid for fast computation of geometric moments. For this design, the 1D and 2D all-pole digital filter design procedure using the Z-transform properties is described. It also describes the recurrence equations for the desired filter outputs. The work shows the digital filter design used in [3] and [4], and shows the implementation of the proposed architecture. Finally, the computation results of the proposed method and the method used in [3] is illustrated.
- **Section 5: Conclusions.** The presentation is concluded, summarizing the contents of the work and discussing possibilities which may be open for the future research.

2. Moment functions

Geometric Moments (GMs) and complex moments (CMs) are the simplest among moment functions, with the kernel function defined as a product of the pixel power coordinates. These type of moments are non-orthogonal and because of this problem, the inverse GMs or CMs formulation are not possible. On the other hand, the orthogonal moment functions have been used widley for image analysis. The orthogonal moment functions are based on the orthogonal polynomials such as Legendre, Zernike, Tchebichef, Krawtchouk and so on. All these moment functions play an important role in continuous or discrete domains of polynomials range.

2.1. Geometric and complex moments

GMs are defined with the basis kernel $\{x^p y^q\}$. The $(p+q)$ th order two-dimensional GMs are denoted by m_{pq}, and can be expressed as

$$m_{pq} = \iint_{-\infty}^{+\infty} x^p y^q f(x, y) dx dy \tag{1}$$

where $f(x, y)$ is the image intensity function. GMs of low orders have an intuitive meaning, m_{00} is a "mass" of the image (for binary images, m_{00} is an area of the object), m_01 / m_00 and m_10 / m_00 define the *center of gravity* or *centroid* of the image. Second-order moments m_{02} and m_{20} describe the "distribution of mass" of the image with respect to the coordinate axes. In mechanics they are called the *moments of inertia*. Another popular mechanical quantity, the *radius of gyration* with respect to an axis, can also be expressed in terms of moments as $\sqrt{m_{02} / m_{00}}$ and $\sqrt{m_{20} / m_{00}}$, respectively.

Another popular choice of the polynomial basis $(x + iy)^p (x - iy)^q$ where i is the imaginary unit, leads to *complex moments*

$$c_{pq} = \iint_{-\infty}^{+\infty} (x + iy)^p (x - iy)^q f(x, y) dxdy \tag{2}$$

GMs and CMs carry the same amount of information. Each CM can be expressed in terms of GMs as

$$c_{pq} = \sum_{k=0}^{p} \sum_{l=0}^{q} \binom{p}{k} \binom{q}{l} (-1)^{q-l} i^{p+q-k-j} m_{k+j, p+q-k-j} \tag{3}$$

and vice versa

$$m_{pq} = \sum_{k=0}^{p} \sum_{l=0}^{q} \binom{p}{k} \binom{q}{l} \frac{(-1)^{q-l}}{2^{p+q} i^q} c_{k+l, p+q-k-l} \tag{4}$$

CMs are introduced because they behave favorably under image rotation. This property can be advantageously employed when constructing invariants with respect to rotation.

2.2. Orthogonal moments

If the polynomial basis $\{p_{kj}(x, y)\}$ is orthogonal, i.e. if its elements satisfy the condition of orthogonality

$$\iint_G P_{pq}(x, y) P_{mn}(x, y) dxdy = 0 \tag{5}$$

or weighted orthogonality

$$\iint_G w(x, y) P_{pq}(x, y) P_{mn}(x, y) dxdy = 0 \tag{6}$$

for any indexes $p \neq m$ or $q \neq n$, we speak about *orthogonal (OG) moments.* G is the area of orthogonality.

In theory, all polynomial bases of the same degree are equivalent because they generate the same space of functions. Any moment with respect to a certain basis can be expressed in terms of moments with respect to any other basis. From this point of view, OG moments of any type are equivalent to geometric moments.

However, a significant difference appears when considering stability and computational issues in a discrete domain. Standard powers are nearly dependent both for small and large values of the exponent and increase rapidly in range as the order increases. This leads to correlated geometric moments and to the need for high computational precision. Using lower precision results in unreliable computation of geometric moments. OG moments can capture the image features in an improved, non-redundant way. They also have the advantage of requiring lower computing precision because we can evaluate them using recurrent relations, without expressing them in terms of standard powers.

Unlike geometric moments, OG moments are coordinates of *f* in the polynomial basis in the common sense used in linear algebra. Thanks to this, the image reconstruction from OG moments can be performed easily as

$$f(x, y)=\sum_{j}\sum_{k} M_{kj} P_{kj}(x, y) \tag{7}$$

2.3. Continuous moments

Some of orthogonal moments are in terms of a continuous variable such as Legendre, Zernike and Gaussian-Hermite moments. All of these continuous moments depend on the same continuous polynomials. Here, we will discuss about such these polynomial functions and moments.

2.3.1. Legendre moments

There have been many works describing the use of the Legendre moments in image processing, e.g. references [29] and [30], among many others. The *Legendre moments* are defined as

$$\lambda_{pq}=\frac{(2p+1)(2q+1)}{4}\iint_{-1}^{+1} P_p(x)P_q(y)f(x, y)dxd;\; p, q=0,\; 1,\; 2,\; \ldots \tag{8}$$

where $P_n(x)$ is the n th degree *Legendre polynomial* (expression by the so-called Rodrigues' formula)

$$P_p(x)=\frac{1}{2^p p!}\frac{d^p}{dx^p}(x^2-1)^p \tag{9}$$

and the image $f(x, y)$ is mapped into the square $(-1, 1)\times(-1, 1)$. The Legendre polynomials of low degrees expressed in terms of x^p are

$$\begin{aligned} &P_0(x)=1,\\ &P_1(x)=x,\\ &P_2(x)=\tfrac{1}{2}(3x^2-1),\\ &P_3(x)=\tfrac{1}{2}(5x^3-3x),\\ &P_4(x)=\tfrac{1}{8}(35x^4-30x^2+3). \end{aligned} \tag{10}$$

The relation of orthogonality is

$$\int_{-1}^{1} P_p(x) P_q(x) dx = \frac{2}{2q+1} \delta_{pq} \tag{11}$$

The recurrence relation, which can be used for efficient computation of the Legendre polynomials, is

$$\begin{aligned} &P_0(x) = 1, \\ &P_1(x) = x, \\ &P_{p+1}(x) = \frac{2p+1}{p+1} x P_p(x) - \frac{p}{p+1} P_{p-1}(x) \end{aligned} \tag{12}$$

2.3.2. Zernike moments

Zernike moments (ZMs) were introduced into image analysis about 30 years ago by Teague [9] who used ZMs to construct rotation invariants. He used the fact that the ZMs keep their magnitude under arbitrary rotation. He also showed that the Zernike invariants of the second and third orders are equivalent to the Hu invariants when expressed in terms of geometric moments. He presented the invariants up to the eighth order in explicit form but no general rule about how to derive them was given. Later, Wallin [31] described an algorithm for the formation of rotation invariants of any order. Numerical properties and possible applications of ZMs in image processing among others. ZMs of the nth order with repetition l are defined as

$$A_{pq} = \frac{p+1}{\pi} \int_0^{2\pi} \int_0^1 V_{pq}^*(r, \theta) f(r, \theta) r dr d\theta, \quad p = 0, 1, 2, \ldots \; q = -p, -p+2, \ldots, p \tag{13}$$

i.e. the difference $n - |l|$ is always even. The asterisk means the complex conjugate. The *Zernike polynomials* are defined as products

$$V_{pq}(r, \theta) = R_{pq}(r) e^{iq\theta} \tag{14}$$

where the radial part is

$$R_{pq}(r) = \sum_{k=|q|, |q|+2, \ldots}^{p} B_{pqk} r^k \tag{15}$$

The coefficients

$$B_{pqk} = \frac{(-1)^{\frac{p-k}{2}} \left(\frac{p+k}{2}\right)!}{\left(\frac{p-k}{2}\right)! \left(\frac{k+q}{2}\right)! \left(\frac{k-q}{2}\right)!} \tag{16}$$

can be used for conversion from geometric moments,

$$A_{pq} = \frac{p+1}{\pi} \sum_{k=|q|,|q|+2,\ldots}^{p} \sum_{j=0}^{\frac{k-|q|}{2}} \sum_{l=0}^{|q|} \binom{\frac{k-|q|}{2}}{j} \binom{|q|}{l} w^{l} B_{pqk} m_{k-2j-l,2j+l} \tag{17}$$

where

$$w = \begin{cases} -i \ ; & q>0 \\ i \ ; & q \leq 0 \end{cases} \tag{18}$$

The Zernike polynomials satisfy the relation of orthogonality

$$\int_0^{2\pi}\int_0^{1} V_{pq}^{*}(r,\theta)V_{kl}(r,\theta)r dr d\theta = \frac{\pi}{p+1}\delta_{kp}\delta_{lq} \tag{19}$$

The recurrence relation for the radial part is

$$R_{pq}(r) = \frac{2pr}{p+q} R_{p-1,q-1}(r) - \frac{p-q}{p+q} R_{p-2.q}(r) \tag{20}$$

Computation of the Zernike polynomials by this formula must commence with

$R_{pp}(r) = R_{p,-p}(r) = r^{p} \ , \ p=0, \ 1, \ \ldots$

2.4. Discrete moments

There is a group of orthogonal polynomials defined directly on a series of points and therefore they are especially suitable for digital images. Some of these polynomials such as Tchebichef, Krawtchouk, Hahn, dual Hahn and Meixner polynomials.

2.4.1. Tchebichef moments

The 2D TMs of order $(p+q)$ of an image intensity function $f(n, m)$ with size $N \times M$ is defined as

$$T_{pq} = A(p, N)A(q, M)\sum_{n=0}^{N-1}\sum_{m=0}^{M-1} t_p(n)t_q(m)f(n, m) \tag{21}$$

where $t_p(n)$ is the p th order orthogonal discrete Tchebichef polynomial defined by [8]

$$t_p(n) = p!\sum_{k=0}^{p}(-1)^{p-k}\binom{N-1-k}{p-k}\binom{p+k}{p}\binom{n}{k} \tag{22}$$

and

$A(p, N) = \frac{\beta(p, N)}{\rho(p, N)}$

where $\beta(p, N)$ is a normalization factor. The simplest choice of this factor is N^p. The recurrence relation of Tchebichef polynomials with respect to the chosen order is:

$$(p+1)t_{p+1}(n) - (2p+1)(2n - N + 1)t_p(n) + p(N^2 - p^2)t_{p-1}(n) = 0 \tag{23}$$

where $p \geq 1$ and the first two polynomials are $t_0(n) = 1$ and $t_1(n) = 2n - N + 1$. The orthogonality property satisfies the following squared-norm:

$$\rho(p, N) = (2p)!\binom{N+p}{2p+1} \tag{24}$$

2.4.2. Krawtchouk moments

The definition of the nth order classical Krawtchouk polynomial is defined as

$$K_n(x; p, N) = \sum_{k=0}^{n} a_{k,n,p} x^k = {}_2F_1\left(-n, -x; -N; \frac{1}{p}\right) \tag{25}$$

where $x, n = 0, 1, 2, \ldots, N$, $N > 0$, $p \in (0,1)$ and ${}_2F_1(\bullet)$ is the generalized hypergeometric function

$${}_2F_1(a, b; c; z) = \sum_{k=0}^{\infty} \frac{(a)_k (b)_k}{(c)_k} \frac{z^k}{k!} \tag{26}$$

and $(x)_k$ is the Pochhammer symbol given by

$$\begin{aligned} &(x)_k = x(x+1)(x+2)\cdots(x+k-1) \\ &k \geq 1 \text{ and } (x)_0 = 1 \end{aligned} \tag{27}$$

The normalized and weighted Krawtchouk polynomials $\{\bar{K}_n(x; p, N)\}$ are defined as [10]

$$\bar{K}_n(x; p, N) = K_n(x; p, N)\sqrt{\frac{w(x; p, N)}{\rho(n; p, N)}} \tag{28}$$

where the weight function, $w(\bullet)$ and the square norm, $\rho(\bullet)$ are given as

$$w(x; p, N) = \binom{N}{x} p^x (1-p)^{N-x} \tag{29}$$

and

$$\rho(n; p, N) = \left(\frac{p-1}{p}\right)^n \frac{n!}{(-N)_n} \tag{30}$$

The normalized and weighted Krawtchouk polynomials have the following three-term recurrence relation:

$$p(n-N)\bar{K}_{n+1}(x;p,N)=A[p(N-2n)+n-x]\bar{K}_n(x;p,N)-Bn(1-p)\bar{K}_{n-1}(x;p,N) \quad (31)$$

where

$$A=\sqrt{\frac{(1-p)(n+1)}{p(N-n)}}$$

$$B=\sqrt{\frac{(1-p)^2(n+1)n}{p^2(N-n)(N-n+1)}}$$

with

$$\bar{K}_0(x;p,N)=\sqrt{w(x;p,N)}$$

$$\bar{K}_1(x;p,N)=\left(1-\frac{1}{Np}x\right)\sqrt{w(x;p,N)}.$$

The 2D Krawtchouk moment of order $(n+m)$ of an image intensity function $f(x, y)$ with size $N \times M$ is defined as [16]

$$Q_{nm}=\sum_{x=0}^{N-1}\sum_{y=0}^{M-1}\bar{K}_n(x;p_1,N-1)\bar{K}_m(y;p_2,M-1)f(x,y) \quad (32)$$

The orthogonality property leads to the following inverse moment transform

$$f(x,y)=\sum_{n=0}^{N-1}\sum_{m=0}^{M-1}Q_{nm}\bar{K}_n(x;p_1,N-1)\bar{K}_m(y;p_2,M-1) \quad (33)$$

If only the moments of order up to (N_{max}, M_{max}) are computed, then the reconstructed image in (2.30) can be approximated by

$$\tilde{f}(x,y)=\sum_{n=0}^{N_{max}}\sum_{m=0}^{M_{max}}Q_{nm}\bar{K}_n(x;p_1,N-1)\bar{K}_m(y;p_2,M-1) \quad (34)$$

3. Formulation of geometric moments using digital filters

This section first reviews the generation of GMs using digital filter structure as proposed by Hatamian [19]. This is then followed by a review of the improved version used by Wong and Siu [20].

3.1. Hatamian's model

Hatamian proposed the all-pole digital filter structure to compute GMs up to the 16th order. The one dimensional GMs of order p for a N -length sequence $x[n]$ is defined in this model as

$$m_p = \sum_{n=0}^{N-1} n^p x[n] \tag{35}$$

One reason for using digital filters as a moment generator is based on the convolution of the aforementioned sequence with impulse response, $n^p u[n]$, where $u[n]$ is the unit step function. Hence, the output of the digital filter y_p evaluated at the point $n = N - 1$ can be expressed:

$$y_p[N] = \sum_{k=0}^{N-1} x[k](N - k)^p \tag{36}$$

where $x[n]$ is the reversed sequence. Figure 1 shows the structure of a single pole digital filter with transfer function $\frac{1}{z - 1}$, which is equivalent to an accumulator with unity feedback. This accumulator has a delay in the feed-forward path. For p cascaded filters, the corresponding transfer function is given by

$$H_p(z) = \frac{1}{(z - 1)^{p+1}} \tag{37}$$

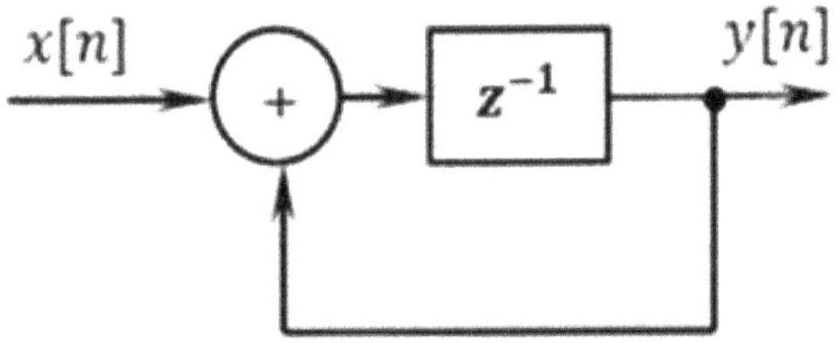

Figure 1. Single-pole filter structure [19] in feedforward path.

The 1D cascaded filters structure is shown in Figure 2, where $H_0(z) = \frac{1}{z - 1}$.

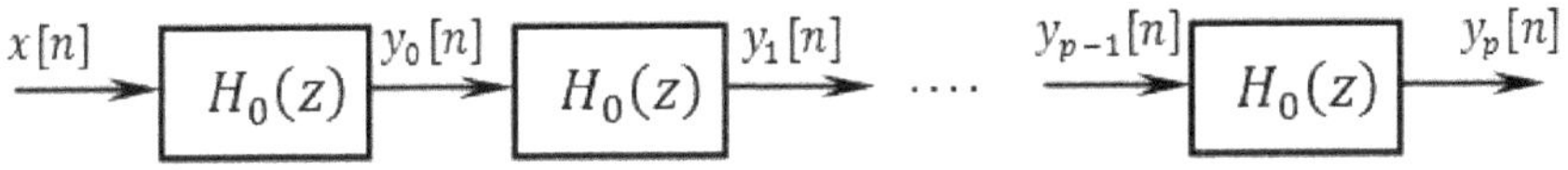

Figure 2. Cascading of single-pole filters for generating of moments up to order p .

The relationship between GMs and the all-pole digital filter outputs are related by the following expression

$$m_p = \sum_{r=0}^{p} C_{p,r} y_r \tag{38}$$

where y_r is the rth digital filter output and $C_{p,r}$ is a matrix of coefficients directly obtained from the impulse responses of the all-pole digital filters as given in [27]:

$$C_{p,r} = \begin{cases} 0 & ; \quad r > p \\ (-1)^p & ; \quad r = 0,\ p \geq 0 \\ rC_{p-1,r-1} - (r+1)C_{p-1,r} & ; \quad r > 0,\ p > 0 \end{cases} \tag{39}$$

For a two dimensional image, the relationship between the GMs and digital filters outputs can be expanded to

$$m_{p,q} = \sum_{r=0}^{p} C_{p,r} \sum_{s=0}^{q} C_{q,s} y_{r,s} \tag{40}$$

However, the all-pole filter structure has delays in the feed-forward path; hence it causes an increase in the computation time.

3.2. Wong and Siu's model

To overcome this problem, Wong and Siu [20] moved the delay unit to the feedback path of the filter, as shown in Figure 3. The transfer function is given as

$$H_p(z) = \left(\frac{1}{1 - z^{-1}}\right)^{p+1} \tag{41}$$

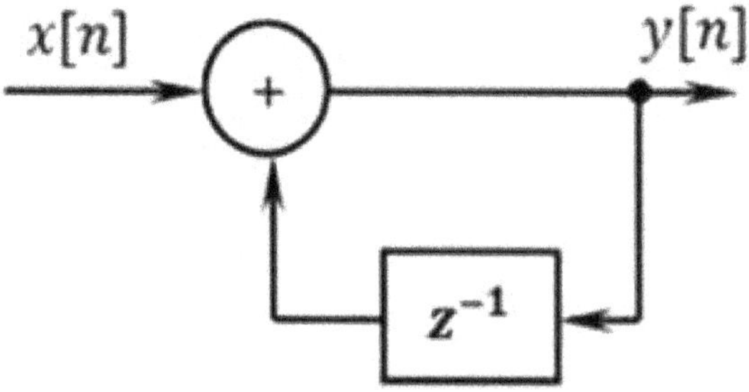

Figure 3. Single-pole filter structure [20] in feedback path.

In this case, however, the GMs are obtained from the digital filter outputs and a different matrix coefficient, as shown below:

$$m_p = \sum_{r=0}^{p} D_{p,r} y_r \tag{42}$$

where the coefficients of $D_{p,r}$ are obtained from the following recurrence formula shown in [23].

$$D_{p,r} = \begin{cases} 0 & ; \quad p>0,\ r=0 \\ 1 & ; \quad p=0,\ r=0 \\ r(D_{p-1.r-1} - D_{p-1,r}) & ; \quad others \end{cases} \tag{43}$$

For a two dimensional image, the relationship between the GMs and the improved digital filters structure outputs are related by

$$m_{p,q} = \sum_{r=0}^{p} D_{p,r} \sum_{s=0}^{q} D_{q,s} y_{r,s} \tag{44}$$

3.3. Proposed method based on the lower output values of digital filters

We begin with the one-dimensional case, where the results can be easily extended to two-dimensional. Consider a digital filter with impulse response $h_p[n] = n^p u[n]$, where $u[n]$ is the unit step function and p is the moment order. Now, assume the input of the digital filter is given as $x[n]$ as mentioned in (3.2). Based on the convolution theorem, the output, $y[n]$ is therefore

$$y_p[n] = x[n] * h_p[n] \tag{45}$$

Using the digital filter as shown in Figure. 3 and changing the unit step function to $u[n+p]$ to accommodate the sampling of the digital filter outputs at earlier instances, we get the following

$$y_p[n] = x[n] * \binom{n+p}{p} u[n+p] = \sum_{k=0}^{n+p} x[k] \binom{n-k+p}{p} \tag{46}$$

Substituting, $n = N - p$ yields

$$y_p[N-p] = \sum_{k=0}^{N} x[k] \binom{N-k}{p} \tag{47}$$

Expanding the binomial equation, and using the Stirling numbers we get

$$y_p[N-p] = \frac{1}{p!}\sum_{k=0}^{N}\sum_{i=0}^{p} s_1(p, i)x[k](N-k)^i \tag{48}$$

where $s_1(p, i)$ are the Stirling numbers of the first kind [28], which satisfy

$$\frac{(N-k)!}{(N-k-p)!} = \sum_{i=0}^{p} s_1(p, i)(N-k)^i \tag{49}$$

Using (3.2), we can rewrite (3.15) in terms of GMs as follows:

$$y_p[N-p] = \frac{1}{p!}\sum_{i=0}^{p} s_1(p, i)m_i \tag{50}$$

Now by taking the inverse of (3.16), the GMs can be obtained in terms of the digital filter outputs thus:

$$m_p = \sum_{r=0}^{p} r!\,s_2(p, r)y_r[N-r] \tag{51}$$

where $s_2(p, r)$ are the Stirling numbers of the second kind [28], and the Stirling numbers of the first and second kind can be considered to be inverses of one another:

$$\sum_{p=0}^{\max\{i,r\}} (-1)^{p-r} s_1(p, i)s_2(r, p) = \delta_{ir} \tag{52}$$

where δ_{ir} is the Kronecker delta.

Notice now, for the p order, it can be shown that the digital filter outputs are sampled at $N-p$, unlike the previous works which were sampled at N or later instances of N [19]-[20]-[21]. As the order $\frac{p}{2}$ is reached, the digital filter output values begin to decrease. This allows the use of low value digital filter outputs for the formulation of GM. The 2D moments can be obtained by expanding the 1D model for the digital filter outputs as follows:

$$m_{p,q} = \sum_{r=0}^{p}\sum_{s=0}^{q} r!\,s!\,s_2(p, r)s_2(q, s)y_{rs}[N-r, N-s] \tag{53}$$

3.4. Experimental studies

A set of experiments were carried out to validate the theoretical framework developed in the previous sections and to evaluate the performance of the proposed structure. This section is divided into 3 parts. In the first subsection, an artificial image of size 4×4 is used to generate GMs up to third order. The computational complexity of three algorithms – the algorithms of [19], [20] and the proposed method – is then analyzed and discussed in the second subsection.

In the third subsection, the speed of the proposed method is compared with the speed achieved using [19] and [20].

3.4.1. *Artificial test image*

rtificial test image of size 4×4 was used to prove the validity of the proposed approach. In this case, the digital filter outputs up to third order were generated. The intensity function of the test image is given in the following matrix:

$$x[m,\ n]=\begin{bmatrix} 111 & 114 & 103 & 116 \\ 109 & 101 & 113 & 108 \\ 106 & 102 & 112 & 104 \\ 107 & 110 & 105 & 115 \end{bmatrix}$$

The difference between the digital filter output values for [20] and proposed structure is shown in Table 1. It is clear that for orders higher than one, the proposed output values are much lower than [20] as the order increases.

Digital filter outputs	Filter structure output values [20]	Proposed structure output values
y_{00}	1736	1736
y_{01}	4326	4326
y_{02}	8651	4325
y_{03}	15148	2172
y_{10}	4358	4358
y_{11}	10870	10870
y_{12}	21750	10880
y_{20}	8742	4384
y_{21}	21811	10941
y_{30}	15331	2205

Table 1. A comparison of filter output values between [20] and the proposed method for an artificial image, $x[m,\ n]$, up to third order.

3.4.2. *Computational complexity*

The advantage of the proposed method lies in the smaller digital filter output values as compared to [19] and [20]. However, it would still be useful to study the computational complexity of these three methods and the direct method in term of the number of additions and multiplications. The proposed method, [19] and [20] consist of two main steps. Digital filter outputs are obtained from the respective digital filter structure. Then, these outputs are linearly combined to compute GMs.

For a grayscale image of size $N \times N$ and GM up to order of s, where $s = (p+q)$, the number of additions and multiplications for the proposed method, [19] and [20] are shown in Table 2.

Algorithm	Additions (Digital filter stages)	Additions (Digital filter outputs to GMs)	Multiplications (Digital filter outputs to GMs)
[19]	$(s+1)\left(N+\frac{s+2}{2}\right)(N+1)$	$\frac{s(s+1)(s+2)(s+7)}{24}$	$\frac{s^4+10s^3+23s^2-34s-24}{24}$
[20]	$(s+1)\left(N+\frac{s+2}{2}\right)N$	$\frac{s(s-1)(s^2+3s+14)}{24}$	$\frac{s(s-1)(s^2+3s+14)}{24}$
Proposed method	$(s+1)\left[N(N+2)-\frac{(s+2)(s-3)}{6}\right]$	$\frac{s(s-1)(s^2+3s+14)}{24}$	$\frac{s(s-1)(s^2+3s+14)}{24}$

Table 2. Complexity analysis of GMs computation using digital filters for image of size $N \times N$ and maximum order of $s = p + q$.

It can be seen that even though the complexity of the linear combination stage for the proposed method is the same as [20], there is saving in the number of additions at the digital filters stage. This can be clearly shown in the example below. For $N = 512$ and $s = 45$, the number of additions needed in the filter stage for [20] is 12612096 while the proposed method just requires 12090594 additions. The summary of the complexity comparison for all the three methods to compute GMs up to 45th order is shown in Table 3.

Algorithm	Additions	Multiplications
[19]	12847524	210795
[20]	12791451	179355
Proposed method	12269949	179355

Table 3. Complexity analysis of GMs computation using digital filters for image of size 512×512 and $s = 45$.

For a 128×128 grayscale image, the advantage of the proposed filter structure as compared to [19] and [20] is clearly depicted in Figure 4.

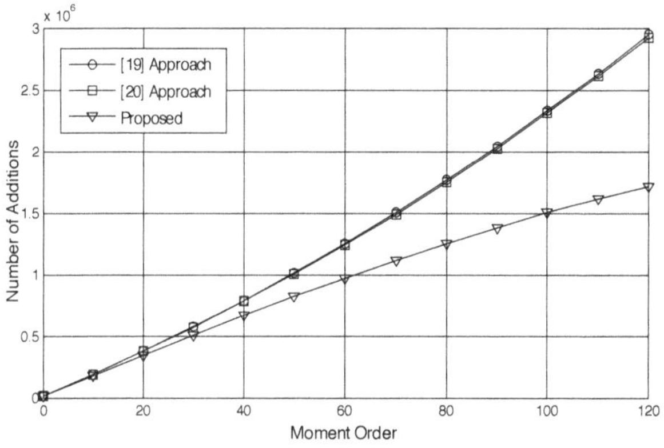

Figure 4. Number of additions for a 128 × 128 grey-scaled image.

3.4.3. Speed performance and comparison studies

The computation speed of the proposed method is compared with [19] and [20]. CPU elapsed time is used in the evaluation process in all the performed numerical experiments. The codes were all written in MATLAB7 and simulations were run with a 3GHz Intel Core2 with 2GB RAM and the average time is used in the discussion. The images used for the experiments were the Pepper and Lena images, as shown in Figure 5.

Table 4 shows the simulation time for GMs computation on the Pepper image of size 128×128 for orders of 5 to 55, in a step of 10. The same experiment was repeated on the Lena image of size 512×512. The simulation times are shown in Table 5.

Moment order ($s = p + q$)	[19]	[20]	Proposed method
5	0.4	0.39	0.388
15	1.12	1.104	1.05
25	2.03	1.97	1.81
35	3.28	3.14	2.82
45	5.15	4.87	4.32
55	7.96	7.48	6.63

Table 4. CPU elapsed time in milliseconds for the 128 × 128 Pepper's test image.

Moment order ($s = p + q$)	[19]	[20]	Proposed method
5	6.03	6.02	6.00
15	16.264	16.222	16.018
25	26.817	26.722	26.13
35	37.866	37.677	36.494
45	49.663	49.325	47.343
55	62.542	61.982	58.99

Table 5. CPU elapsed time in milliseconds for the 512×512 Lena's test image.

The tables clearly show that the proposed method requires less time than [19] and [20] to compute GMs of the same order. They also show that the time saving increases as the order of moments increases. In Figure 6 and Figure 7 we provide a comparison of required CPU time between [19], [20] and the proposed method.

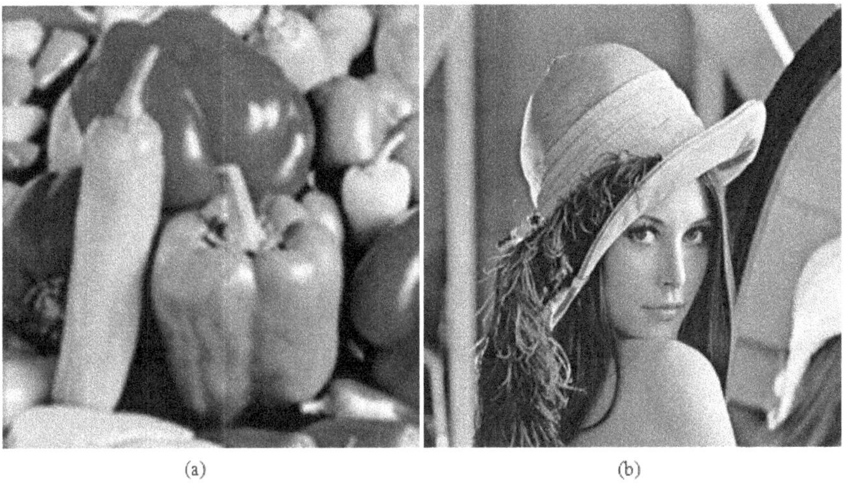

Figure 5. Test images: (a) Peppers and (b) Lena.

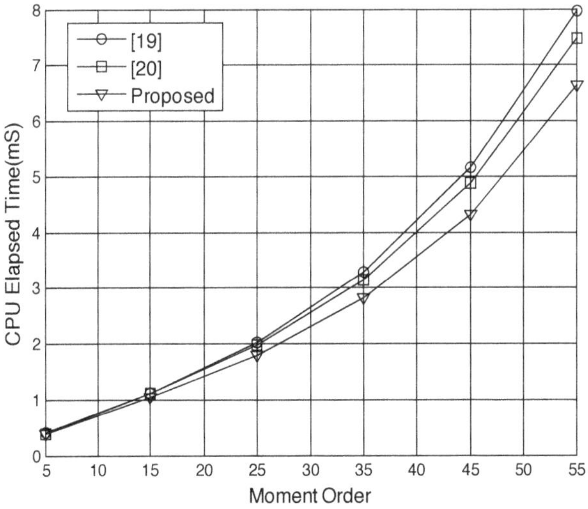

Figure 6. Linear scale of CPU time in seconds for the 128×128 gray-scale peppers image.

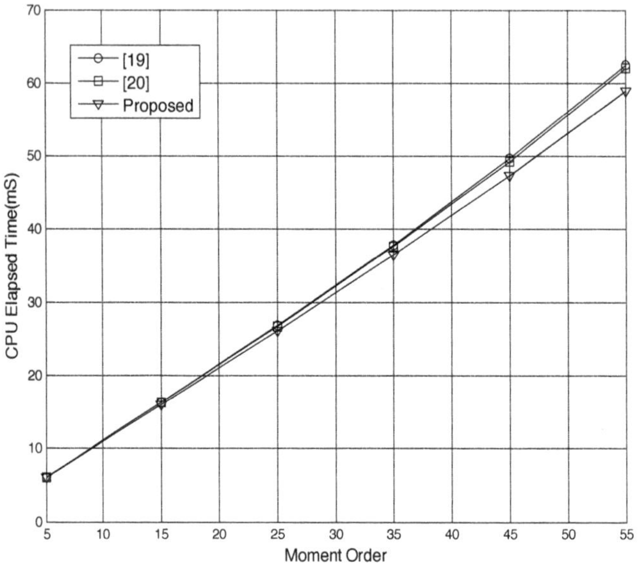

Figure 7. Linear scale of CPU time in seconds for the 512×512 gray-scale Lena image.

4. A Reduced 2D digital filter structures for fast computation of geometric moments

In the previous works, the linear transformation is performed using an external PC to relate the outputs of the digital filter structure to generate the moments. However, finding of a relationship between the geometric or orthogonal moments and the digital filters outputs has been stood as the main meaning of similar works [19]-[20].

In all the above aforementioned literatures, the basic concept involved the computation of geometric and other types of moments using the accumulator grid structure has remain unchanged over the past decades.

First we summarize the designing modality of all-pole digital filters using 1-D and 2-D Z-transforms [32]. We believe that relevance creation manner among the real-time and Z-domain is one of the most essential issues in studying of digital image moments. This section presents the results of a study aimed at the formulation of a new method to reduce the filter resources used in the digital filter structure. This study, though not exhaustive, serves to provide a suitable classification framework that underlies a new approach in digital filter design for the computation of moments.

4.1. Reduced digital filter structure

Unlike the previous researchers who used a 1-D Z-transform to derive a 2-D digital filter structure by cascading the filters both in the rows and columns. However, we use a 2-D definition of Z-transform to obtain the impulse response of the filter which led to a reduced digital filter structure as compared with [19]. The 2-D Z-transform for the image, $f[m, n]$ is given as:

$$F(z_1, z_2)=\sum_{m=1}^{+\infty}\sum_{n=1}^{+\infty} f[m, n]z_1^{-m}z_2^{-n} \tag{54}$$

The impulse response for a 2-D image becomes:

$$h_{p,q}[m, n]=\binom{m+p}{m}u[m]\binom{n+q}{n}u[n] \tag{55}$$

and the transfer function $H_{p,q}(z_1, z_2)$, in the 2-D Z-transform domain for this filter structure is shown as

$$H_{p,q}(z_1, z_2)=\frac{1}{\left(1-z_1^{-1}\right)^{p+1}\left(1-z_2^{-1}\right)^{q+1}} \tag{56}$$

Using the above transfer function, the relationship between the input and the output of the digital filter is:

$$Y_{p,q}(z_1, z_2)=\frac{X(z_1, z_2)}{\left(1-z_1^{-1}\right)^{p+1}\left(1-z_2^{-1}\right)^{q+1}} \tag{57}$$

Based on (4.4), the zero order of the digital filter output is derived as

$$Y_{00}(z_1, z_2)=\frac{X(z_1, z_2)}{\left(1-z_1^{-1}\right)\left(1-z_2^{-1}\right)} \tag{58}$$

Thereafter, a recurrence relationship between the previous and the next outputs of each digital filter for the row and columns as shown in Figure 8 can be obtained:

$$Y_{p+1,q}(z_1, z_2)=\frac{Y_{p,q}(z_1, z_2)}{1-z_1^{-1}} \tag{59}$$

$$Y_{p,q+1}(z_1, z_2)=\frac{Y_{p,q}(z_1, z_2)}{1-z_2^{-1}} \tag{60}$$

By taking the inverse 2-D Z- transform of (4.6) and (4.7), we get the following:

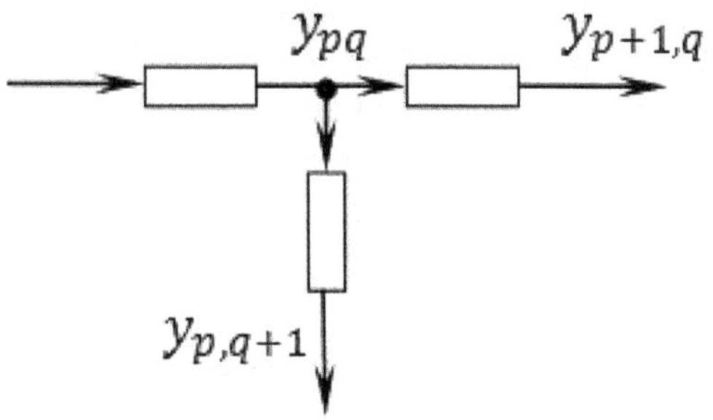

Figure 8. Recurrence relationship between the digital filter outputs of the row and column.

$$y_{p+1,q}[m, n] = y_{p+1,q}[m-1, n] + y_{p,q}[m, n] \tag{61}$$

$$y_{p,q+1}[m, n] = y_{p,q+1}[m, n-1] + y_{p,q}[m, n] \tag{62}$$

The implementation of proposed digital filter structure for generating moments up to $(p+q)$ orders is shown in Figure 9.

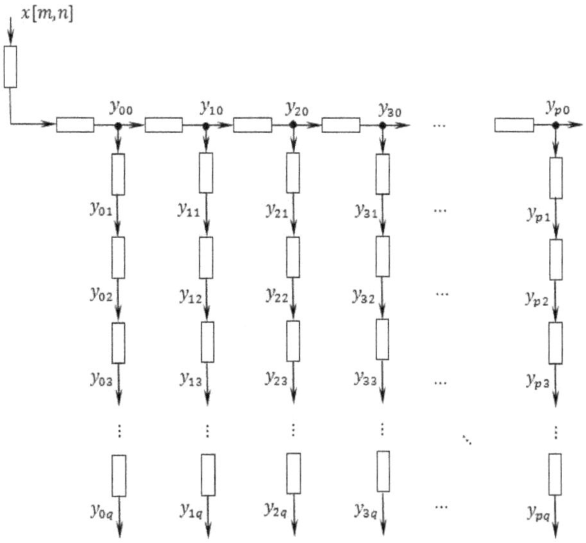

Figure 9. Reduced digital filters for generating the 2-D geometric moments up to $(p+q)$ order.

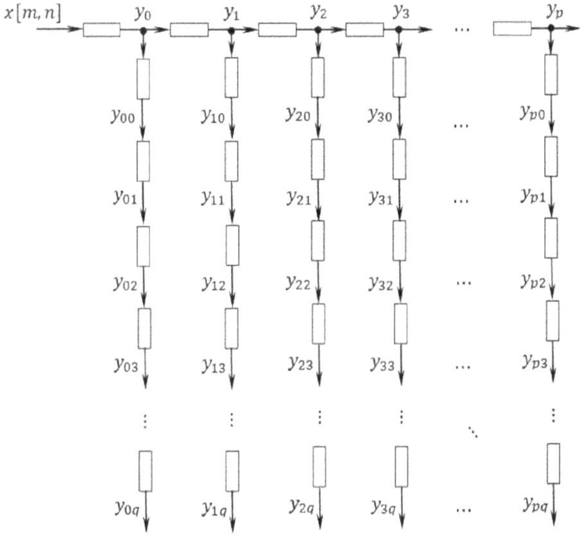

Figure 10. Digital filters for generating the 2-D geometric moments up to $(p+q)$ order used in [2] and [8].

As compared to the digital filter structure used in [19] and [20] as shown in Figure 10, it can be seen that the difference is the prefix, y_{pq} and y_p used in the rows. In the proposed model, the outputs y_{00}, y_{10},..., y_{p0} occur at the row filter, unlike in [19] and [20], they occur at the column filters. Hence, the difference in savings of digital filters used in the proposed method is determined by the maximum p order. for example, in the design of the 40th order geometric moments, where the maximum p order is 40, and thus the number of digital filters used in proposed method will be 40 digital filters less than [19] and [20].

In our proposed method, we begin by showing the relationship between the digital filter outputs and the geometric moments by considering a 1D image. So, the first few outputs of the digital filter are

$$\begin{aligned} y_0[N] &= \sum_{k=1}^{N} x[k]h_0[N-k] = \mu_0 \\ y_1[N] &= \sum_{k=1}^{N} x[k]h_1[N-k] = \sum_{k=1}^{N} x[k](N-k+1) = (N+1)\mu_0 - \mu_1 \\ y_2[N] &= \sum_{k=1}^{N} x[k]h_2[N-k] = \sum_{k=1}^{N} x[k]\tfrac{1}{2}(N-k+1)(N-k+2) = \tfrac{1}{2}(N+1)(N+2)\mu_0 - \tfrac{1}{2}(2N+3)\mu_1 + \tfrac{1}{2}\mu_2 \end{aligned} \tag{63}$$

Hence by solving them, the above geometric moments can be obtained in terms of the digital filters outputs.

$$\begin{aligned}\mu_0 &= y_0\\ \mu_1 &= (N+1)y_0 - y_1\\ \mu_2 &= (N+1)^2 y_0 - (2N+3)y_1 + 2y_2\end{aligned} \tag{64}$$

A matrix notation showing the above relationship between the geometric moments and the digital filter outputs as expressed

$$\mu_P = C_N Y_P \tag{65}$$

where μ_P is the geometric moment vector, Y_P is the digital filter output vector, and C_N is a square matrix that a recurrence relationship between this matrix elements is given as

$$C_{pr} = \begin{cases} 0 & ; \quad p<r \\ (-1)^p p! & ; \quad p=r \\ (N+1)^p & ; \quad r=0 \\ -pC_{p-1,r-1} + p\sum_{k=r}^{p-1} \dfrac{C_{p-1,k}}{2^{k-r+1} \times 3^{\left\lfloor \frac{k-r+1}{2} \right\rfloor}} & ; \quad oth. \end{cases} \tag{66}$$

where, $\lfloor \bullet \rfloor$ represents the *floor function.*

Following the same procedure used in obtaining the 1-D relationship digital filter outputs and geometric moments, it can also be extended for 2-D image. This is done by taking the transpose of the digital filter outputs and the transpose of the matrix, C_N . For a 2-D image of dimensions $\times N$, the geometric moments can be obtained from

$$\mu_{pq} = C_M Y_{pq}^T C_N{}^T \tag{67}$$

Also, the summation forms of (4.12) and (4.14) can be written as:

$$\mu_p = \sum_{r=0}^{p} C_{pr} y_r \tag{68}$$

$$\mu_{pq} = \sum_{r=0}^{p} \sum_{s=0}^{q} C_{pr} C_{qs} y_{sr} \tag{69}$$

4.2. Experimental results

In this subsection, we will begin with an example to determine the geometric moments up to third order for 2-D image using the proposed and existing method [19]. Figure.3. shows the proposed digital filter structure and the structure used in [19]. As can be seen from Figure 11(a) and (b), the proposed filter structure used three less digital filters. The difference between them is maximum p- order. As the number of moment orders increases the savings of the digital filters will be large. For example, the geometric moments used are forty orders, and then the number of digital filters will be reduced by forty as compared with existing methods.

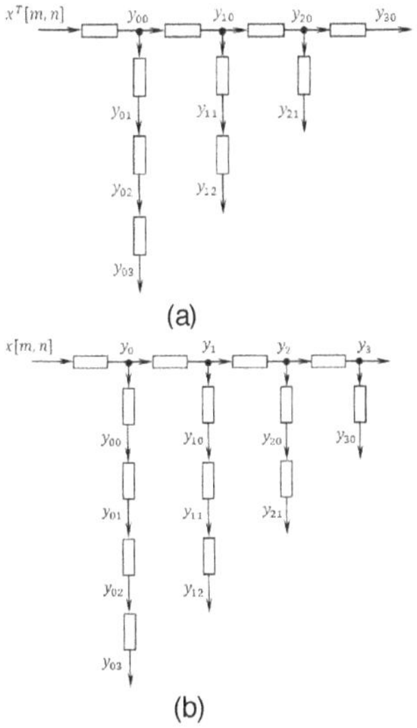

Figure 11. Digital filter structure for generating up to third order geometric moments for 2-D image. (a) proposed model and (b) existing model.

4.2.1. Artifical test image

Artificial test images of small size are used to prove validity of the proposed architecture. To illustrate the workings, an artificial image with size of 2×3 to generate up to third order of geometric moments is considered. The intensity function of the test image is represented by the following input matrix:

$$x[m, n] = \begin{bmatrix} 5 & 3 & 6 \\ 2 & 4 & 1 \end{bmatrix}$$

Consider the dimension of the input image is 2×3, it means the endpoint of the filter outputs, is 6. Therefore, all generated outputs by the row filters are defined at different intervals till 6, where be represented by the Table 6. Table 7 shows the generated outputs by the column filters for two sampled seconds (3,6). The input image must be reversed in this method and denoted as x^R.

n	0	1	2	3	4	5	6
x^R	0	1	4	2	6	3	5
y_0	0	1	5	7	6	9	14
y_1	0	1	6	13	6	15	29
y_2	0	1	7	20	6	21	50
y_3	0	1	8	28	6	27	77

Table 6. State table method used in [8] for obtaining row filter outputs.

n	3	6	n	3	6	n	3	6	n	3	6
y_0	7	14	y_1	13	29	y_2	20	50	y_3	28	77
y_{00}	7	21	y_{10}	13	42	y_{20}	20	70	y_{30}	28	105
y_{01}	7	28	y_{11}	13	55	y_{21}	20	90			
y_{02}	7	35	y_{12}	13	68						
y_{03}	7	42									

Table 7. State table method used in [8] for obtaining column filter outputs.

The highlighted outputs can be collected as Y matrix in (4.12) and the coefficients matrix (D) as defined in (3.9), will generate the third order moments' set of the artificial image as follows:

$$\begin{aligned} &\mu_{00}=21,\ \mu_{01}=42,\ \mu_{02}=98,\ \mu_{03}=252 \\ &\mu_{10}=28,\ \mu_{11}=55,\ \mu_{12}=125 \\ &\mu_{20}=42,\ \mu_{21}=81 \\ &\mu_{30}=70 \end{aligned} \tag{70}$$

If we derive the moments of the same example with the proposed method, the endpoint of the procedure is also 6. However, the results of the digital filter outputs are listed in Tables 8 and 9. In this case, according to Figure 11(a), it is well-known that the number of the digital filter is deduced up to maximum p- order. Furthermore, in this method it does not need to reverse of the input sequence.

n	0	1	2	3	4	5	6
x^T	0	5	2	3	4	6	1
y	0	5	7	3	7	6	7
y^T	0	5	3	6	7	7	7
y_{00}	0	5	8	14	7	14	21
y_{10}	0	5	13	27	7	21	42
y_{20}	0	5	18	45	7	28	70
y_{30}	0	5	23	68	7	35	105

Table 8. State table of proposed method for obtaining row filter outputs.

n	3	6	n	3	6	n	3	6	n	3	6
y_{00}	14	21	y_{10}	27	42	y_{20}	45	70	y_{30}	68	105
y_{01}	14	35	y_{11}	27	69	y_{21}	45	115			
y_{02}	14	49	y_{12}	27	96						
y_{03}	14	63									

Table 9. State table of proposed method for obtaining column filter outputs.

Based on (4.13) and (4.14), we can derive the digital filter outputs' matrix (Y) and the coefficients matrices (C_M , C_N), respectively. Then, the set of third order moments are obtained as (4.17).

4.2.2. Real test image

The images used for the experiments were the Pepper (128×128) and Lena (512×512) images as shown in Figure 12(a) and (b). As shown in Figure 13(a) and (b) the number of additions for digital filter processes are compared in [19] and the proposed method.

(a) (b)

Figure 12. Test images (a) Pepper and (b) Lena.

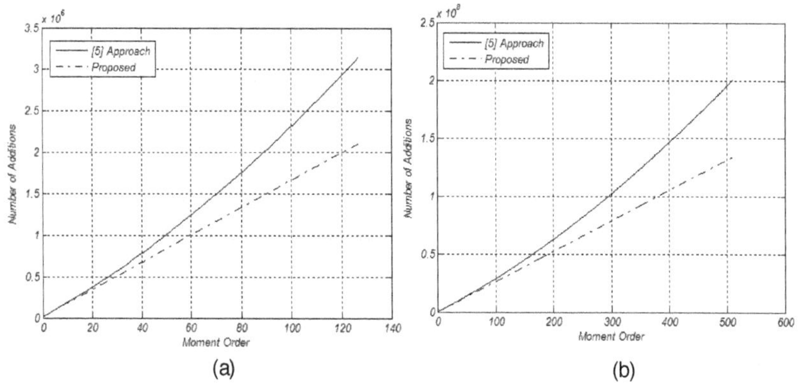

Figure 13. Comparison of the number of additions in digital filter part for [19] and proposed method for (a) Pepper image (b) Lena image.

5. Conclusions

Orthogonal moments are widely used in image analysis and as pattern features in pattern classification. The computation of orthogonal moments can be achieved via geometric moments. Hatamian introduced cascaded digital filters, where each filter operates as an accumulator to generate geometric moments.

One of the weaknesses in using the outputs of the cascaded digital filters to generate the GMs is that the filter outputs increase exponentially as the orders of the moments increase. This work proposes a new formulation to solve this problem by sampling at earlier instances of $N,\ N-1,\ \ldots.N+p$, where p is the maximum moment order for an $N \times N$ image. This step then paved the way to use a set of lower digital filter output values. The work demonstrates the efficacy and validity of the new algorithm in two ways: one, by comparing its complexity of computation with the complexity of two state of the art models proposed by Katoulos and Andreadis [21] and Wong and Siu [20] and two, by carrying out a set of experiments on speed of computation comparing the results obtained using the proposed method with those obtained using existing methods. A number of findings indicate the superiority of the proposed method: (1) A savings of as much as 45% is achieved for the proposed method in the number of additions as the moment order approaches N when compared with the existing methods and (2) this leads to less computational time for the proposed method to derive the GMs. This work has focused on the software implications of the algorithm. If the proposed method is implemented in FPGA or ASIC based platforms, a great savings in terms of bit-widths will be realized.

Also we showed a reduced method where the number of digital filters used in the generating of geometric moments is reduced by p -order when compared with the existing methods. The proposed method is modeled using the 2D Z-transform, and the theoretical framework for the proposed reduced digital filter structure is developed and the experimental results validate the performance.

Author details

Barmak Honarvar Shakibaei Asli and Raveendran Paramesran

Elecrtical Engineering Department, University of Malaya, KulaLumpur, Malaysia

References

[1] M. K. Hu. Visual pattern recognition by moment invariants, IRE Trans. Information Theory 8: 1962, 179-187.

[2] H. S. Hsu. Moment preserving edge detection and its application to image data compression, Opt. Eng. 32: 1993, 1596–1608.

[3] M. I. Heywood. Fractional central moment method for moment-invariant object classification, Proc. Inst. Elect. Eng. 142: 1995, 213–219.

[4] S. Ghosal and R. Mehrotra. Orthogonal moment operators for subpixel edge detection, Pattern Recognition 26: 1993, 295–306.

[5] S. O. Belkasim. Pattern recognition with moment invariants—A comparative study and new results, Pattern Recognition 24: 1991, 1117–1138.

[6] J. Flusser. Pattern recognition by affine moment invariants, Pattern Recognition 26: 1993, 167–174.

[7] J.F. Boyce and W.J. Hossack. Moment invariants for pattern recognition, Pattern Recognition Letters 1: 1983, 451-456.

[8] Dong Xu and Hua Li. Geometric moment invariant. Pattern Recognition 41: 2008, 240-249.

[9] M.R. Teague. Image analysis via the general theory of moments, J. Optical Society of America 70: 1980, 920-930.

[10] A. Goshtasby. Template matching in rotated images, IEEE Transaction on Pattern Analysis and Machine Intelligence PAMI-7: 1985, 338–344.

[11] V. Markandey and R. J. P. Figureueiredo. Robot sensing techniques based on high dimensional moment invariants and tensors, IEEE Transactions on Robotics and Automation 8: 1992, 186–195.

[12] Y.S. Kim, W.Y. Kim. Content-based trademark retrieval system using visually salient feature, Image and Vision Computing 16: 1998, 931–939.

[13] C.W. Chong, P. Raveendran, R. Mukundan. A comparative analysis of algorithms for fast computation of Zernike moments, Pattern Recognition, 36: 2003, 731–742.

[14] Yap P. T., Paramesran R., Ong S. H. Image analysis using Hahn moments, IEEE Transaction on Pattern Analysis and Machine Intelligence PAMI-11: 2007, 2057–2062.

[15] R. Mukundan, S. H. Ong, and P. A. Lee. Image analysis by Tchebichef moments, IEEE Transaction on Image Processing 10: 2001, 1357–1364.

[16] P. T. Yap, R. Paramesran, and S. H. Ong. Image analysis by Krawtchouk moments, IEEE Transaction on Image Processing 12: 2003, 1367–1377.

[17] G.Wang and S.Wang. Recursive computation of Chebyshev moment and its inverse transform, Pattern Recognition 39: 2006, 47–56.

[18] Guojun Zhang, Zhu Luo, Bo Fu, Bo Li, Jiaping Liao, Xiuxiang Fan, Zheng Xi. A symmetry and bi-recursive algorithm of accurately computing Krawtchuk moments, Pattern Recognition Letters 31: 2010, 548-554.

[19] M. Hatamian. A real-time two-dimensional moment generating algorithm and its single chip implementation, IEEE Transaction on Acoustic, Speech and Signal Processing, ASSP-34: 1986, 546–553.

[20] Wong. W.-H and Siu. W.-C. Improved digital filter structure for the fast moments computation, Proceedings of IEE on Vision, Image and Signal Processing, 146: 1999, 73-79.

[21] L. Kotoulas and I. Andreadis. Fast computation of Chebyshev moments, IEEE Transaction on Circuits and System for Video Technology 16: 2006, 884–888.

[22] L. Kotoulas and I. Andreadis. Real-time computation of Zernike moments, IEEE Transaction on Circuits and System for Video Technology 15: 2005, 801–809.

[23] M. Al-Rawi, Fast zernike moments. Journal on Real-Time Image Processing 3: 2008, 86–96.

[24] H. S. Kim and H Lee. Invariant image watermark using Zernike moments, IEEE Transaction on Circuits and System for Video Technology 13: 2003, 766–775.

[25] S. P. Prismall, M. S. Nixon, and J. N. Carter. On moving object reconstruction by moments, in 13th British Machine Vision Conference, 2002, pp: 73–82.

[26] G. Amayeh, G. Bebis, A. Erol, and M. Nicolescu. Peg-free hand shape verification using high order Zernike moments, in Proceeding Conference on Computer Vision and Pattern Recognition Workshop, 2006, pp: 17–22.

[27] M. Al-Rawi, Y. Jie. Practical fast computation of Zernike moments, Journal of Computer Science and Technology 17: 2002, 181–188.

[28] Hayes MH. Schaums's outline of theory and problems of digital signal processing, McGraw Hill, 1999, New York.

[29] Zhang, H., Shu, H., Luo, L. and Dillenseger, J. L. A Legendre orthogonal moment based 3D edge operator, Science in China Series G: Physics, Mechanics and Astronomy, vol. 48, no. 1, 2005, pp. 1–13.

[30] R. Mukundan and K. R. Ramakrishnan, Fast computation of Legendre and Zernike moments, Pattern Recognition 28, 1995, 1433–1442.

[31] Wallin, A. and Kübler, O. Complete sets of complex Zernikemoment invariants and the role of the pseudoinvariants, IEEE Transactions Pattern Analysis and Machine Intelligence, vol. 17, no. 11, 1995, pp. 1106–10.

[32] Barmak Honarvar, Raveendran Paramesran, Kim Han-Thung, Kah-Hyong Chang. A reduced 2-D digital filter structure for fast implemebntation of geometric moments, International Conference on Computer and Electrical Engineering 4th (ICCEE), Singapore, 2011.

[33] Chern-Loon Lim, Barmark Honarvar, Kim Han Thung, Raveendran Paramesran. Fast computation of exact Zernike moments using cascaded digital filters. Journal Information Sciences, Vol. 181, No.17, 2011, pp 3638-3651.

[34] Kah-Hyong Chang, Raveendran Paramesran, Barmak Honarvar Shakibaei Asli and Chern-Loon Lim. Efficient Hardware Accelerators for the Computation of Tchebichef Moments, IEEE Transaction on Circuit, System, and Video Technology. Vol 22, No. 3, 2012, pp 414-425.

Chapter 7

A Study on a Filter Bank Structure With Rational Scaling Factors and Its Applications

Fumio Itami

Additional information is available at the end of the chapter

1. Introduction

Filter banks decompose signals into multiple sub-bands to perform various processes and reconstruct the original signals [1]-[8]. It has been demonstrated that sub-band processing with filter banks improves the performance of numerous image processing applications such as image recognition, watermarking, image coding,and so on [9]-[14].

For example, image recognition accuracy is improved with sub-band processing. A watermark is typically embedded into a middle frequency band in order to make the scheme not only robust to compression but also secure against attacks on the watermark. Thus, it is highly desirable that analysis filters have excellent frequency characteristics to appropriately decompose signals in such applications.

In sub-band coding, on the other hand, we usually focus more on various visual issues proper to compression such as checkerboard artifacts and blocking effects, rather than the frequency characteristics of the analysis filters. Coding gain is also one of the significant issues in image coding. Therefore, various constraints are imposed on the filters to deal with such problems in the design. Accordingly, the design is usually performed with non-linear optimization, which increases its design complexity.

Multi-resolution processing, which is an extension of sub-band processing, is also in demand in a number of applications. Multi-resolution schemes such as wavelet transform not only decompose the frequencies of signals but also yield various lower resolution versions of the original signals by gradually reducing the sampling rate with filter banks.

Meanwhile, image scaling has become highly required processing in the context of the diversification of the means of displaying images [15]-[17]. Image scaling changes the resolution of images more flexibly than multi-resolution processing, although it does not decompose the frequency components of signals. Typically it consists of up/down samplers and a filter which suppresses the imaging or aliasing components caused by the samplers.

Under the circumstances, we discuss complex processes where sub-band processing and scaling are carried out sequentially in this chapter. For instance, when watermarked images are scaled, or scaling is applied to decoded images, a filter bank and a scaling structure are designed separately and implemented sequentially. However, they have the structural similarity in that they involve samplers and filters. This implies that one can integrate them in order to implement them with lower implementation cost.

Therefore, this chapter provides a profound discussion on a filter bank structure to yield an effective scheme in computation in such complex processing. We introduce a simple filter bank structure for computational simplicity which directly synthesizes the analysis decomposed signals to produce a scaled signal, not to reconstruct the original signal as in traditional perfect reconstruction filter banks.

First, we discuss the frequency decomposition characteristics of the filter bank to show how to obtain arbitrarily scaled signals. We briefly clarified the band-width of the synthesis filters to obtain the scaled signal from the analysis part of the filter bank in [18]. However, the discussion is not sufficient since the behavior of the aliasing components, which can significantly affect the filter bank performance, is not clarified. This chapter discusses the issue so that the feature of the filter bank will be manifested. It is shown that the aliasing problem is similar to that of perfect reconstruction filter banks.

Next, theoretical conditions for addressing the above aliasing problem are given through the input-output relation of the filter bank. We also discussed the strategy to make the filter bank equivalent to the direct scaling structure by deriving such theoretical conditions in [18]. However, we observed the number of conditions is proportional to that of aliasing components, which can lead to time-consuming design or complicated optimization. Moreover, we also know the conditions are usually not satisfied when all the filters have linear phase, so that the derived filter banks do not have excellent performance.

In this chapter, we discuss the conditions from a viewpoint of the aliasing components to verify that they are the same regardless of the aliasing components, so that the number of equations to solve is significantly reduced. A design procedure with the conditions is also discussed. We mention that the procedure is applicable to various image processing mentioned above. In addition, it is demonstrated through simulation results that scaled images' quality is comparable to that of the sequential structure even though not all the filters have linear phase characteristics. Finally, we discuss potential issues and advantages in making use of the scheme as well as traditional ones in practical image processing.

2. Image scaling

Image-scaling which changes the resolution of images has become highly required processing in the context of the diversification of the means of displaying images. In such context, various image processing such as image coding, watermarking and recognition is also applied before image scaling, in which a filter bank is often accompanied by a scaling structure sequentially in practice since filter banks are used in many image processing applications with their sub-band decomposition performance.

Typically, the direct scaling structure consists of up/down samplers and a filter which suppresses the imaging or aliasing components caused by the samplers. Similarly, the filter bank structure also involves the samplers and filters. This indicates that one can integrate both the structure in order to implement them with lower computational complexity than that of their sequential structure shown in Figure 1.

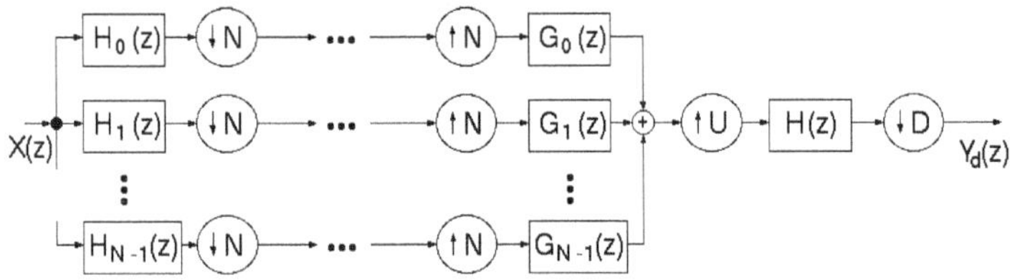

Figure 1. The typical sequential structure.

For this, we provide a study on a simple filter bank structure depicted in Figure 2 which is a simplified version of the structure given in [18]. Please note that the sampling rate at the output in both the structures Figure 1 and Figure 2 is the same when $R = U$ and $S = D$. In addition, the number of channels in both the systems is also the same when $N = D$. Thus, we assume this relation in this chapter.

The aim of this chapter is to demonstrate that the computational complexity of Figure 2 for scaling is reduced compared to that of Figure 1, while its scaling performance is comparable to that of Figure 1.

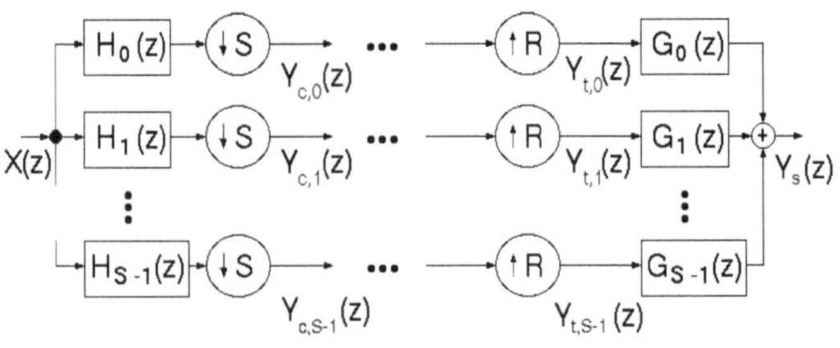

Figure 2. The introduced filter bank structure.

3. A discussion on the computational complexity of the filter bank for scaling

First, we discuss the computation cost of the introduced filter bank drawn in Figure 2 as well as the sequential structure shown in Figure 1.

The sequential structure decomposes the input signal into sub-band frequency components first with down-sampling by the factor D . Next, the decomposed signals are synthesized after up-sampling by D . Such sub-band processing with the filter bank is followed by the direct scaling structure, where further up-sampling by U is carried out. Thus, the sampling rate increases with two successive up-sampling processes. After this, the sampling rate decreases with down-sampling by D .

On the other hand, the introduced filter bank structure also decomposes the input into multiple sub-band signals with down-sampling by D first, and then synthesizes the signals with up-sampling by U .

Accordingly, we see that the sequential structure requires two down-sampling and up-sampling processes, while the introduced filter bank structure is carried out with only one down-sampling and up-sampling process. Such a structural difference can cause different computation time for scaling signals.

If we assume that both the systems have the same length of the analysis filters, then the difference in length between both the systems will be in their synthesis part. The difference will also cause different computation time. We also discuss these concerns with the design and simulation in both of the systems in detail.

4. A discussion on the filter bank structure for scaling

4.1. The frequency decomposition and synthesis characteristics of the filter bank

Next, we discuss the performance of the frequency decomposition and synthesis with the filter bank. Figure 3(a) shows an original signal $X(z)$ where the bold line represents π [rad] on the frequency axis. Figure 3(b) shows the frequency decomposition characteristics with the analysis low-pass filter $H_0(z)$ followed by down-sampling by the factor D . We also show the counterpart of the high-pass filter $H_{D\text{-}1}(z)$ and down-samplingin Figure 3(c). The frequency components colored with dots represent the aliasing components. Please note that these decomposition characteristics in the analysis part are the same as those of traditional (perfect reconstruction) filter banks.

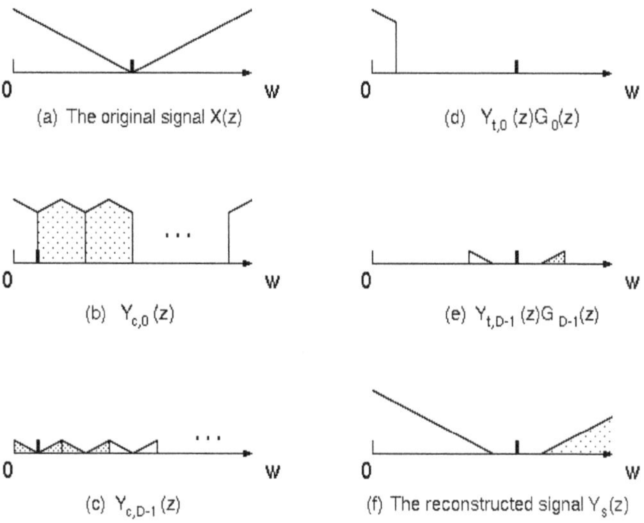

Figure 3. The frequency decomposition and synthesis in the filter bank

The analysis signals such as Figure 3(b) and (c) are described in the z domain as

$$Y_{c,i}(z)=\frac{1}{D}\sum_{r=0}^{D-1} X\left(W^r z^{\frac{1}{D}}\right)H_i\left(W^r z^{\frac{1}{D}}\right)(i=0,1,\ \cdots,\ D-1) \tag{1}$$

where

$$W=e^{-j\frac{2\pi}{D}} \tag{2}$$

These analysis signals are up-sampled by the factor U and then filtered with the filters $G_i(z)$ in the synthesis part, which is illustrated with Figure 3(d) and (e) in the case of $G_0(z)$ and $G_{D-1}(z)$, respectively. Please notice that the pass band of the filters $G_i(z)$ is not the same as-that of $H_i(z)$ since the factor U and D are not the same either to yield scaled signals.

These signals in the synthesis part are represented as

$$Y_{t,i}(z)=\frac{1}{D}\sum_{r=0}^{D-1} X\left(W^r z^{\frac{U}{D}}\right)H_i\left(W^r z^{\frac{U}{D}}\right)(i=0,1,\ \cdots,\ D-1) \tag{3}$$

The eventual scaled signal, which is the output of the filter bank, is drawn in the Figure 3(f), and also represented as

$$Y_s(z)=\frac{1}{D}\sum_{i=0}^{D-1} Y_{t,i}(z)G_i(z) \tag{4}$$

In this manner, one can obtain a scaled signal by using the filter bank, and the scaled signal is completely reconstructed when the ideal filters are used in the filter bank. However, it is not possible to use such filters, as in traditional filter banks, and the aliasing components affect the scaling performance with the filter bank.

Therefore, we provide a discussion on the aliasing components in the next, which manifests the scaling behavior of the filter bank.

4.2. The aliasing components of the filter bank

The output signal of the filter bank is rewritten as

$$Y_s(z)=\sum_{i=0}^{D-1} G_i(z)\frac{1}{D}\sum_{r=0}^{D-1} X\left(W^r z^{\frac{U}{D}}\right)H_i\left(W^r z^{\frac{U}{D}}\right) \tag{5}$$

When $U=D$, that is, a perfect reconstruction filter bank is used, then the pass-band of the synthesis filters is the same as that of the analysis one. In this case, "serious" aliasing caused by the "mismatch" of the band of the filters is avoided, and only minor aliasing caused by the non-ideal filters exists in the adjacent areas between the aliasing componentsin the frequency domain.

On the other hand, when $U \neq D$, which is necessary to obtain an arbitrarily scaled signal, the pass-band of the synthesis ones is not the same as that of the analysis ones, and the serious aliasing occurs. Actually, it can be observed that the pass-band of $G_i(z)$ overlaps with the aliasing components in the above equation and also in Figure 3.

However, it can also be verified in the frequency domain that such serious aliasing is essential and necessary to obtain a scaled signal, since the aliasing components themselves form the scaled signal in part in thiscase.

In this sense, one can conclude that even though the input signal is scaled, only minor aliasing between the aliasing components caused by the non-ideal filters is taken into account to design the filter banks as well as the traditional filter banks.

We also compare the direct scaling structure in the sequential structure shown in Figure 1 with the filter bank of Figure 2 in terms of aliasing. In the direct scaling structure, up-sampling by the factor of U is followed by down-sampling by D. Hence, when $U>D$, which means the input signal is extended, then down-sampling is not critical sampling, that is, oversampling. We illustrate this with Figure 4.

On the other hand, the filter bank structure performs down-sampling by D first, which means critical sampling. This implies that the aliasing caused by down-sampling in Figure 1 is less than that of Figure 2. Thus, it is also concluded that the filter bank structures have more aliasing than the direct scaling structures do when the scaling factors are larger than 1.

These are possible issues to address when we design the filter banks that serve as scaling systems.

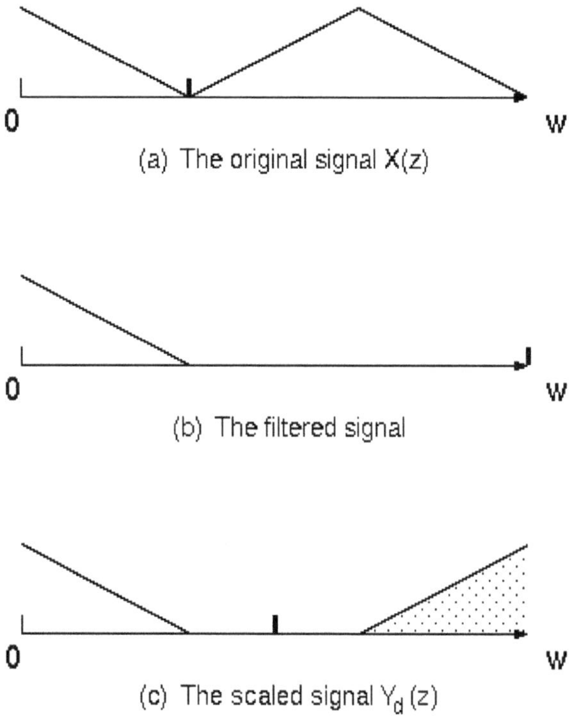

Figure 4. The frequency characteristics of the direct scaling structure

5. The theoretical conditions for the improvement of the scaling performance

Typically, traditional perfect reconstruction filter banks are designed to satisfy perfect reconstruction conditions completely or approximately so that aliasing will be eliminated. Similarly, we provide theoretical conditions for the introduced filter banks to reduce aliasing.

We see, from Figure 3 and 4, that aliasing in Figure 3 is reduced if the eventual signal shown in Figure 3(f) is the same as Figure 4(c). In other words, it is desired that the introduced filter banks precisely work as the direct scaling structures. Therefore, a design scheme for the filter banks that approximately or exactly serve as the direct scaling structures is provided in this section.

5.1. The input-output relation in the filter banks and the direct scaling structures in the frequency domain

First, we give the relation between the input and output of the filter banks and the direct scaling structures. The output signal of the direct scaling structure in the sequential structure shown in Figure 1 is written as

$$Y_d(z) = \frac{1}{D}\sum_{r=0}^{D-1} X\left(W^{Ur} z^{\frac{U}{D}}\right) H\left(W^{r} z^{\frac{1}{D}}\right). \tag{6}$$

Please note that we usually use a perfect reconstruction filter bank in Figure 1, therefore, it is assumed that the output of the filter bank is the same as its input.

On the other hand, the output signal $Y_s(z)$ of the filter bank shown in Figure 2 is represented as the equation (4).

If the signal $Y_d(z)$ is equal to the output signal $Y_s(z)$, then

$$X\left(W^{Ur} z^{\frac{U}{D}}\right) H\left(W^{r} z^{\frac{1}{D}}\right) = X\left(W^{r} z^{\frac{U}{D}}\right) \sum_{i=0}^{D-1} H_i\left(W^{r} z^{\frac{U}{D}}\right) G_i(z) \ (r = 0, 1, \cdots, D-1) \tag{7}$$

in which

$$X\left(W^{Ur} z^{\frac{U}{D}}\right) = X\left(W^{r} z^{\frac{U}{D}}\right) \tag{8}$$

hold in some cases without any manipulation, or holds in the other cases by changing the order of r on the left and right side of (8). This depends on the values of U and D. Therefore, the conditions (7) are proportional tothe number of aliasing components $D - 1$, and the parameters in the conditions are $H(z)$, $H_i(z)$ and $G_i(z)$. Moreover, it is observed that the conditions are mostly difficult to satisfy when all the filters have linear phase characteristics, i.e., the symmetry of the coefficients. Therefore, we give a profound discussion on the conditions to address these issues in the next.

5.2. A discussion on the conditions

First, we discuss the equations (8) to verify that they hold. We assume that $U > D$, but the following discussion is straightforwardly extended to the case where $U < D$. Since the equation

$$W^{r} = W^{nD+r} \tag{9}$$

hold, the equations

$$Ur = nD + r \tag{10}$$

need to hold in order to satisfy the equations (8). n is an arbitrary integer number.

If $U = D+1$, then $Ur = rD + r$, which leads to (10). Thus, the equations (8) hold in this case. When $U \neq D+1$, (10) does not hold. Hence, we replace r on the right side of (10) with $r' = r(U-D)$. Therefore, $nD + r' = Ur + D(n-r)$, which also leads to

$$Ur = nD + r'. \tag{11}$$

If we replace r' with

$$r'' = mod(r', D) \tag{12}$$

then we also obtain

$$Ur = nD + r''. \tag{13}$$

$mod(a, b)$ represents the reminder in the case where a is divided by b .Therefore, the equations

$$X\left(W^{Ur} z^{\frac{U}{D}}\right) = X\left(W^{r''} z^{\frac{U}{D}}\right) \tag{14}$$

hold. We see that the equations (14) also hold when $U = D+1$, i.e., (14) are equivalent to (8) when $U = D+1$.

Hence, the equations

$$H\left(W^{r} z^{\frac{1}{D}}\right) = \sum_{i=0}^{D-1} H_i\left(W^{r''} z^{\frac{U}{D}}\right) G_i(z) \quad (r = 0, 1, \cdots, D-1) \tag{15}$$

need to hold in order to satisfy the conditions (7).

Next, we verify the equations (15) are the same regardless of r . When $r'' = r = 0$, the equations (15) are rewritten with $h_{i,j}$, $g_{i,j}$ and h_j partly (every D-th equation) as

$$\begin{cases} h_{qD+p} = h_{0,qD+p} g_{0,lU+k} + h_{0,(q+1)D+p} g_{0,(l-1)U+k} + \cdots + h_{1,qD+p} g_{1,lU+k} + \cdots \\ h_{qD+p+1} = h_{0,qD+p+1} g_{0,lU+k-1} + h_{0,(q+1)D+p+1} g_{0,(l-1)U+k-1} + \cdots + h_{1,qD+p+1} g_{1,lU+k-1} + \cdots \\ h_{qD+p+2} = h_{0,qD+p+2} g_{0,lU+k-2} + h_{0,(q+1)D+p+2} g_{0,(l-1)U+k-2} + \cdots + h_{1,qD+p+2} g_{1,lU+k-2} + \cdots \\ h_{qD+p+3} = h_{0,qD+p+3} g_{0,lU+k-3} + h_{0,(q+1)D+p+3} g_{0,(l-1)U+k-3} + \cdots + h_{1,qD+p+3} g_{1,lU+k-3} + \cdots \\ \vdots \end{cases} \tag{16}$$

where $h_{i,j}$, $g_{i,j}$ and h_j are the coefficients of the filter $H_i(z)$, $G_i(z)$ and $H(z)$, respectively, and the order of the rows except the 0-th row on the right side of (16), however, changes depending on the factors D and U.

Here we define the following vector

$$v_r^{''}=\begin{bmatrix} W^{cr^{''}} & W^{2cr^{''}} & \cdots & W^{(D-1)cr^{''}} \end{bmatrix} \tag{17}$$

in which c is a constant, and the $p^{''}$-th row on the right side of (16) is multiplied by the $p^{''}$-th element $W^{p^{''}cr^{''}}$ in each $r^{''}$. For example, the $p^{''}$-th row on the right side of (16) is multiplied by the coefficient $W^{p^{''}c}$ when $r^{''}=1$.

Moreover, the $p^{''}$-th row on the right side includes $h_{i,p+qD}$, since the coefficient $h_{i,p+qD}$ has the delay $z^{\frac{U}{D}(p+qD)}$, where $p^{''}$ has the relation to p, which is

$$p^{''}=mod(U(p+qD),\ D) \tag{18}$$

If we denote p which satisfies $p^{''}=1$ as p_s, then $c=p_s$, since the first row on the right side, which includes h_{i,p_s+qD}, is multiplied by the coefficient W^{p_s} when $r^{''}=1$. In this manner, the constant c is determined.

On the other hand, we also define the following vector

$$v_r=\begin{bmatrix} W^{r} & W^{2r} & \cdots & W^{(D-1)r} \end{bmatrix} \tag{19}$$

where the $p^{''}$-th row on the left side of (16) is multiplied by the $p^{''}$-th element $W^{p^{''}r}$ in each r. If $r=p_s$ when $r^{''}=1$, then $v_r^{''}=v_r$ since $c=p_s$. In this case, the equations (16) are the same as those in $r=r^{''}=0$. In fact, the equation $r=p_s$ holds when $r^{''}=1$, since (18) is equivalent to (12). In other words, the relation between $p^{''}$ and p is the same as that between $r^{''}$ and r. Furthermore, the equations $v_r^{''}=v_r$ hold in all the $r^{''}$ and r under (12). Therefore, we see that the equations (16) are the same regardless of r or $r^{''}$.

However, it is observed that equations are usually not satisfied when all the filters have linear phase characteristics. Accordingly, we do not impose the symmetry constraints on all the filters. We also see that the filter in the desired direct scaling structure needs to have some delays in order to satisfy the equations. The performance of such filter banks is examined in the next section.

6. Design and Simulation Results

The equations derived in the previous section have unknown parameters $H_i(z)$, $G_i(z)$, and $H(z)$. The filter $H(z)$ in the direct scaling structure should be known, or designed, in advance. Thus, $H_i(z)$ and $G_i(z)$ are obtained by solving the equations with a pre-designed desired $H(z)$ so that the filter bank structure will be exactly or approximately equivalent to the direct scaling structure.

The concrete design procedure depends on applications the filter banks are utilized in. For example, several constraints should be imposed on the analysis filters $H_i(z)$ for the filter banks to have non DC leakage, high coding gain, etc., when the filter banks are exploited in sub-band image coding.

The analysis filters should have high attenuation in the stop band, or excellent frequency decomposition performance, when the filter banks are employed for watermarking, since the watermarks are often embedded only into one narrow sub-band that is robust to attacks.

These tell that the analysis filters typically suffer from severe constraints. In this section, we survey the case where the analysis and synthesis filters are separately designed, i.e., only $G_i(z)$ are determined so that the equations are satisfied with not only a pre-designed $H(z)$ but also $H_i(z)$, in order not to lay much burden on $H_i(z)$. On the other hand, we do not impose the symmetry constraints on $G_i(z)$ to satisfy the equations more precisely.

With this design policy, we here design filter banks. As an example, two-channel and three-channel filter banks are designed and explored. In the two-channel case, the scaling factor is 3/2, and in the three channel one, it is 2/3. The length of the linear phase analysis filters is 6 in the two-channel case, and 8 in the three-channel one. The length of the obtained (non-linear) synthesis filters is 9, and 6, respectively. The length of the linear phase filter in the direct scaling structure is 3, and 2, respectively, and it has third and half band amplitude characteristics, respectively. This is summarized in Table1.

Please note that in the introduced filter bank structures, the direct scaling filters are not required for scaling since the filter banks perform not only sub-band processing but also scaling. The direct scaling filters are used only to design the filter banks. The round brackets "(·)" are used in this sense in Table 1. It is observed that the direct scaling filters have some delays so that the conditions will be satisfied.

The number of channels	The scaling Factors	The length of the analysis, synthesis and direct scaling filters		
2	3/2	6	9	(3)
3	2/3	8	6	(2)

Table 1. The length of the designed filters in the introduced filter banks.

For comparison, we also design nearly perfect reconstruction filter banks in the two-channel and three-channel case, where the length of the analysis filters is 6, and 8, respectively. Please note that the sequential structure consists of the perfect reconstruction filter bank and the same direct scaling structure (except the delays). This is also shown in Table2.In the sequential structures, the coefficients of the analysis filters partially have a few zero-values to satisfy the perfect reconstruction conditions. The sequential structures also require direct scaling filters to perform the scaling process.

The number of channels	The scaling Factors	The length of the analysis, synthesis and direct scaling filters		
2	3/2	6	6	3
3	2/3	8	9	2

Table 2. The length of the designed filters in the sequential structures

The elapsed computation time required for signal scaling in both the sequential structure and the introduced filter bank is also compared. The computation is carried out on a computer with a 2.2 GHz processor. In this comparison, we examine the elapsed time with various scaling factors and length of the filters.

For example, the elapsed time is 0.2030, and 0.2660 in the introduced structure and the sequential one designed above, respectively, when $D=2$, $U=3$ and the input image size is 256×256 .

In addition, Table 3 shows the elapsed time[s] when $D=5$, $U=4, 6, 9, 11$, the length of the analysis filters in both systems is 10, and 30, the length of the filter in the direct scaling structure is equal to the factor U , and the input image size is 256×256 .

The factor *U*	The elapsed time[s]		The length of the synthesis filters	
	Proposed	Sequential	Proposed	Sequential
4	0.375, 0.406(0.281)	0.437, 0.500(0.359)	8, 24	10, 30
6	0.390, 0.453(0.297)	0.453, 0.531(0.390)	12, 36	10, 30
9	0.437, 0.578(0.375)	0.484, 0.563(0.421)	18, 54	10, 30
11	0.453, 0.641(0.391)	0.500, 0.593(0.485)	22, 66	10, 30

Table 3. Comparison of the elapsed time[s] between the proposed and sequential schemes

It is observed that the elapsed time is proportional to the factor U and also the length of the filters, and that the elapsed time of the introduced filter banks is more than that of the sequential structures when the synthesis filters are long with $U = 9, 11$. However, this drawback is improved if each branch of the filter banks is completely carried out in parallel, i.e., the computation time of the filter banks is the same as that of each branch in the filter banks.

We also explored the elapsed time of both systems in this case, which is also shown in the round brackets "(·)", where the length of the analysis filters is also 30, in Table 3.

We also explore the performance of image-scaling in the two-channel and three-channel cases. Several well-known test images (256×256) are used as the input to the filter banks. Table 4 shows the PSNR[dB] of the output of the introduced filter banks against that of the sequential structures. It is observed that the obtained scaled images with the introduced filter banks are comparable to that of the sequential structures. However, it is not certain whether the proposed approach can yield the optimal solution for image-scaling. A detailed discussion on this issue is future work.

The number of channels	LENA	BARBARA	BOAT	Building	Lighthouse
2	147.60	148.88	145.40	146.64	146.56
3	122.80	123.96	121.02	122.12	121.82

Table 4. The PSNR[dB] of the introduced filter banks against the sequential structures.

The introduced filter banks as well as the sequential structures can be applicable to practical image processing applications. In this simulation, we examine the performance of both the systems in watermarking, where it is assumed that the watermark is embedded into a subband of the original image and then the watermarked image is rescaled, and that the watermark is extracted from the rescaled image subsequently. We illustrate both the systems in watermarking with Figure 5 and Figure 6 in which the scaling factor is 3/2.

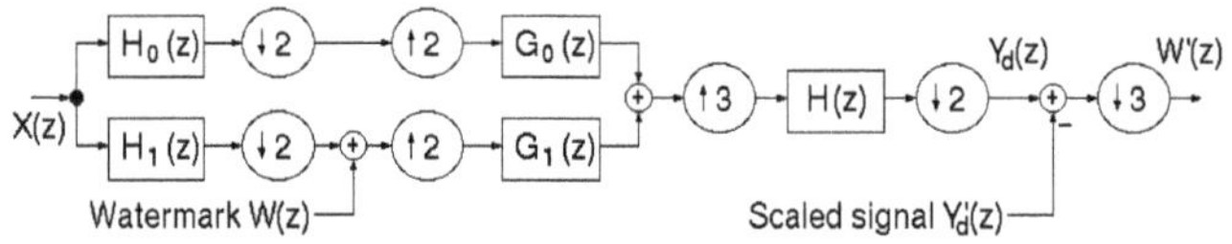

Figure 5. The sequential structure employed for watermarking.

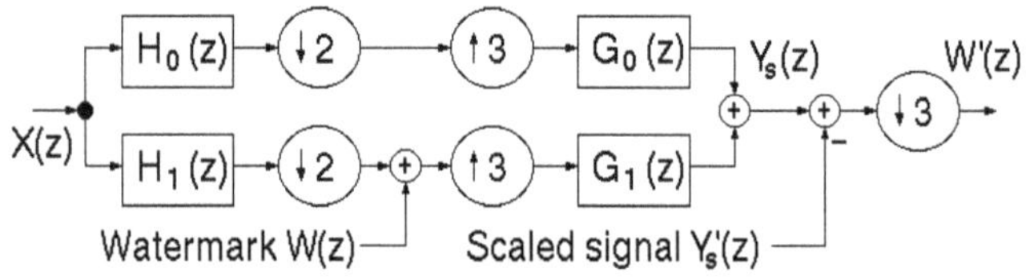

Figure 6. The introduced filter bank structure for watermarking.

Please notice that the subsequent extracting stage in both the systems requires down-sampling and also the use of a scaled version of the original image, in order to extract the watermark embedded in a sub-band of the original image in this simulation. The only difference between both of the systems is that the sub-band signals are synthesized and then rescaled sequentially in Figure 5 while they are rescaled in each sub-band and synthesized in Figure 6. This also holds for other image processing applications such as image coding, recognition, etc.

To demonstrate this, we use several original test images (256×256) and bi-level watermark images (128×128) shown in Figure 7 and Figure 8 in both the structures, and the watermark is embedded into the high frequency band of the original image because it is known that the low frequency one is visually susceptible to the watermarking process.

Figure 7. The bi-level image used as Watermark1.

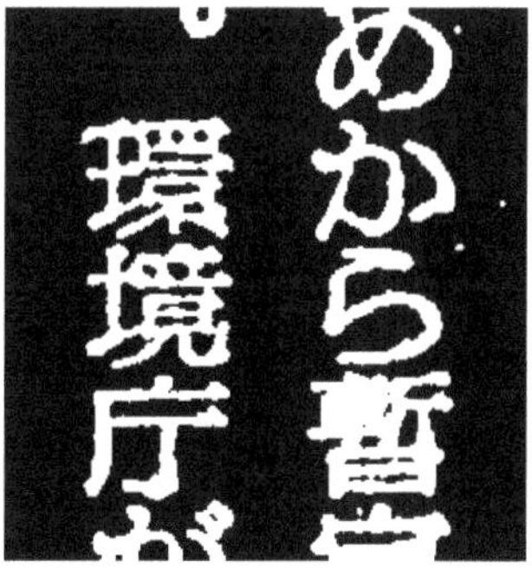

Figure 8. The bi-level image used as Watermark2.

Table 5 represents the PSNR[dB] of the extracted watermark against the original watermark in both the systemsin which the original image "LENA" is used. It is observed that the obtained watermarks with the introduced filter banks are comparable to those of the sequential structures, while the computation time of the proposed filter bank is less than that of the sequential structures as shown above.

Watermark1		Watermark2	
Proposed structure	Sequential structure	Proposed structure	Sequential structure
34.38	34.56	34.66	34.88

Table 5. The PSNR[dB] of the extracted watermark against the original watermark.

We, however, see that the water marks are not extracted with very high PSNR in both the systems, which is because re-sampling or filtering is carried out in the image-scaling and watermark-extracting stage.

Typically, watermarks can also be deteriorated in quality or eliminated by attacks such as filtering, scaling, etc. Thus, they are more sensitive to such attacks than usual watermarking (without scaling) is. A detailed discussion is also the future work.

7. Conclusion

This chapter has discussed a filter bank structure which directly synthesizes the analysis decomposed signals to produce a scaled signal, which leads to computational simplicity in complex processing such as sub-band processing followed by scaling.

First, we have discussed the frequency decomposition characteristics of the filter bank to show how to obtain arbitrarily scaled signals, where the behavior of the aliasing components has also been manifested and addressed, since it can significantly affect the filter bank performance as in traditional perfect reconstruction filter banks.

Next, theoretical conditions for designing the filter bank that is equivalent to the direct scaling structure have been given with the input-output relation of the filter bank, where it has been mentioned that the number of conditions is proportional to that of aliasing components, and that the conditions are mostly not satisfied when all the filters have linear phase. Therefore, we have discussed the conditions, so that they are the same regardless of the aliasing components, which means that the number of equations to solve is significantly reduced.

In addition, it has been demonstrated through simulation results that the computation time required for signal scaling with the proposed scheme is usually less than that of the sequential structures, and that the quality of scaled images is comparable to that of the sequential structures even though not all the filters have linear phase characteristics.

Finally, we have discussed potential issues and advantages in employing the proposed scheme as well as traditional ones in practical image processing such as watermarking.

The findings in this research are that the derived solution by using the proposed approach yields comparable scaling and watermarking performance to that of the usual sequential structures that consist of filter banks and also the direct scaling processes, while the computation time required for scaling is less than that of the sequential scheme and the design procedure is considerabley simplified.

Author details

Fumio Itami*

Address all correspondence to: itami@sit.ac.jp

Faculty of Engineering, Saitama Institute of Technology, Japan

References

[1] Vaidyanathan, P. P. (1993). Multirate Systems and Filter Banks. *Englewood Cliffs. NJ:Prentice-Hall.*

[2] Vaidyanathan, P. P., & Kirac, A. (1998). Results on optimal biorthogonal filter banks. *IEEE Transactions on Circuits Syst. II Aug.*, 45.

[3] Soman, A. K., Vaidyanathan, P. P., & Nguyen, T. Q. (1993). Linear-phase orthogonal filter banks. *IEEE Transactions on Signal Processing Dec.*, 41.

[4] Strang, G., & Nguyen, T. (1996). Wavelets and Filter Banks. *Wellesley, MA: Wellesley-Cambridge.*

[5] Vetterli, M., & Kovacevic, J. (1995). Wavelets and Subband Coding. *Englewood Cliffs NJ: Prentice-Hall.*

[6] Vetterli, M., & Herley, C. (1992). Wavelets and filter banks: Theory and design. *IEEE Transactions on Signal Processing Sep.*, 40, 2207-2232.

[7] Kovacevic, J., & Vetterli, M. (1993). Perfect Reconstruction Filter Banks with Rational Sampling Factors. *IEEE Transactions on Signal Processing Jun.*, 41, 2047-2066.

[8] Gopinath, R. A., & Burrus, C. S. (1994). On Upsampling, Downsampling, and Rational Sampling Rate Filter Banks. *IEEE Transactions on Signal Processing Apr.*, 42, 812-824.

[9] Wang, Y., Doherty, J., & Van Dyck, R. A. (2002). Wavelet-Based Watermarking Algorithm for Ownership Verification of Digital Images. *IEEE TransactionsonImage Processing Feb.*, 11, 77-88.

[10] Lin, C. Y., Wu, M., Bloom, J. A., Cox, I. J., Miller, M. L., & Lui, Y. M. (2001). Rotation, scale, and translation resilient watermarking for images. *IEEE Transactions on Image Processing May*, 10.

[11] Hernandez, J., Amado, M., & Perez-Gonzalez, F. (2000). DCT-domain watermarking techniques for still images, detector performance analysis and a new structure. *IEEE Transactions on Image Processing Jan.*, 9, 55-68.

[12] Pitas, I. (1998). A method for watermark casting on digital image. *IEEE Transactions on Circuits Syst. Video Technol. Oct.*, 8, 775-780.

[13] Woods, J. W., & O'neil, S. (1986). Subband Coding of Images. *IEEE Transactionson Acoustics, Speech, and Signal Processing Oct.*, 34 , 1278-1288.

[14] Oraintara, S., Tran, T.D., & Nguyen, T.Q. (2003). A Class of Regular Biorthogonal Linear-PhaseFilterbanks: Theory, Structure, andApplication in Image Coding. *IEEE Transactionson Signal Processing Dec.*, 513220-3235.

[15] Unser, M., Aldroubi, A., & Eden, M. (1993). B-Spline Signal Processing: Part I-Theory. *IEEE Transactions on Signal Processing Feb.*, 41, 821-833.

[16] Unser, M., Aldroubi, A., & Eden, M. (1993). B-Spline Signal Processing: Part II-Efficient Design and Applications. *IEEE Transactions on Signal Processing Feb.*, 41, 834-848.

[17] Yang, S., & Nguyen, T. Q. (2002). Interpolated Mth-Band Filters for Image Size Conversion. *IEEE Transactions on Signal Processing Dec.*, 50.

[18] Itami, F., Watanabe, E., & Nishihara, A. (2006). Multirate Filter Bank-based Conversion of Image Resolution. *Proc. of IEEE, Asia Pacific Conference on Circuits and Systems Dec.*, 1234-1237.

Chapter 8

Analytical Approach for Synthesis of Minimum L_2-Sensitivity Realizations for State-Space Digital Filters

Shunsuke Yamaki, Masahide Abe and
Masayuki Kawamata

Additional information is available at the end of the chapter

1. Introduction

On the fixed-point implementation of digital filters, undesirable finite-word-length (FWL) effects arise due to the coefficient truncation and arithmetic roundoff. These FWL effects must be reduced as small as possible because such effects may cause serious degradation of characteristic of digital filters. L_2-sensitivity is one of the evaluation functions which evaluate the coefficient quantization effects of state-space digital filters [1–14]. The L_2-sensitivity minimization is quite beneficial technique for the synthesis of high-accuracy digital filter structures, which achieves quite low-coefficient quantization error.

To the L_2-sensitivity minimization problem, Yan *et al.* [1] and Hinamoto *et al.* [2] proposed solutions using iterative calculations. Both of the solutions in [1] and [2] try to solve nonlinear equations by successive approximation. Their solutions do not guarantee that the L_2-sensitivity surely converges to the minimum L_2-sensitivity since their solutions are not analytical solutions. It is necessary to derive some analytical solutions to the L_2-sensitivity minimization problem in order to guarantee that their conventional solutions surely derive the minimum L_2-sensitivity.

This chapter presents analytical approach for synthesis of the minimum L_2-sensitivity realizations for state-space digital filters. In Section 3, we derive closed form solutions to the L_2-sensitivity minimization problem for second-order digital filters [12, 13]. This problem can be converted into the problem to find the solution to fourth-degree polynomial equation of constant coefficients, which can be algebraically solved in closed form. Next, we reveal that the L_2-sensitivity minimization problem can be solved analytically for arbitrary

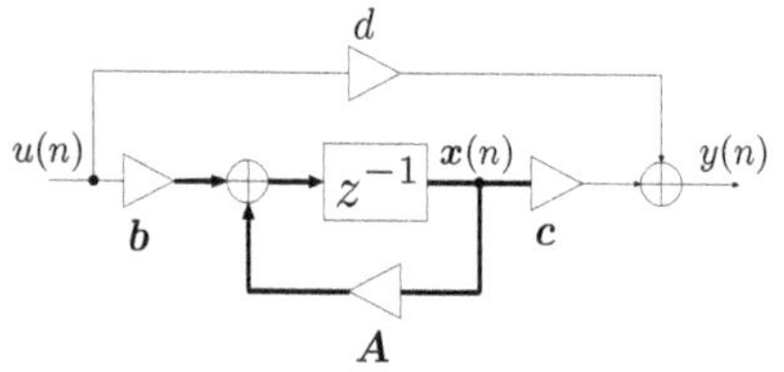

(a) Nth-order digital filter

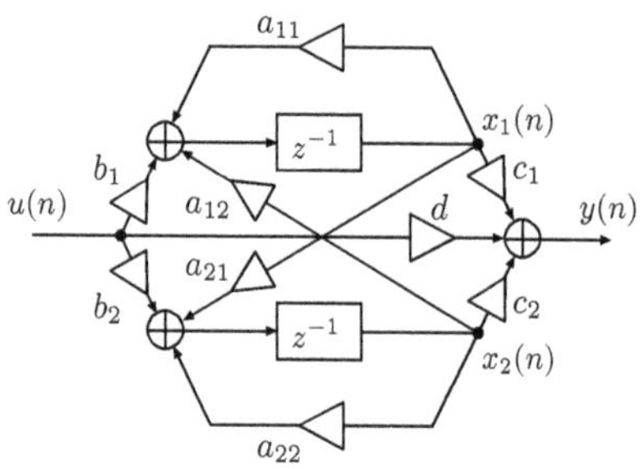

(b) Second-order digital filter

Figure 1. Block diagram of a state-space digital filter.

filter order if second-order modes are all equal [14] in Section 4. We derive a general expression of the transfer function of digital filters with all second-order modes equal. We show that the general expression is obtained by a frequency transformation on a first-order prototype FIR digital filter. Furthermore, we show the absence of limit cycles of the minimum L_2-sensitivity realizations [11] in Section 5. The minimum L_2-sensitivity realization without limit cycles can be synthesized by selecting an appropriate orthogonal matrix in the coordinate transformation matrix.

2. Preliminaries

This section gives the preliminaries in order to lay groundwork for the main topics of this chapter, which appear in later sections. In Subsection 2.1, we begin with introduction of state-space digital filters. Subsection 2.2 provides the introduction of L_2-sensitivity. Subsection 2.3 explains coordinate transformations, the operation for changing the structures of state-space digital filters under the transfer function invariant. Subsection 2.4 formulates the L_2-sensitivity minimization problem.

2.1. State-space digital filters

It is beneficial to introduce the state-space representation for the synthesis of high accuracy digital filters. For a given Nth-order transfer function $H(z)$, a state-space digital filter can be described by the following state-space equations:

$$\boldsymbol{x}(n+1) = \boldsymbol{A}\boldsymbol{x}(n) + \boldsymbol{b}u(n) \tag{1}$$

$$y(n) = \boldsymbol{c}\boldsymbol{x}(n) + du(n) \tag{2}$$

where $\boldsymbol{x}(n) \in \Re^{N\times 1}$ is a state variable vector, $u(n) \in \Re$ is a scalar input, $y(n) \in \Re$ is a scalar output, and $\boldsymbol{A} \in \Re^{N\times N}$, $\boldsymbol{b} \in \Re^{N\times 1}$, $\boldsymbol{c} \in \Re^{1\times N}$, $d \in \Re$ are real constant matrices called coefficient matrices. The block diagram of the state-space digital filter $(\boldsymbol{A}, \boldsymbol{b}, \boldsymbol{c}, d)$ is shown in Fig. 1(a). In case of second-order digital filters, the block diagram in Fig. 1(a) can be rewritten as shown in Fig. 1(b). The transfer function $H(z)$ is described in terms of the coefficient matrices $(\boldsymbol{A}, \boldsymbol{b}, \boldsymbol{c}, d)$ as

$$H(z) = \boldsymbol{c}(z\boldsymbol{I} - \boldsymbol{A})^{-1}\boldsymbol{b} + d. \tag{3}$$

In this chapter, the state-space representation $(\boldsymbol{A}, \boldsymbol{b}, \boldsymbol{c}, d)$ is assumed to be a minimal realization of $H(z)$, that is, the state-space representation $(\boldsymbol{A}, \boldsymbol{b}, \boldsymbol{c}, d)$ is controllable and observable. The transfer function $H(z)$ in Eq. (3) can be rewritten as

$$H(z) = \boldsymbol{c}\frac{\text{adj}(z\boldsymbol{I} - \boldsymbol{A})}{\det(z\boldsymbol{I} - \boldsymbol{A})}\boldsymbol{b} + d. \tag{4}$$

We know from the above equation that the poles of $H(z)$ are the solutions of the characteristic equation $\det(z\boldsymbol{I} - \boldsymbol{A}) = 0$, that is, the eigenvalues of the coefficient matrix $\boldsymbol{A}$. Since we assume that the transfer function $H(z)$ is stable, all absolute values of the eigenvalues of the coefficient matrix A are less than unity. It follows from Eq. (4) that the absolute values of poles of $H(z)$ are less than unity.

2.2. L_2-sensitivity

The L_2-sensitivity is one of the measurements which evaluate coefficient quantization errors of digital filters. The L_2-sensitivity of the filter $H(z)$ with respect to the realization $(\boldsymbol{A}, \boldsymbol{b}, \boldsymbol{c}, d)$ is defined by

$$\begin{aligned} S(\boldsymbol{A}, \boldsymbol{b}, \boldsymbol{c}) =& \sum_{k=1}^{N}\sum_{l=1}^{N}\frac{1}{2\pi}\int_0^{2\pi}\left|\frac{\partial H(e^{j\omega})}{\partial a_{kl}}\right|^2 d\omega \\ &+ \sum_{k=1}^{N}\frac{1}{2\pi}\int_0^{2\pi}\left|\frac{\partial H(e^{j\omega})}{\partial b_k}\right|^2 d\omega + \sum_{l=1}^{N}\frac{1}{2\pi}\int_0^{2\pi}\left|\frac{\partial H(e^{j\omega})}{\partial c_l}\right|^2 d\omega \\ =& \left\|\frac{\partial H(z)}{\partial \boldsymbol{A}}\right\|_2^2 + \left\|\frac{\partial H(z)}{\partial \boldsymbol{b}}\right\|_2^2 + \left\|\frac{\partial H(z)}{\partial \boldsymbol{c}}\right\|_2^2 \end{aligned} \tag{5}$$

where $\|\cdot\|_2$ denotes the L_2-norm. The derivatives of the transfer function $H(z)$ with respect to the coefficient matrices are described by

$$\frac{\partial H(z)}{\partial \boldsymbol{A}} = \boldsymbol{G}^T(z)\boldsymbol{F}^T(z), \quad \frac{\partial H(z)}{\partial \boldsymbol{b}} = \boldsymbol{G}^T(z), \quad \frac{\partial H(z)}{\partial \boldsymbol{c}} = \boldsymbol{F}^T(z) \tag{6}$$

where $\boldsymbol{F}(z)$ and $\boldsymbol{G}(z)$ are defined by

$$\boldsymbol{F}(z) = (z\boldsymbol{I} - \boldsymbol{A})^{-1}\boldsymbol{b} \tag{7}$$

$$\boldsymbol{G}(z) = \boldsymbol{c}(z\boldsymbol{I} - \boldsymbol{A})^{-1} \tag{8}$$

respectively. Substituting Eqs. (6) into Eq. (5), the L_2-sensitivity can be rewritten as

$$\begin{aligned} S(A,b,c) &= \left\| \frac{\partial H(z)}{\partial A} \right\|_2^2 + \left\| \frac{\partial H(z)}{\partial b} \right\|_2^2 + \left\| \frac{\partial H(z)}{\partial c} \right\|_2^2 \\ &= \left\| G^T(z) F^T(z) \right\|_2^2 + \left\| G^T(z) \right\|_2^2 + \left\| F^T(z) \right\|_2^2 . \end{aligned} \tag{9}$$

We can express the L_2-sensitivity $S(A,b,c)$ by using complex integral as [1]

$$\begin{aligned} S(A,b,c) = &\mathrm{tr}\left[\frac{1}{2\pi j} \oint_{|z|=1} F(z)G(z)(F(z)G(z))^{\dagger} \frac{dz}{z} \right] \\ &+ \mathrm{tr}\left[\frac{1}{2\pi j} \oint_{|z|=1} G^{\dagger}(z)G(z) \frac{dz}{z} \right] + \mathrm{tr}\left[\frac{1}{2\pi j} \oint_{|z|=1} F(z)F^{\dagger}(z) \frac{dz}{z} \right] . \end{aligned} \tag{10}$$

Applying Parseval's relation to Eq. (10), Hinamoto *et al.* expressed the L_2-sensitivity in terms of the general Gramians such as [2]

$$S(A,b,c) = \mathrm{tr}(W_0)\mathrm{tr}(K_0) + \mathrm{tr}(W_0) + \mathrm{tr}(K_0) + 2\sum_{i=1}^{\infty} \mathrm{tr}(W_i)\mathrm{tr}(K_i). \tag{11}$$

The general controllability Gramian K_i and the general observability Gramian W_i in Eq. (11) are defined as the solutions to the following Lyapunov equations:

$$K_i = AK_iA^T + \frac{1}{2}\left(A^i bb^T + bb^T (A^T)^i \right) \tag{12}$$

$$W_i = A^T W_i A + \frac{1}{2}\left(c^T c A^i + (A^T)^i c^T c \right) \tag{13}$$

for $i = 0, 1, 2, \cdots$, respectively. The general controllability and observability Gramians are natural expansions of the controllability and observability Gramians, respectively. Letting $i = 0$ in Eqs. (12) and (13), we have the Lyapunov equations for the controllability Gramian K_0 and the observability Gramian W_0 as follows:

$$K_0 = AK_0A^T + bb^T \tag{14}$$

$$W_0 = A^T W_0 A + c^T c. \tag{15}$$

The controllability Gramian K_0 and the observability Gramian W_0 are positive definite symmetric, and the eigenvalues $\theta_i^2 (i = 1, \cdots, N)$ of the matrix product $K_0 W_0$ are all positive.

Second-order modes are defined by the square roots of the eigenvalues θ_i's as follows [3, 5]:

$$(\theta_1, \cdots, \theta_N) = \sqrt{\text{Eigenvalues of } \boldsymbol{K}_0 \boldsymbol{W}_0}. \tag{16}$$

In the field of digital signal processing, the controllability and observability Gramians are also called the covariance and noise matrices of the filter $(\boldsymbol{A}, \boldsymbol{b}, \boldsymbol{c}, d)$, respectively.

2.3. Coordinate transformations

It is well known that the number of state-space realizations of a transfer function is infinite since choice of the state variable vector $\boldsymbol{x}(n)$ is not unique. We can change the filter structures of state-space digital filters by the operation called *coordinate transformation* under the transfer function invariant.

Let $\boldsymbol{T}$ be a nonsingular $N \times N$ real matrix. If a coordinate transformation defined by

$$\bar{\boldsymbol{x}}(n) = \boldsymbol{T}^{-1} \boldsymbol{x}(n) \tag{17}$$

is applied to a filter structure $(\boldsymbol{A}, \boldsymbol{b}, \boldsymbol{c}, d)$, we obtain a new filter structure which has the following coefficient matrices:

$$(\overline{\boldsymbol{A}}, \overline{\boldsymbol{b}}, \overline{\boldsymbol{c}}, \overline{d}) = (\boldsymbol{T}^{-1} \boldsymbol{A} \boldsymbol{T}, \boldsymbol{T}^{-1} \boldsymbol{b}, \boldsymbol{c} \boldsymbol{T}, d). \tag{18}$$

It should be noted that the coordinate transformation does not affect the transfer function $H(z)$, that is,

$$\begin{aligned} \overline{H}(z) &= \overline{\boldsymbol{c}}(z\boldsymbol{I} - \overline{\boldsymbol{A}})^{-1} \overline{\boldsymbol{b}} + \overline{d} \\ &= \boldsymbol{c}(z\boldsymbol{I} - \boldsymbol{A})^{-1} \boldsymbol{b} + d \\ &= H(z). \end{aligned} \tag{19}$$

It implies that there exist infinite filter structures for a given transfer function $H(z)$ since nonsingular $N \times N$ matrices exist infinitely. Therefore, one can synthesize infinite filter structures by the coordinate transformation with keeping the transfer function invariant.

Under the coordinate transformation by the nonsingular matrix $\boldsymbol{T}$, the general controllability Gramian $\boldsymbol{K}_i$ and the general observability Gramian $\boldsymbol{W}_i$ are transformed into $\overline{\boldsymbol{K}}_i$ and $\overline{\boldsymbol{W}}_i$ given by

$$(\overline{\boldsymbol{K}}_i, \overline{\boldsymbol{W}}_i) = (\boldsymbol{T}^{-1} \boldsymbol{K}_i \boldsymbol{T}^{-T}, \boldsymbol{T}^T \boldsymbol{W}_i \boldsymbol{T}) \tag{20}$$

respectively. Letting $i = 0$ in Eqs. (20) yields

$$(\overline{K}_0, \overline{W}_0) = (T^{-1}K_0T^{-T}, T^TW_0T) \tag{21}$$

From Eqs. (21), we have

$$\overline{K}_0\overline{W}_0 = T^{-1}K_0W_0T \tag{22}$$

which shows that $\overline{K}_0\overline{W}_0$ has the same eigenvalues of K_0W_0. Thus, the second order modes defined by Eq. (16) are invariant under the coordinate transformations. It implies that second-order modes depends on only the transfer function.

2.4. L_2-sensitivity minimization problem

The value of L_2-sensitivity depends on not only the transfer function $H(z)$ but also the coordinate transformation matrix T. The L_2-sensitivity of the filter $(T^{-1}AT, T^{-1}b, cT, d)$ can be expressed in terms of the complex integral as

$$\begin{aligned} &S(T^{-1}AT, T^{-1}b, cT) \\ &= \mathrm{tr}\left[\frac{1}{2\pi j}\oint_{|z|=1} T^{-1}F(z)G(z)TT^T(F(z)G(z))^{\dagger}T^{-T}\frac{dz}{z}\right] \\ &+\mathrm{tr}\left[\frac{1}{2\pi j}\oint_{|z|=1} T^TG^{\dagger}(z)G(z)T\frac{dz}{z}\right] + \mathrm{tr}\left[\frac{1}{2\pi j}\oint_{|z|=1} T^{-1}F(z)F^{\dagger}(z)T^{-T}\frac{dz}{z}\right] \end{aligned} \tag{23}$$

or in terms of the general Gramians as

$$\begin{aligned} &S(T^{-1}AT, T^{-1}b, cT) \\ &= \mathrm{tr}(T^TW_0T)\mathrm{tr}(T^{-1}K_0T^{-T}) + \mathrm{tr}(T^TW_0T) + \mathrm{tr}(T^{-1}K_0T^{-T}) \\ &+2\sum_{i=1}^{\infty}\mathrm{tr}(T^TW_iT)\mathrm{tr}(T^{-1}K_iT^{-T}). \end{aligned} \tag{24}$$

The L_2-sensitivity (23) can be expressed as the function of the positive definite symmetric matrix P as follows [1]:

$$\begin{aligned} S(P) &= \mathrm{tr}\left[\frac{1}{2\pi j}\oint_{|z|=1} F(z)G(z)P(F(z)G(z))^{\dagger}P^{-1}\frac{dz}{z}\right] \\ &+\mathrm{tr}\left[\frac{1}{2\pi j}\oint_{|z|=1} G^{\dagger}(z)G(z)P\frac{dz}{z}\right] + \mathrm{tr}\left[\frac{1}{2\pi j}\oint_{|z|=1} F(z)F^{\dagger}(z)P^{-1}\frac{dz}{z}\right] \end{aligned} \tag{25}$$

where $P = TT^T$. Similarly, the L_2-sensitivity (24) can be expressed as the function of the positive definite symmetric matrix P as follows [2]:

$$\begin{aligned} S(\boldsymbol{P}) = & \operatorname{tr}(\boldsymbol{W}_0\boldsymbol{P})\operatorname{tr}(\boldsymbol{K}_0\boldsymbol{P}^{-1}) + \operatorname{tr}(\boldsymbol{W}_0\boldsymbol{P}) + \operatorname{tr}(\boldsymbol{K}_0\boldsymbol{P}^{-1}) \\ & +2\sum_{i=1}^{\infty} \operatorname{tr}(\boldsymbol{W}_i\boldsymbol{P})\operatorname{tr}(\boldsymbol{K}_i\boldsymbol{P}^{-1}) \end{aligned} \tag{26}$$

where $\boldsymbol{P} = \boldsymbol{T}\boldsymbol{T}^T$. The problem we consider here is to derive *the optimal positive definite symmetric matrix* $\boldsymbol{P}_{\text{opt}}$, which gives the global minimum of $S(\boldsymbol{P})$ as follows:

$$\left.\frac{\partial S(\boldsymbol{P})}{\partial \boldsymbol{P}}\right|_{\boldsymbol{P}=\boldsymbol{P}_{\text{opt}}} = \mathbf{0}. \tag{27}$$

If one can obtain the optimal positive definite symmetric matrix $\boldsymbol{P}_{\text{opt}}$, the optimal coordinate transformation matrix $\boldsymbol{T}_{\text{opt}}$ is given by

$$\boldsymbol{T}_{\text{opt}} = \boldsymbol{P}_{\text{opt}}^{\frac{1}{2}}\boldsymbol{U} \tag{28}$$

where $\boldsymbol{U}$ is an *arbitrary* orthogonal matrix. It implies that the minimum L_2-sensitivity realizations exist infinitely for a given digital filter $H(z)$. The minimum L_2-sensitivity realizations have freedom for orthogonal transformations.

3. Analytical solutions to the L_2-sensitivity minimization problem for second-order digital filters

This section proposes analytical synthesis of the minimum L_2-sensitivity realizations for second-order digital filters. We propose closed form solutions to the L_2-sensitivity minimization problem of second-order state-space digital filters. The proposed closed form solutions can greatly save the computation time and guarantee that the L_2-sensitivity obtained by the iterative algorithm of the conventional method surely converges to the theoretical minimum. We show that the L_2-sensitivity is expressed by a simple linear combination of exponential functions, and we can obtain the minimum L_2-sensitivity realization by solving a fourth degree polynomial equation of constant coefficients in closed form without iterative calculations [12, 13].

3.1. Problem formulation

We adopt the balanced realization $(\boldsymbol{A}_\text{b}, \boldsymbol{b}_\text{b}, \boldsymbol{c}_\text{b}, d_\text{b})$ as the initial realization to synthesize the minimum L_2-sensitivity realization. The coefficient matrices $(\boldsymbol{A}_\text{b}, \boldsymbol{b}_\text{b}, \boldsymbol{c}_\text{b}, d_\text{b})$ satisfy symmetric properties as follows:

$$\boldsymbol{A}_\text{b}^T = \boldsymbol{\Sigma}\boldsymbol{A}_\text{b}\boldsymbol{\Sigma},\ \boldsymbol{c}_\text{b}^T = \boldsymbol{\Sigma}\boldsymbol{b}_\text{b} \tag{29}$$

where $\boldsymbol{\Sigma}$ is a signature matrix defined as follows:

$$\boldsymbol{\Sigma} = \mathrm{diag}(\sigma_1, \cdots, \sigma_N),\ \sigma_i = \pm 1\ (i = 1, \cdots, N). \tag{30}$$

We exploit the symmetric properties of the balanced realization in order to simplify the L_2-sensitivity formulation and minimization in the following discussion. Under this condition, the L_2-sensitivity $S(\boldsymbol{P})$ in Eq. (26) is rewritten as

$$\begin{aligned} S(\boldsymbol{P}) = & \,\mathrm{tr}(\boldsymbol{W}_0^{(b)}\boldsymbol{P})\mathrm{tr}(\boldsymbol{K}_0^{(b)}\boldsymbol{P}^{-1}) + \mathrm{tr}(\boldsymbol{W}_0^{(b)}\boldsymbol{P}) + \mathrm{tr}(\boldsymbol{K}_0^{(b)}\boldsymbol{P}^{-1}) \\ & + 2\sum_{i=1}^{\infty} \mathrm{tr}(\boldsymbol{W}_i^{(b)}\boldsymbol{P})\mathrm{tr}(\boldsymbol{K}_i^{(b)}\boldsymbol{P}^{-1}) \end{aligned} \tag{31}$$

and thus, the L_2-sensitivity minimization problem is formulated as follows:

$$\min_{\boldsymbol{P}} S(\boldsymbol{P}) \text{ in Eq. (31)} \tag{32}$$

where $\boldsymbol{P}$ is an arbitrary positive definite symmetric matrix.

We derive the optimal positive definite symmetric matrix $\boldsymbol{P}_{\mathrm{opt}}$ which gives the global minimum of the L_2-sensitivity $S(\boldsymbol{P})$ in Eq. (31).

3.2. Second-order digital filters

Consider second-order digital filters with complex conjugate poles given by

$$H(z) = \frac{\alpha}{z - \lambda} + \frac{\alpha^*}{z - \lambda^*} + d \tag{33}$$

where (λ, λ^*) are *complex conjugate* poles, α is a *complex* scalar, and d is a *real* scalar [1]. We define scalar parameters P, Q, and R as follows:

$$P = \frac{|\alpha|}{1 - |\lambda|^2} \tag{34}$$

$$R + jQ = \frac{\alpha}{1 - \lambda^2} \tag{35}$$

which can be calculated directly from the transfer function $H(z)$. The closed form expression of the balanced realization of the filter $H(z)$ is given as follows [15]:

$$\left[\begin{array}{c|c} \boldsymbol{A}_{\mathrm{b}} & \boldsymbol{b}_{\mathrm{b}} \\ \hline \boldsymbol{c}_{\mathrm{b}} & d_{\mathrm{b}} \end{array}\right] = \left[\begin{array}{cc|c} \lambda_{\mathrm{r}} - \dfrac{\kappa - \kappa^{-1}}{2}\lambda_{\mathrm{i}} & \dfrac{\kappa + \kappa^{-1}}{2}\lambda_{\mathrm{i}} & \mu_1 + \mu_2 \\ -\dfrac{\kappa + \kappa^{-1}}{2}\lambda_{\mathrm{i}} & \lambda_{\mathrm{r}} + \dfrac{\kappa - \kappa^{-1}}{2}\lambda_{\mathrm{i}} & \mu_1 - \mu_2 \\ \hline \mu_1 + \mu_2 & -(\mu_1 - \mu_2) & d \end{array}\right] \tag{36}$$

[1] For also second-order digital filters, we can derive analytical solutions to the L_2-sensitivity minimization problem [13].

where

$$\begin{cases} \lambda = \lambda_r + j\lambda_i,\ \alpha = \alpha_r + j\alpha_i,\ \kappa = \sqrt{\dfrac{P+Q}{P-Q}}, \\ \mu_1 = \sqrt{\dfrac{\kappa(|\alpha| - \alpha_i)}{2}},\ \mu_2 = \sqrt{\dfrac{|\alpha| + \alpha_i}{2\kappa}}\mathrm{sign}(\alpha_r). \end{cases} \tag{37}$$

Using the parameters P, Q, and R, the controllability Gramian $\boldsymbol{K}_0^{(b)}$ and the observability Gramian $\boldsymbol{W}_0^{(b)}$ of the balanced realization $(\boldsymbol{A}_b, \boldsymbol{b}_b, \boldsymbol{c}_b, d_b)$ can be expressed as follows:

$$\boldsymbol{K}_0^{(b)} = \boldsymbol{W}_0^{(b)} = \boldsymbol{\Theta} \tag{38}$$

$$\begin{aligned} \boldsymbol{\Theta} &= \mathrm{diag}(\theta_1, \theta_2) \\ &= \mathrm{diag}(\sqrt{P^2 - Q^2} + R, \sqrt{P^2 - Q^2} - R). \end{aligned} \tag{39}$$

3.3. Property of the positive definite symmetric matrix P

In this subsection, we consider the property of the positive definite symmetric matrix $\boldsymbol{P}$. The following two theorems lead a symmetric property of the optimal positive definite symmetric matrix $\boldsymbol{P}_{\mathrm{opt}}$ [1].

Theorem 1. [9] *L_2-sensitivity $S(\boldsymbol{P})$ has the unique global minimum, which is achieved by a positive definite symmetric matrix $\boldsymbol{P}_{\mathrm{opt}}$ satisfying*

$$\left.\frac{\partial S(\boldsymbol{P})}{\partial \boldsymbol{P}}\right|_{\boldsymbol{P}=\boldsymbol{P}_{\mathrm{opt}}} = \mathbf{0}. \tag{40}$$

□

Theorem 2. [1] *If a positive definite symmetric matrix $\boldsymbol{P}_{\mathrm{opt}}$ satisfies*

$$\left.\frac{\partial S(\boldsymbol{P})}{\partial \boldsymbol{P}}\right|_{\boldsymbol{P}=\boldsymbol{P}_{\mathrm{opt}}} = \mathbf{0} \tag{41}$$

then the positive definite symmetric matrix $\boldsymbol{\Sigma}\boldsymbol{P}_{\mathrm{opt}}^{-1}\boldsymbol{\Sigma}$ also satisfies

$$\left.\frac{\partial S(\boldsymbol{P})}{\partial \boldsymbol{P}}\right|_{\boldsymbol{P}=\boldsymbol{\Sigma}\boldsymbol{P}_{\mathrm{opt}}^{-1}\boldsymbol{\Sigma}} = \mathbf{0} \tag{42}$$

for the signature matrix $\boldsymbol{\Sigma}$ which satisfies Eq. (29). □

The derivative $\partial S(\boldsymbol{P})/\partial \boldsymbol{P}$ is given by differentiating $S(\boldsymbol{P})$ in Eq. (31) with respect to $\boldsymbol{P}$ as

$$\begin{aligned}\frac{\partial S(\boldsymbol{P})}{\partial \boldsymbol{P}} &= (1+\mathrm{tr}(\boldsymbol{K}_0^{(\mathrm{b})}\boldsymbol{P}^{-1}))\boldsymbol{W}_0^{(\mathrm{b})} + 2\sum_{i=1}^{\infty}\mathrm{tr}(\boldsymbol{K}_i^{(\mathrm{b})}\boldsymbol{P}^{-1})\boldsymbol{W}_i^{(\mathrm{b})} \\ &\quad -\boldsymbol{P}^{-1}\left((1+\mathrm{tr}(\boldsymbol{W}_0^{(\mathrm{b})}\boldsymbol{P}))\boldsymbol{K}_0^{(\mathrm{b})} + 2\sum_{i=1}^{\infty}\mathrm{tr}(\boldsymbol{W}_i^{(\mathrm{b})}\boldsymbol{P})\boldsymbol{K}_i^{(\mathrm{b})}\right)\boldsymbol{P}^{-1}. \end{aligned} \tag{43}$$

From Theorem 1 and Theorem 2, it is proved that the optimal positive definite symmetric matrix $\boldsymbol{P}_{\mathrm{opt}}$ has the following symmetric property [1]:

$$\boldsymbol{P}_{\mathrm{opt}} = \boldsymbol{\Sigma}\boldsymbol{P}_{\mathrm{opt}}^{-1}\boldsymbol{\Sigma} \tag{44}$$

for the signature matrix $\boldsymbol{\Sigma}$ which satisfies Eq. (29). We will thus search the optimal positive definite symmetric matrix $\boldsymbol{P}_{\mathrm{opt}}$ among the positive definite symmetric matrices $\boldsymbol{P}$ which satisfy

$$\boldsymbol{P} = \boldsymbol{\Sigma}\boldsymbol{P}^{-1}\boldsymbol{\Sigma}. \tag{45}$$

3.4. Closed form expression of the positive definite symmetric matrix $\boldsymbol{P}$

In this subsection, we consider the case of second-order digital filters and give the closed form expression of the positive definite symmetric matrix $\boldsymbol{P}$. When we restrict ourselves to the second-order case of state-space digital filters, we can give an closed form expression of the positive definite symmetric matrix $\boldsymbol{P}$ which satisfies Eq. (45), considering the form of the signature matrix $\boldsymbol{\Sigma}$ classified into the following two cases:

$$\begin{cases}\boldsymbol{\Sigma} = \pm\mathrm{diag}(1,1) = \pm\boldsymbol{I} \\ \boldsymbol{\Sigma} = \pm\mathrm{diag}(1,-1). \end{cases} \tag{46}$$

For each case, we next consider the closed form expression of the positive definite symmetric matrix $\boldsymbol{P}$ which satisfies Eq. (45).

In case of $\boldsymbol{\Sigma} = \pm\boldsymbol{I}$, Eq. (44) yields $\boldsymbol{P}_{\mathrm{opt}} = \boldsymbol{P}_{\mathrm{opt}}^{-1}$, that is, $\boldsymbol{P}_{\mathrm{opt}} = \boldsymbol{I}$. It means that the minimum L_2-sensitivity realization can be synthesized without any coordinate transformation to the balanced realization, that is, the initial realization. In other words, the minimum L_2-sensitivity realization is the balanced realization as follows:

$$(\boldsymbol{A}_{\mathrm{opt}}, \boldsymbol{b}_{\mathrm{opt}}, \boldsymbol{c}_{\mathrm{opt}}, d_{\mathrm{opt}}) = (\boldsymbol{A}_{\mathrm{b}}, \boldsymbol{b}_{\mathrm{b}}, \boldsymbol{c}_{\mathrm{b}}, d_{\mathrm{b}}). \tag{47}$$

Thus, we need no more discussion on this case since the minimum L_2-sensitivity realization is already achieved as the balanced realization.

On the other hand, in case of $\boldsymbol{\Sigma} = \pm\mathrm{diag}(1,-1)$, the authors have derived the closed form expression of a positive definite symmetric matrix $\boldsymbol{P}$ which satisfies Eq. (45) as follows:

$$\boldsymbol{P} = \begin{bmatrix} \cosh(p) & \sinh(p) \\ \sinh(p) & \cosh(p) \end{bmatrix} \tag{48}$$

where p is a real scalar variable [12, 13].

3.5. Closed form expression of the L_2-sensitivity $S(\boldsymbol{P})$

In this subsection, the closed form expression of the L_2-sensitivity of second-order digital filters is given. We give the closed form expression of the L_2-sensitivity $S(\boldsymbol{P})$ in Eq. (31). We first express the general Gramians of the balanced realization $(\boldsymbol{A}_\mathrm{b}, \boldsymbol{b}_\mathrm{b}, \boldsymbol{c}_\mathrm{b}, d_\mathrm{b})$ as follows:

$$\begin{aligned} \boldsymbol{K}_i^{(\mathrm{b})} &= \frac{1}{2}\left(\boldsymbol{A}_\mathrm{b}^i \boldsymbol{K}_0^{(\mathrm{b})} + \boldsymbol{K}_0^{(\mathrm{b})} (\boldsymbol{A}_\mathrm{b}^T)^i\right) \\ &= \frac{1}{2}\left(\boldsymbol{A}_\mathrm{b}^i \boldsymbol{\Theta} + \boldsymbol{\Theta}(\boldsymbol{A}_\mathrm{b}^T)^i\right) \end{aligned} \tag{49}$$

$$\begin{aligned} \boldsymbol{W}_i^{(\mathrm{b})} &= \frac{1}{2}\left(\boldsymbol{W}_0^{(\mathrm{b})} \boldsymbol{A}_\mathrm{b}^i + (\boldsymbol{A}_\mathrm{b}^T)^i \boldsymbol{W}_0^{(\mathrm{b})}\right) \\ &= \frac{1}{2}\left(\boldsymbol{\Theta}\boldsymbol{A}_\mathrm{b}^i + (\boldsymbol{A}_\mathrm{b}^T)^i \boldsymbol{\Theta}\right). \end{aligned} \tag{50}$$

We express the L_2-sensitivity $S(\boldsymbol{P})$ by substituting Eqs. (49) and (50) into Eq. (31) as follows:

$$S(\boldsymbol{P}) = \mathrm{tr}(\boldsymbol{\Theta}\boldsymbol{P})\mathrm{tr}(\boldsymbol{\Theta}\boldsymbol{P}^{-1}) + \mathrm{tr}(\boldsymbol{\Theta}\boldsymbol{P}) + \mathrm{tr}(\boldsymbol{\Theta}\boldsymbol{P}^{-1}) + 2\sum_{i=1}^{\infty} \mathrm{tr}(\boldsymbol{\Theta}\boldsymbol{A}_\mathrm{b}^i\boldsymbol{P})\mathrm{tr}(\boldsymbol{\Theta}(\boldsymbol{A}_\mathrm{b}^T)^i\boldsymbol{P}^{-1}). \tag{51}$$

The L_2-sensitivity $S(\boldsymbol{P})$ in Eq. (51) can be expressed more simply. Exploiting the symmetric property of coefficient matrix $\boldsymbol{A}_\mathrm{b}$ and $\boldsymbol{P}$ given in Eqs. (29) and (45) respectively, we can rewrite the L_2-sensitivity $S(\boldsymbol{P})$ as

$$S(\boldsymbol{P}) = 2\mathrm{tr}(\boldsymbol{\Theta}\boldsymbol{P}) - (\mathrm{tr}(\boldsymbol{\Theta}\boldsymbol{P}))^2 + 2\sum_{i=0}^{\infty}\left(\mathrm{tr}(\boldsymbol{\Theta}\boldsymbol{A}_\mathrm{b}^i\boldsymbol{P})\right)^2. \tag{52}$$

In order to give the closed form expression of the L_2-sensitivity $S(\boldsymbol{P})$ in Eq. (52), it is necessary to derive the closed form expressions of matrices $\boldsymbol{A}_\mathrm{b}$, $\boldsymbol{\Theta}$, and $\boldsymbol{P}$. The closed form expressions of matrices $\boldsymbol{A}_\mathrm{b}$ and $\boldsymbol{\Theta}$ are given in Eqs. (36) and (39), respectively. The closed form expression of the positive definite symmetric matrix $\boldsymbol{P}$ is given in Eq. (48). Substituting the closed form expressions of matrices $\boldsymbol{A}_\mathrm{b}$, $\boldsymbol{\Theta}$, and $\boldsymbol{P}$ into Eq. (52) gives the closed form expression of the L_2-sensitivity $S(p)$ as

$$S(\boldsymbol{P}) = S(p) = \sum_{n=-2}^{2} s_n e^{np}. \tag{53}$$

It is remarkable that Eq. (53) is a simple linear combination of exponential functions which does not contain infinite summations. These coefficients s_n's are easily computed directly from the transfer function $H(z)$ [12, 13].

3.6. Synthesis of minimum L_2-sensitivity realizations

The parameter p which minimizes $S(p)$ in Eq. (53) can be derived by solving the following equation with respect to p:

$$\frac{\partial S(p)}{\partial p} = \sum_{n=-2}^{2} n s_n e^{np} = 0. \tag{54}$$

Letting $\beta = e^p$ gives

$$\sum_{n=-2}^{2} n s_n \beta^n = 0. \tag{55}$$

The above equation is a fourth-degree polynomial equation with respect to β of constant coefficients. In 1545, G. Cardano states in his book entitled *Ars Magna* (its translated edition is [16]) that there exists the formula of solutions for fourth-degree polynomial equations. Therefore, Eq. (55) can be solved analytically. Eq. (55) has four solutions, from which the positive real solution $\beta_{\text{opt}} = e^{p_{\text{opt}}}$ is adopted to derive the optimal positive definite symmetric matrix $\boldsymbol{P}_{\text{opt}}$ as

$$\begin{aligned}\boldsymbol{P}_{\text{opt}} &= \begin{bmatrix} \cosh(p_{\text{opt}}) & \sinh(p_{\text{opt}}) \\ \sinh(p_{\text{opt}}) & \cosh(p_{\text{opt}}) \end{bmatrix} \\ &= \frac{1}{2}\begin{bmatrix} \beta_{\text{opt}} + \beta_{\text{opt}}^{-1} & \beta_{\text{opt}} - \beta_{\text{opt}}^{-1} \\ \beta_{\text{opt}} - \beta_{\text{opt}}^{-1} & \beta_{\text{opt}} + \beta_{\text{opt}}^{-1} \end{bmatrix}.\end{aligned} \tag{56}$$

The diagonalization of the optimal positive definite symmetric matrix $\boldsymbol{P}_{\text{opt}} = \boldsymbol{T}_{\text{opt}}\boldsymbol{T}_{\text{opt}}^T$ is given by

$$\begin{aligned}\boldsymbol{P}_{\text{opt}} &= \begin{bmatrix} \frac{1}{\sqrt{2}} & -\frac{1}{\sqrt{2}} \\ \frac{1}{\sqrt{2}} & \frac{1}{\sqrt{2}} \end{bmatrix} \begin{bmatrix} \beta_{\text{opt}} & 0 \\ 0 & \beta_{\text{opt}}^{-1} \end{bmatrix} \begin{bmatrix} \frac{1}{\sqrt{2}} & \frac{1}{\sqrt{2}} \\ -\frac{1}{\sqrt{2}} & \frac{1}{\sqrt{2}} \end{bmatrix} \\ &\equiv \boldsymbol{R}^T \boldsymbol{B}_{\text{opt}} \boldsymbol{R}.\end{aligned} \tag{57}$$

Once the optimal positive definite symmetric matrix $\boldsymbol{P}_{\text{opt}}$ is derived, the optimal coordinate transformation matrix $\boldsymbol{T}_{\text{opt}}$ is calculated as

$$\boldsymbol{T}_{\text{opt}} = \boldsymbol{P}_{\text{opt}}^{\frac{1}{2}} \boldsymbol{U} \tag{58}$$

where $\boldsymbol{U}$ is an arbitrary orthogonal matrix. We have to note that the optimal coordinate transformation matrix $\boldsymbol{T}_{\text{opt}}$ is not unique because of the non-uniqueness of matrix $\boldsymbol{U}$. By letting $\boldsymbol{U} = \boldsymbol{I}$ in Eq. (58), for instance, one of the optimal coordinate transformation matrices $\boldsymbol{T}_{\text{opt}} = \boldsymbol{P}_{\text{opt}}^{1/2}$ is given by

$$\begin{aligned}\boldsymbol{T}_{\text{opt}} &= \boldsymbol{P}_{\text{opt}}^{\frac{1}{2}} \\ &= \boldsymbol{R}^T \boldsymbol{B}_{\text{opt}}^{\frac{1}{2}} \boldsymbol{R} \\ &= \begin{bmatrix} \frac{1}{\sqrt{2}} & -\frac{1}{\sqrt{2}} \\ \frac{1}{\sqrt{2}} & \frac{1}{\sqrt{2}} \end{bmatrix} \begin{bmatrix} \beta_{\text{opt}}^{\frac{1}{2}} & 0 \\ 0 & \beta_{\text{opt}}^{-\frac{1}{2}} \end{bmatrix} \begin{bmatrix} \frac{1}{\sqrt{2}} & \frac{1}{\sqrt{2}} \\ -\frac{1}{\sqrt{2}} & \frac{1}{\sqrt{2}} \end{bmatrix} \\ &= \frac{1}{2} \begin{bmatrix} \beta_{\text{opt}}^{\frac{1}{2}} + \beta_{\text{opt}}^{-\frac{1}{2}} & \beta_{\text{opt}}^{\frac{1}{2}} - \beta_{\text{opt}}^{-\frac{1}{2}} \\ \beta_{\text{opt}}^{\frac{1}{2}} - \beta_{\text{opt}}^{-\frac{1}{2}} & \beta_{\text{opt}}^{\frac{1}{2}} + \beta_{\text{opt}}^{-\frac{1}{2}} \end{bmatrix}. \end{aligned} \tag{59}$$

We can give a geometrical interpretation of the above optimal coordinate transformation matrix $\boldsymbol{T}_{\text{opt}}$. In Eq. (59), $\boldsymbol{R}$ is an orthogonal matrix, which means $\pi/4$[rad] rotation of the coordinate axes, $\boldsymbol{B}_{\text{opt}}$ is a positive definite diagonal matrix, which means a simple scaling of each coordinate axis. Eq. (59) shows that the minimum L_2-sensitivity realization is obtained by the following operations for the coordinate axes of the initial balanced realization; $-\pi/4$[rad] rotation, simple scaling, and $+\pi/4$[rad] rotation. Finally, the minimum L_2-sensitivity realization $(\boldsymbol{A}_{\text{opt}}, \boldsymbol{b}_{\text{opt}}, \boldsymbol{c}_{\text{opt}}, d_{\text{opt}})$ is synthesized as follows:

$$\left[\begin{array}{c|c} \boldsymbol{A}_{\text{opt}} & \boldsymbol{b}_{\text{opt}} \\ \hline \boldsymbol{c}_{\text{opt}} & d_{\text{opt}} \end{array}\right] = \left[\begin{array}{c|c} \boldsymbol{T}_{\text{opt}}^{-1} \boldsymbol{A}_{\text{b}} \boldsymbol{T}_{\text{opt}} & \boldsymbol{T}_{\text{opt}}^{-1} \boldsymbol{b}_{\text{b}} \\ \hline \boldsymbol{c}_{\text{b}} \boldsymbol{T}_{\text{b}} & d_{\text{b}} \end{array}\right] \tag{60}$$

and the minimum L_2-sensitivity $S_{\min}$, which is achieved by $p_{\text{opt}} = \log(\beta_{\text{opt}})$, is expressed by substituting p_{opt} into Eq. (53) as

$$S_{\min} = \sum_{n=2}^{2} s_n \beta_{\text{opt}}^n. \tag{61}$$

3.7. Numerical examples

We present numerical examples to demonstrate the validity of the proposed method. Consider a second-order digital filter $H(z)$ with complex conjugate poles, of which transfer function is given by

$$\begin{aligned} H(z) &= \frac{\alpha}{z-\lambda} + \frac{\alpha^*}{z-\lambda^*} + d \\ &= \frac{0.0396 + 0.0793z^{-1} + 0.0396z^{-2}}{1 - 1.3315z^{-1} + 0.49z^{-2}} \end{aligned} \tag{62}$$

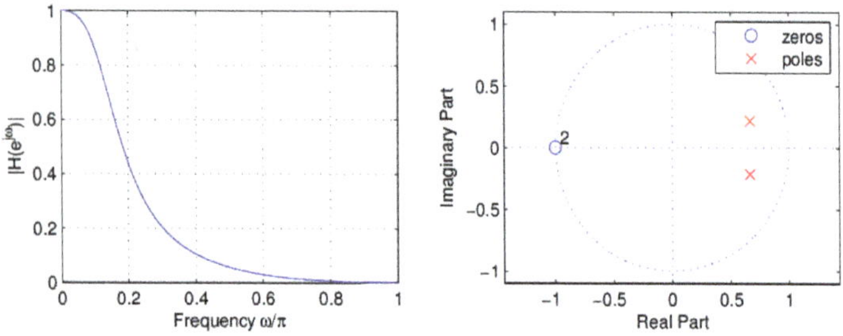

Figure 2. Frequency responses and zero-pole configurations of $H(z)$.

where $\lambda = 0.7\exp(j0.1\pi)$, $\alpha = 1.6657 - j6.3055$, and $d = 1$. The frequency responses and zero-pole configurations of digital filter $H(z)$ are shown in Fig. 2. From the transfer function $H(z)$, parameters P, Q, and R are calculated as

$$P = 0.5068,\ Q = -0.2947,\ R = 0.25. \tag{63}$$

The coefficients s_n's are computed as follows:

$$(s_{-2}, s_{-1}, s_0, s_1, s_2) = (0.3345, 0.8246, 0.8987, 0.8246, 0.7951). \tag{64}$$

We solve the following fourth-degree polynomial equation to derive the optimal solution $\beta_{\rm opt}$:

$$\sum_{n=-2}^{2} n s_n \beta^n = 0. \tag{65}$$

The fourth-degree polynomial equation above has the following four solutions:

$$\beta = 0.8568, -0.6960, -0.3396 \pm j0.7682. \tag{66}$$

We adopt $\beta_{\rm opt} = 0.8568$, which is a positive real scalar, to derive the optimal positive definite symmetric matrix $\boldsymbol{P}_{\rm opt}$. We can derive the minimum L_2-sensitivity realization $(\boldsymbol{A}_{\rm opt}, \boldsymbol{b}_{\rm opt}, \boldsymbol{c}_{\rm opt}, d_{\rm opt})$ in closed form as follows:

$$\left[\begin{array}{c|c} \boldsymbol{A}_{\rm opt} & \boldsymbol{b}_{\rm opt} \\ \hline \boldsymbol{c}_{\rm opt} & d_{\rm opt} \end{array}\right] = \left[\begin{array}{cc|c} 0.7810 & 0.2451 & 0.4751 \\ -0.2451 & 0.5505 & 0.3061 \\ \hline 0.4751 & -0.3061 & 0.0396 \end{array}\right] \tag{67}$$

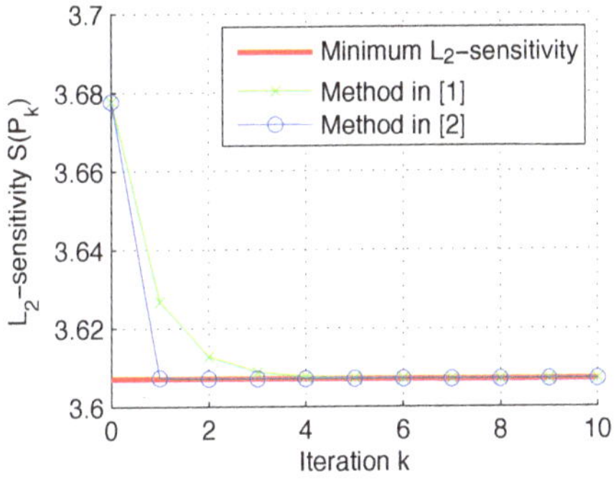

Figure 3. Minimum L_2-sensitivity and convergence behaviors of L_2-sensitivity in conventional methods [1] and [2].

of which L_2-sensitivity $S_{\min}$ is

$$S_{\min} = \sum_{n=-2}^{2} s_n \beta_{\text{opt}}^n = 3.6070. \tag{68}$$

Figure 3 shows the comparison of our proposed method with the iterative methods reported in [1] and [2], where the initial realization is the balanced realization. Our proposed method achieves the minimum L_2-sensitivity by only solving a fourth-degree polynomial equation without iterative calculations, while both of the methods in [1] and [2] require many iterative calculations to achieve the minimum L_2-sensitivity. Furthermore, our proposed method can guarantee that the L_2-sensitivity surely converges to the theoretical minimum when using conventional methods in [1] and [2].

4. Analytical solutions to the L_2-sensitivity minimization problem for digital filters with all second-order modes equal

This section reveals that the L_2-sensitivity minimization problem can be solved analytically if second-order modes are all equal. Furthermore, we clarify the general expression of the transfer functions of digital filters with all second-order modes equal [14].

4.1. Analytical synthesis of the minimum L_2-sensitivity realizations

We have discovered that there exist some digital filters whose minimum L_2-sensitivity realization is equal to the balanced realization. Such digital filters satisfy a sufficient condition summarized in the following theorem:

Theorem 3. *If all the second-order modes $\theta_i (i = 1, \cdots, N)$ of a digital filter $H(z)$ are equal, then*

$$(\boldsymbol{A}_{\text{opt}}, \boldsymbol{b}_{\text{opt}}, \boldsymbol{c}_{\text{opt}}, d_{\text{opt}}) = (\boldsymbol{A}_{\text{b}}, \boldsymbol{b}_{\text{b}}, \boldsymbol{c}_{\text{b}}, d_{\text{b}}) \tag{69}$$

that is, the minimum L_2-sensitivity realization is equal to the balanced realization. □

Proof: The general Gramians of the balanced realization are given by Eqs. (49) and (50), respectively. If all the second-order modes $\theta_i (i = 1, \cdots, N)$ satisfies

$$\theta_i = \theta \ (i = 1, \cdots, N) \tag{70}$$

the controllability and observability Gramians are expressed as

$$\boldsymbol{K}_0^{(\mathrm{b})} = \boldsymbol{W}_0^{(\mathrm{b})} = \mathrm{diag}(\theta, \cdots, \theta) = \theta \boldsymbol{I}. \tag{71}$$

Substituting Eq. (71) into Eqs. (49) and (50), the general Gramians are given by

$$\boldsymbol{K}_i^{(\mathrm{b})} = \boldsymbol{W}_i^{(\mathrm{b})} = \frac{1}{2}\theta(\boldsymbol{A}_{\mathrm{b}}^i + (\boldsymbol{A}_{\mathrm{b}}^T)^i). \tag{72}$$

We can express the general Gramians as $\boldsymbol{K}_i^{(\mathrm{b})} = \boldsymbol{W}_i^{(\mathrm{b})} = \boldsymbol{\Theta}_i$, which is defined by

$$\boldsymbol{\Theta}_i = \frac{1}{2}\theta(\boldsymbol{A}_{\mathrm{b}}^i + (\boldsymbol{A}_{\mathrm{b}}^T)^i) \ \ (i = 0, 1, \cdots). \tag{73}$$

Substituting $\boldsymbol{K}_i^{(\mathrm{b})} = \boldsymbol{W}_i^{(\mathrm{b})} = \boldsymbol{\Theta}_i$ into Eq. (43) yields

$$\begin{aligned} \frac{\partial S(\boldsymbol{P})}{\partial \boldsymbol{P}} = & (1 + \mathrm{tr}(\boldsymbol{\Theta}_0 \boldsymbol{P}^{-1}))\boldsymbol{\Theta}_0 + 2\sum_{i=1}^{\infty} \mathrm{tr}(\boldsymbol{\Theta}_i \boldsymbol{P}^{-1})\boldsymbol{\Theta}_i \\ & - \boldsymbol{P}^{-1}\left((1 + \mathrm{tr}(\boldsymbol{\Theta}_0 \boldsymbol{P}))\boldsymbol{\Theta}_0 + 2\sum_{i=1}^{\infty} \mathrm{tr}(\boldsymbol{\Theta}_i \boldsymbol{P})\boldsymbol{\Theta}_i \right) \boldsymbol{P}^{-1}. \end{aligned} \tag{74}$$

It is obvious that

$$\left. \frac{\partial S(\boldsymbol{P})}{\partial \boldsymbol{P}} \right|_{\boldsymbol{P}=\boldsymbol{I}} = 0 \tag{75}$$

which means that the minimum L_2-sensitivity realization can be synthesized without any coordinate transformation to the balanced realization, that is, the initial realization. Therefore, it is proved that the minimum L_2-sensitivity is equal to the balanced realization. □

4.2. Class of digital filters with all second-order modes equal

In the previous subsection, we revealed that the L_2-sensitivity minimization problem can be solved analytically if second-order modes are all equal. We next clarify the class of digital filters with all second-order modes equal. We have newly derived a general expression of the transfer function of Nth-order digital filters with all second-order modes equal.

4.2.1. General expression

In Ref. [14], we have newly derived a general expression of the transfer function of Nth-order digital filters with all second-order modes equal.

Corollary 1. *Let the second-order modes of an Nth-order digital filter $H(z)$ be $\theta_i (i = 1, \cdots, N)$. The second-order modes of the transfer function $\theta H(z)$ are given by $|\theta|\theta_i (i = 1, \cdots, N)$, where θ is a nonzero real scalar.*

Theorem 4. *The transfer function of an Nth-order digital filter $H(z)$ with all second-order modes equal such as*

$$\theta_1 = \theta_2 = \cdots = \theta_{N-1} = \theta_N \tag{76}$$

can be expressed as the following form:

$$H(z) = \theta H_{AP}(z) + \rho \tag{77}$$

where θ is a nonzero real scalar, ρ is a real scalar, and $H_{AP}(z)$ is an Nth-order all-pass digital filter. The second-order modes of the digital filter $H(z)$ are given by

$$\theta_i = |\theta| \ (i = 1, \cdots, N). \tag{78}$$

□

4.2.2. Frequency transformation

Furthermore, it is remarkable that the transfer function $H(z)$ in Eq. (77) is generally obtained by the frequency transformation on a first-order FIR prototype filter using an Nth-order all-pass digital filter.

Remark 1. *An Nth-order digital filter $H(z)$ in Eq. (77) is obtained by the frequency transformation such as*

$$H(z) = H_P(z)|_{z^{-1} \leftarrow H_{AP}(z)} \tag{79}$$

where $H_P(z) = \theta z^{-1} + \rho$ is a first-order prototype FIR digital filter and $H_{AP}(z)$ is an Nth-order all-pass digital filter. □

The variable substitution in Eq. (79) represents the frequency transformation, where $H_P(z)$ is the prototype digital filter. Block diagrams of the prototype digital filter $H_P(z)$ and transformed digital filter $H(z)$ are shown in Fig. 4. These figures show that a block diagram of the transformed digital filter $H(z)$ can be obtained by simple substitution of the all-pass digital filter $H_{AP}(z)$ into the unit delay z^{-1}. Theorem 4 and Remark 1 give us the class of digital filters with all second-order modes equal.

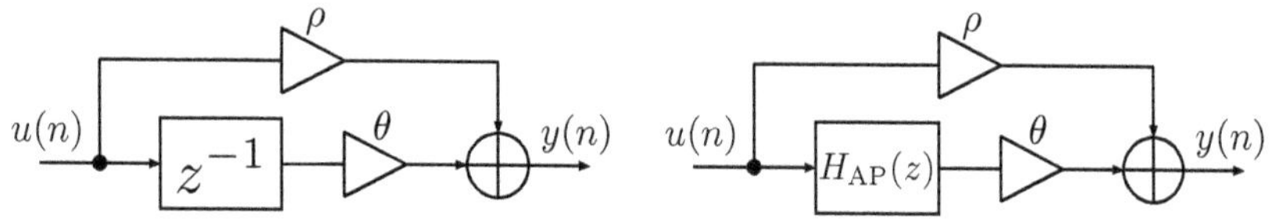

(a) First-order prototype digital filter $H_P(z)$

(b) Nth-order transformed digital filter $H(z)$

Figure 4. Block diagrams of prototype digital filter $H_P(z)$ and transformed digital filter $H(z)$.

4.3. Examples of digital filters with all second-order modes equal

There are many types of digital filters with all second-order modes equal. We can design various digital filters by setting variables θ, ρ, and $H_{AP}(z)$.

4.3.1. The unit delay

The unit delay is the simplest example for digital filter with all second-order modes equal. It is obvious that letting $\theta = 1$, $\rho = 0$, and $H_{AP}(z) = z^{-1}$ in Eq. (77) yields the unit delay z^{-1}.

4.3.2. First-order digital filters

Any first-order digital filter can be expressed in the form of Eq. (77). Consider a first-order IIR digital filter given by

$$H_{IIR}(z) = \frac{b_0 + b_1 z^{-1}}{1 + a_1 z^{-1}} \tag{80}$$

where b_0 and b_1 are numerator coefficients, a_1 is a denominator coefficient. One can easily show that Eq. (80) can be rewritten as the form of Eq. (77) where

$$\theta = \frac{b_1 - a_1 b_0}{1 - a_1^2}, \ \rho = \frac{b_0 - a_1 b_1}{1 - a_1^2}, \ H_{AP}(z) = \frac{a_1 + z^{-1}}{1 + a_1 z^{-1}}. \tag{81}$$

4.3.3. All-pass digital filters

It is obvious that all-pass digital filters are included in the class of digital filters expressed as Eq. (77). The transfer function $H(z)$ in Eq. (77) is an all-pass digital filter when we let $\theta = 1$ and $\rho = 0$.

4.3.4. Multi-notch comb digital filters

The transfer function of an Nth-order multi-notch comb digital filter is given by

$$H_{\mathrm{MN}}(z) = \frac{1+\alpha}{2}\frac{1-z^{-N}}{1-\alpha z^{-N}}. \tag{82}$$

This filter has N notches at the frequency $2\pi k/N$[rad] for $k = 1, \cdots, N$. One can easily show that Eq. (82) can be rewritten as the form of Eq. (77) where

$$\theta = \frac{1}{2},\ \rho = \frac{1}{2},\ H_{\mathrm{AP}}(z) = \frac{\alpha - z^{-1}}{1-\alpha z^{-1}}. \tag{83}$$

4.4. Numerical examples

This subsection gives numerical examples of synthesis of the minimum L_2-sensitivity realizations for various types of digital filters with all second-order modes equal.

4.4.1. First-order FIR digital filters

Consider a first-order FIR digital filter $H_{\mathrm{FIR}}(z)$ given by

$$H_{\mathrm{FIR}}(z) = 0.5 + 0.5z^{-1} \tag{84}$$

of which frequency magnitude and phase responses are shown in Fig. 5 (a). The second-order mode of the digital filter $H_{\mathrm{FIR}}(z)$ is $\theta = 0.5$. The balanced realization (A_b, b_b, c_b, d_b), which is equal to the minimum L_2-sensitivity realization, of $H_{\mathrm{FIR}}(z)$ is derived as

$$\left[\begin{array}{c|c} A_b & b_b \\ \hline c_b & d_b \end{array}\right] = \left[\begin{array}{c|c} 0 & 0.7071 \\ \hline 0.7071 & 0.5 \end{array}\right] \tag{85}$$

and controllability Gramian $K_0^{(b)}$ and observability Gramians $W_0^{(b)}$ are calculated as

$$K_0^{(b)} = W_0^{(b)} = 0.5. \tag{86}$$

4.4.2. First-order IIR digital filters

Consider a first-order IIR digital filter $H_{\mathrm{IIR}}(z)$ given by

$$H_{\mathrm{IIR}}(z) = \frac{0.25 + 0.25z^{-1}}{1 - 0.5z^{-1}} \tag{87}$$

of which frequency magnitude and phase responses are shown in Fig. 5 (b). The second-order mode of the digital filter $H_{\mathrm{IIR}}(z)$ is $\theta = 0.5$. The balanced realization

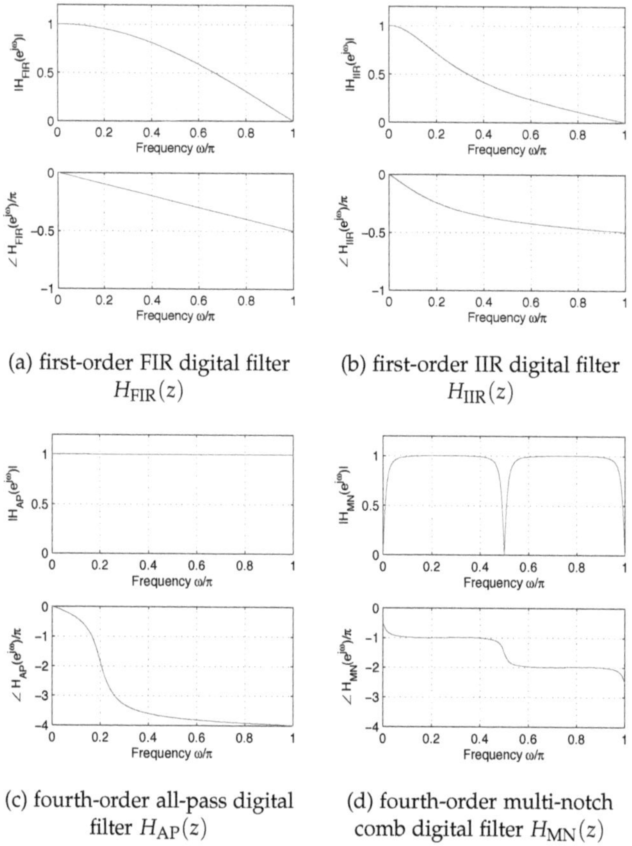

(a) first-order FIR digital filter $H_{\mathrm{FIR}}(z)$

(b) first-order IIR digital filter $H_{\mathrm{IIR}}(z)$

(c) fourth-order all-pass digital filter $H_{\mathrm{AP}}(z)$

(d) fourth-order multi-notch comb digital filter $H_{\mathrm{MN}}(z)$

Figure 5. Frequency magnitude and phase responses of digital filters with all second-order modes equal.

$(\boldsymbol{A}_{\mathrm{b}}, \boldsymbol{b}_{\mathrm{b}}, \boldsymbol{c}_{\mathrm{b}}, d_{\mathrm{b}})$, which is equal to the minimum L_2-sensitivity realization, of $H_{\mathrm{IIR}}(z)$ is derived as

$$\left[\begin{array}{c|c} \boldsymbol{A}_{\mathrm{b}} & \boldsymbol{b}_{\mathrm{b}} \\ \hline \boldsymbol{c}_{\mathrm{b}} & d_{\mathrm{b}} \end{array}\right] = \left[\begin{array}{c|c} 0.5 & 0.6124 \\ \hline 0.6124 & 0.25 \end{array}\right] \tag{88}$$

and controllability Gramian $\boldsymbol{K}_0^{(\mathrm{b})}$ and observability Gramians $\boldsymbol{W}_0^{(\mathrm{b})}$ are calculated as

$$\boldsymbol{K}_0^{(\mathrm{b})} = \boldsymbol{W}_0^{(\mathrm{b})} = 0.5. \tag{89}$$

4.4.3. All-pass digital filters

Consider a fourth-order all-pass digital filter $H_{AP}(z)$ given by

$$H_{AP}(z) = \frac{0.5184 - 1.9805z^{-1} + 3.3350z^{-2} - 2.7507z^{-3} + z^{-4}}{1 - 2.7507z^{-1} + 3.3350z^{-2} - 1.9805z^{-3} + 0.5184z^{-4}} \tag{90}$$

of which poles $\lambda_p (p = 1, 2, 3, 4)$ are given by

$$\begin{cases} \lambda_1 = 0.9\exp(j0.2\pi), \ \lambda_2 = 0.9\exp(-j0.2\pi), \\ \lambda_3 = 0.8\exp(j0.2\pi), \ \lambda_4 = 0.8\exp(-j0.2\pi) \end{cases} \tag{91}$$

and of which frequency magnitude and phase responses are shown in Fig. 5 (c). The second-order modes θ_i $(i = 1, 2, 3, 4)$ of the all-pass digital filter $H_{AP}(z)$ are given by

$$(\theta_1, \theta_2, \theta_3, \theta_4) = (1, 1, 1, 1). \tag{92}$$

The balanced realization $(\boldsymbol{A}_b, \boldsymbol{b}_b, \boldsymbol{c}_b, d_b)$, which is equal to the minimum L_2-sensitivity realization, of $H_{AP}(z)$ is derived as

$$\left[\begin{array}{c|c} \boldsymbol{A}_b & \boldsymbol{b}_b \\ \hline \boldsymbol{c}_b & d_b \end{array}\right] = \left[\begin{array}{cccc|c} 0.8144 & -0.1106 & 0.2499 & -0.5039 & -0.0903 \\ 0.1599 & 0.4698 & -0.7114 & -0.1105 & -0.4853 \\ -0.1616 & 0.2997 & 0.6211 & 0.1062 & -0.6978 \\ 0.5318 & 0.0448 & 0.0006 & 0.8453 & 0.0252 \\ \hline 0.0481 & 0.8217 & 0.2137 & -0.0894 & 0.5184 \end{array}\right] \tag{93}$$

and controllability Gramian $\boldsymbol{K}_0^{(b)}$ and observability Gramians $\boldsymbol{W}_0^{(b)}$ are calculated as

$$\boldsymbol{K}_0^{(b)} = \boldsymbol{W}_0^{(b)} = \text{diag}(1, 1, 1, 1). \tag{94}$$

4.4.4. Multi-notch comb digital filters

Consider a fourth-order multi-notch comb digital filter $H_{MN}(z)$ given by

$$H_{MN}(z) = \frac{0.9073 - 0.9073z^{-4}}{1 - 0.8145z^{-4}} \tag{95}$$

of which poles $\lambda_p (p = 1, 2, 3, 4)$ are given by

$$\begin{cases} \lambda_1 = 0.95, \ \lambda_2 = -0.95, \\ \lambda_3 = j0.95, \ \lambda_4 = -j0.95 \end{cases} \tag{96}$$

and of which frequency magnitude and phase responses are shown in Fig. 5 (d). The second-order modes θ_i $(i = 1,2,3,4)$ of the multi-notch comb digital filter $H_{\mathrm{MN}}(z)$ are given by

$$(\theta_1, \theta_2, \theta_3, \theta_4) = (0.5, 0.5, 0.5, 0.5). \tag{97}$$

The balanced realization $(\boldsymbol{A}_{\mathrm{b}}, \boldsymbol{b}_{\mathrm{b}}, \boldsymbol{c}_{\mathrm{b}}, d_{\mathrm{b}})$, which is equal to the minimum L_2-sensitivity realization, of $H_{\mathrm{MN}}(z)$ is derived as

$$\left[\begin{array}{c|c} \boldsymbol{A}_{\mathrm{b}} & \boldsymbol{b}_{\mathrm{b}} \\ \hline \boldsymbol{c}_{\mathrm{b}} & d_{\mathrm{b}} \end{array}\right] = \left[\begin{array}{cccc|c} 0 & 0 & 0 & 0.8145 & 0.4102 \\ 1 & 0 & 0 & 0 & 0 \\ 0 & 1 & 0 & 0 & 0 \\ 0 & 0 & 1 & 0 & 0 \\ \hline 0 & 0 & 0 & -0.4102 & 0.9073 \end{array}\right] \tag{98}$$

and controllability Gramian $\boldsymbol{K}_0^{(\mathrm{b})}$ and observability Gramians $\boldsymbol{W}_0^{(\mathrm{b})}$ are calculated as

$$\boldsymbol{K}_0^{(\mathrm{b})} = \boldsymbol{W}_0^{(\mathrm{b})} = \mathrm{diag}(0.5, 0.5, 0.5, 0.5). \tag{99}$$

5. Absence of limit cycles in the minimum L_2-sensitivity realizations

This section proves the absence of limit cycles of the minimum L_2-sensitivity realization from the viewpoint of the controllability and observability Gramians. The minimum L_2-sensitivity realizations have freedom for orthogonal transformations. In other words, minimum L_2-sensitivity realizations are not unique. We select the minimum L_2-sensitivity realization without limit cycles among these minimum L_2-sensitivity realizations. The controllability and observability Gramians of the selected minimum L_2-sensitivity realization satisfy a sufficient condition for the absence of limit cycles [11].

5.1. Theoretical proof of the absence of limit cycles

For high-order digital filters, we synthesize the minimum L_2-sensitivity realization by the successive approximation methods in [1] or [2], for examples. For second-order digital filters, we can synthesize the minimum L_2-sensitivity realization by the closed form solutions proposed in Section 3. For both cases, we can construct the minimum L_2-sensitivity realization *without limit cycles*.

We begin by reviewing the procedure to synthesize the minimum L_2-sensitivity. We solve the L_2-sensitivity minimization problem in (32) adopting the balanced realization $(\boldsymbol{A}_{\mathrm{b}}, \boldsymbol{b}_{\mathrm{b}}, \boldsymbol{c}_{\mathrm{b}}, d_{\mathrm{b}})$ as an initial realization. We obtain the optimal positive definite symmetric matrix $\boldsymbol{P}_{\mathrm{opt}}$. In case of high-order digital filters, we can derive the optimal positive definite symmetric matrix $\boldsymbol{P}_{\mathrm{opt}}$ by successive approximation method in [1] or [2], for example. In case of second-order digital filters, we can derive the optimal positive definite symmetric matrix $\boldsymbol{P}_{\mathrm{opt}}$ analytically as proposed in Section 3.

We can give the diagonalization of the matrix $\boldsymbol{P}_{\text{opt}}$ as follows:

$$\boldsymbol{P}_{\text{opt}} = \boldsymbol{R}^T \boldsymbol{B}_{\text{opt}} \boldsymbol{R}. \tag{100}$$

Since the matrix $\boldsymbol{P}_{\text{opt}}$ is positive definite symmetric, it can be diagonalized by an orthogonal matrix $\boldsymbol{R}$, and $\boldsymbol{B}_{\text{opt}}$ is a positive definite diagonal matrix. The optimal coordinate transformation matrix $\boldsymbol{T}_{\text{opt}}$ is given by

$$\begin{aligned} \boldsymbol{T}_{\text{opt}} &= \boldsymbol{P}_{\text{opt}}^{\frac{1}{2}} \boldsymbol{U} \\ &= \boldsymbol{R}^T \boldsymbol{B}_{\text{opt}}^{\frac{1}{2}} \boldsymbol{R} \boldsymbol{U}. \end{aligned} \tag{101}$$

In the above expression, $\boldsymbol{U}$ is an arbitrary orthogonal matrix. It means that the minimum L_2-sensitivity realizations exist infinitely for a given digital filter $H(z)$. The minimum L_2-sensitivity realizations have freedom for orthogonal transformations. We show that the minimum L_2-sensitivity realization does not generate limit cycles if we specify the orthogonal matrix as $\boldsymbol{U} = \boldsymbol{R}^T$, which yields

$$\tilde{\boldsymbol{T}}_{\text{opt}} = \boldsymbol{R}^T \boldsymbol{B}_{\text{opt}}^{\frac{1}{2}}. \tag{102}$$

Theorem 5. *The minimum L_2-sensitivity realization $(\tilde{\boldsymbol{A}}_{\text{opt}}, \tilde{\boldsymbol{b}}_{\text{opt}}, \tilde{\boldsymbol{c}}_{\text{opt}}, \tilde{d}_{\text{opt}})$, obtained by the coordinate transformation by $\tilde{\boldsymbol{T}}_{opt}$ such as*

$$(\tilde{\boldsymbol{A}}_{\text{opt}}, \tilde{\boldsymbol{b}}_{\text{opt}}, \tilde{\boldsymbol{c}}_{\text{opt}}, \tilde{d}_{\text{opt}}) = (\tilde{\boldsymbol{T}}_{\text{opt}}^{-1} \boldsymbol{A}_{\text{b}} \tilde{\boldsymbol{T}}_{\text{opt}}, \tilde{\boldsymbol{T}}_{\text{opt}}^{-1} \boldsymbol{b}_{\text{b}}, \boldsymbol{c}_{\text{b}} \tilde{\boldsymbol{T}}_{\text{opt}}, d_{\text{b}}) \tag{103}$$

does not generate limit cycles. □

Proof: Under the coordinate transformation by $\tilde{\boldsymbol{T}}_{\text{opt}}$ in Eq. (102), the controllability Gramian $\tilde{\boldsymbol{K}}_0^{(\text{opt})}$ and the observability Gramian $\tilde{\boldsymbol{W}}_0^{(\text{opt})}$ of the minimum L_2-sensitivity realization $(\tilde{\boldsymbol{A}}_{\text{opt}}, \tilde{\boldsymbol{b}}_{\text{opt}}, \tilde{\boldsymbol{c}}_{\text{opt}}, \tilde{d}_{\text{opt}})$ are expressed as

$$\begin{aligned} \tilde{\boldsymbol{K}}_0^{(\text{opt})} &= \tilde{\boldsymbol{T}}_{\text{opt}}^{-1} \boldsymbol{K}_0^{(\text{b})} \tilde{\boldsymbol{T}}_{\text{opt}}^{-T} \\ &= \boldsymbol{B}_{\text{opt}}^{-\frac{1}{2}} \boldsymbol{R} \boldsymbol{\Theta} \boldsymbol{R}^T \boldsymbol{B}_{\text{opt}}^{-\frac{1}{2}} \end{aligned} \tag{104}$$

$$\begin{aligned} \tilde{\boldsymbol{W}}_0^{(\text{opt})} &= \tilde{\boldsymbol{T}}_{\text{opt}}^{T} \boldsymbol{W}_0^{(\text{b})} \tilde{\boldsymbol{T}}_{\text{opt}} \\ &= \boldsymbol{B}_{\text{opt}}^{\frac{1}{2}} \boldsymbol{R} \boldsymbol{\Theta} \boldsymbol{R}^T \boldsymbol{B}_{\text{opt}}^{\frac{1}{2}} \end{aligned} \tag{105}$$

where $\boldsymbol{\Theta} = \text{diag}(\theta_1, \cdots, \theta_N)$. From Eqs. (104) and (105), we can derive the relation between the controllability and observability Gramians as follows:

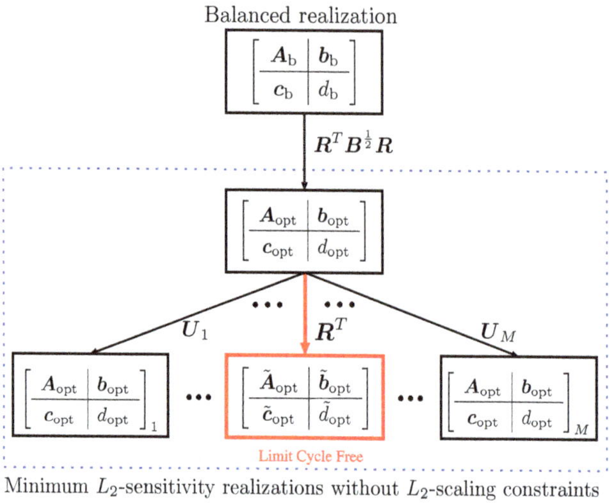

Figure 6. Synthesis of the minimum L_2-sensitivity realization which does not generate limit cycles.

$$\tilde{W}_0^{(\mathrm{opt})} = B_{\mathrm{opt}} \tilde{K}_0^{(\mathrm{opt})} B_{\mathrm{opt}}. \tag{106}$$

Eq. (106) is equivalent to a sufficient condition for the absence of limit cycles proposed in Ref. [6]. Therefore, the minimum L_2-sensitivity realization $(\tilde{A}_{\mathrm{opt}}, \tilde{b}_{\mathrm{opt}}, \tilde{c}_{\mathrm{opt}}, \tilde{d}_{\mathrm{opt}})$ does not generate limit cycles. □

Theorem 5 shows that we can synthesize the minimum L_2-sensitivity realization without limit cycles by choosing appropriate orthogonal matrix U. Fig. 6 shows the synthesis procedure of the minimum L_2-sensitivity realization which does not generate limit cycles. The coefficient matrices of the minimum L_2-sensitivity realization without limit cycles $(\tilde{A}_{\mathrm{opt}}, \tilde{b}_{\mathrm{opt}}, \tilde{c}_{\mathrm{opt}}, \tilde{d}_{\mathrm{opt}})$ are given by

$$\left[\begin{array}{c|c} \tilde{A}_{\mathrm{opt}} & \tilde{b}_{\mathrm{opt}} \\ \hline \tilde{c}_{\mathrm{opt}} & \tilde{d}_{\mathrm{opt}} \end{array}\right] = \left[\begin{array}{c|c} B_{\mathrm{opt}}^{-\frac{1}{2}} R A_{\mathrm{b}} R^T B_{\mathrm{opt}}^{\frac{1}{2}} & B_{\mathrm{opt}}^{-\frac{1}{2}} R b_{\mathrm{b}} \\ \hline c_{\mathrm{b}} R^T B_{\mathrm{opt}}^{\frac{1}{2}} & d \end{array}\right]. \tag{107}$$

5.2. Numerical examples

We present numerical examples to demonstrate the validity of our proposed method. We synthesize the minimum L_2-sensitivity realizations of second-order and fourth-order digital filters which do not generate limit cycles.

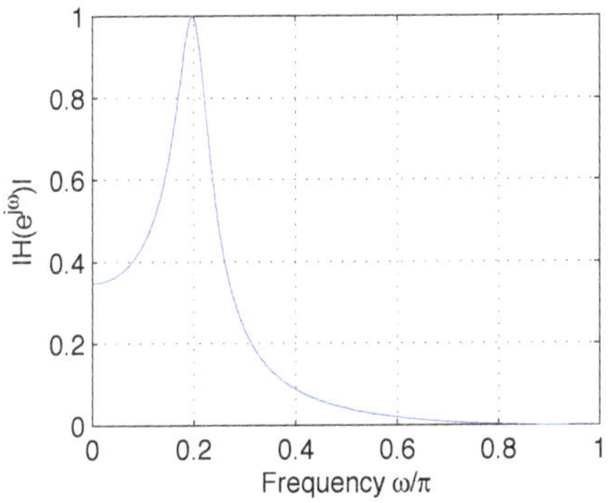

Figure 7. Frequency Response of digital filter $H(z)$ in Eq. (108).

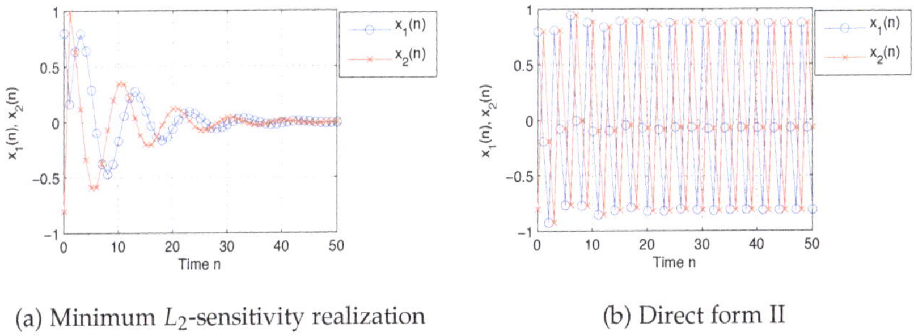

Figure 8. Zero-input responses of $H(z)$ in Eq. (108).

5.2.1. *Second-order digital filters*

Consider a second-order narrow-band band-pass digital filter $H(z)$ given by

$$H(z) = \frac{0.0316 + 0.0602z^{-1} + 0.0316z^{-2}}{1 - 1.4562z^{-1} + 0.81z^{-2}}. \tag{108}$$

The poles of the transfer function $H(z)$ in Eq. (108) are $0.9\exp(\pm j0.2\pi)$, which are very close to the unit circle. The frequency response of the digital filter $H(z)$ in Eq. (108) is shown in Fig. 7. The coefficient matrices of the minimum L_2-sensitivity realization which is free of limit cycle is derived by Eq. (107) as follows:

$$\left[\begin{array}{c|c}\tilde{A}_{\rm opt} & \tilde{b}_{\rm opt} \\ \hline \tilde{c}_{\rm opt} & \tilde{d}_{\rm opt}\end{array}\right] = \left[\begin{array}{cc|c} 0.7281 & 0.5229 & 0.4146 \\ -0.5351 & 0.7281 & -0.1282 \\ \hline 0.1282 & -0.4146 & 0.0316 \end{array}\right]. \tag{109}$$

The controllability Gramian $K_0^{(\rm opt)}$ and the observability Gramian $W_0^{(\rm opt)}$ are given as follows:

$$\tilde{K}_0^{(\rm opt)} = \begin{bmatrix} 0.5100 & -0.0870 \\ -0.0870 & 0.4901 \end{bmatrix} \tag{110}$$

$$\tilde{W}_0^{(\rm opt)} = \begin{bmatrix} 0.4901 & -0.0870 \\ -0.0870 & 0.5100 \end{bmatrix}. \tag{111}$$

We have to note that the controllability Gramian $\tilde{K}_0^{(\rm opt)}$ and the observability Gramian $\tilde{W}_0^{(\rm opt)}$ satisfy the sufficient condition of the absence of limit cycles given in Eq. (106) with

$$B_{\rm opt} = {\rm diag}(0.9803, 1.0201). \tag{112}$$

Therefore, $(\tilde{A}_{\rm opt}, \tilde{b}_{\rm opt}, \tilde{c}_{\rm opt}, \tilde{d}_{\rm opt})$ is the minimum L_2-sensitivity realization without limit cycles.

We demonstrate the absence of limit cycles in the minimum L_2-sensitivity realization by observing its zero-input response. We calculate the zero-input responses of the minimum L_2-sensitivity realization and the dierct form II, setting the initial state as $x(0) = [0.8\ \ -0.8]^T$. We let the dynamic range of signals to be $[-1, 1)$ and adopt two's complement as the overflow characteristic. The zero-input responses are shown in Fig. 8(a) and 8(b). We assume that each filter coefficient and signal have 16[bits] fixed-point representation, of which lower 14[bits] are fractional bits. In this numerical example, the overflow of the state variables occurs in both cases. It is desirable that the effect of the overflow is decreasing since the digital filter $H(z)$ in Eq. (108) is stable. For the minimum L_2-sensitivity realization synthesized by our proposed method, the state variables $x_1(n)$ and $x_2(n)$ converge to zero after the overflow, as shown in Fig. 8(a). Therefore, there are no limit cycles. On the other hand, for the direct form II, a large-amplitude autonomous oscillation is observed as shown in Fig. 8(b). Therefore, the direct form II generates the limit cycles.

5.2.2. *High-order digital filters*

We can demonstrate the validity of the proposed method for also high-order digital filters. Consider a fourth-order band-pass digital filter $H(z)$ given by

$$H(z) = \frac{0.0178 - 0.0252z^{-1} + 0.0173z^{-2} - 0.0252z^{-3} + 0.0178z^{-4}}{1 - 2.6977z^{-1} + 3.5410z^{-2} - 2.3340z^{-3} + 0.7497z^{-4}} \tag{113}$$

The frequency response of the digital filter $H(z)$ in Eq. (113) is shown in Fig. 9. We obtain the limit cycle free minimum L_2-sensitivity realization $(\tilde{A}_{\rm opt}, \tilde{b}_{\rm opt}, \tilde{c}_{\rm opt}, \tilde{d}_{\rm opt})$ by successive approximation method:

$$\left[\begin{array}{c|c} \tilde{A}_{opt} & \tilde{b}_{opt} \\ \hline \tilde{c}_{opt} & \tilde{d}_{opt} \end{array}\right] = \left[\begin{array}{cccc|c} 0.6028 & 0.6394 & 0.1512 & 0.0655 & 0.0344 \\ -0.6360 & 0.7461 & -0.0655 & -0.0297 & -0.0153 \\ -0.0806 & -0.0283 & 0.6028 & 0.6360 & 0.3950 \\ 0.0283 & 0.0118 & -0.6394 & 0.7461 & -0.1423 \\ \hline 0.3950 & 0.1423 & 0.0344 & 0.0153 & 0.0178 \end{array}\right]. \tag{114}$$

The controllability Gramian $\tilde{K}_0^{(opt)}$ and the observability Gramian $\tilde{W}_0^{(opt)}$ are given as follows:

$$\tilde{K}_0^{(opt)} = \begin{bmatrix} 0.3531 & -0.0004 & 0.2470 & 0.0155 \\ -0.0004 & 0.3563 & -0.0157 & 0.2470 \\ 0.2470 & -0.0157 & 0.5309 & 0.0006 \\ 0.0155 & 0.2470 & 0.0006 & 0.5264 \end{bmatrix} \tag{115}$$

$$\tilde{W}_0^{(opt)} = \begin{bmatrix} 0.5309 & -0.0006 & 0.2470 & 0.0157 \\ -0.0006 & 0.5264 & -0.0155 & 0.2470 \\ 0.2470 & -0.0155 & 0.3531 & 0.0004 \\ 0.0157 & 0.2470 & 0.0004 & 0.3563 \end{bmatrix}. \tag{116}$$

We have to note that the controllability Gramian $\tilde{K}_0^{(opt)}$ and the observability Gramian $\tilde{W}_0^{(opt)}$ satisfy the sufficient condition of the absence of limit cycles given in Eq. (106) with

$$B_{opt} = \mathrm{diag}(1.2261, 1.2155, 0.8156, 0.8227). \tag{117}$$

Therefore, $(\tilde{A}_{opt}, \tilde{b}_{opt}, \tilde{c}_{opt}, \tilde{d}_{opt})$ is the minimum L_2-sensitivity realization without limit cycles.

We demonstrate the absence of limit cycles in the minimum L_2-sensitivity realization by observing its zero-input response. We calculate the zero-input responses of the minimum L_2-sensitivity realization and the direct form II, setting the initial state as $x(0) = [0.9\ 0.9\ 0.9\ 0.9]^T$. We let the dynamic range of signals to be $[-1, 1)$ and adopt two's complement as the overflow characteristic. The zero-input responses are shown in Fig. 10(a) and 10(b). We assume that each filter coefficient and signal have 16[bits] fixed-point representation, of which lower 13[bits] are fractional bits. In this numerical example, the overflow of the state variables occurs in both cases. Also in this case, we can confirm that the minimum L_2-sensitivity realization does not generate limit cycles. On the other hand, a large-amplitude autonomous oscillation is observed in the zero-input response of the direct form II.

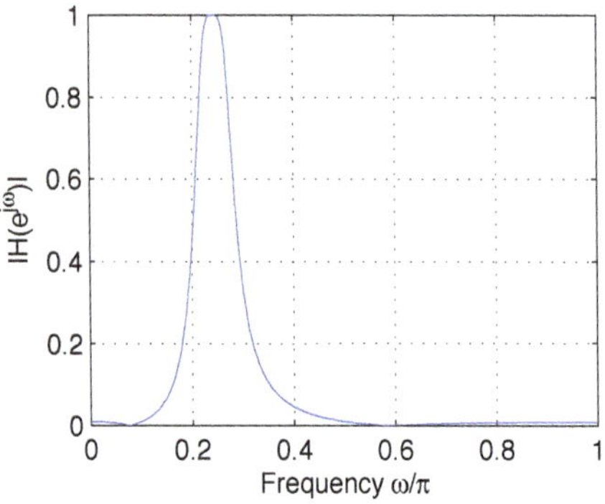

Figure 9. Frequency Response of digital filter $H(z)$ in Eq. (113).

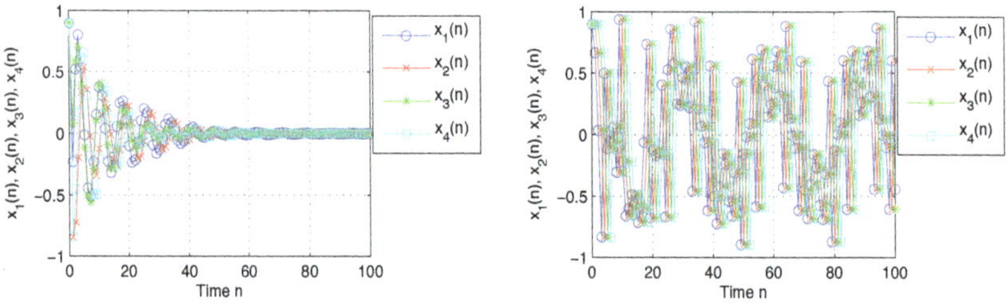

(a) Minimum L_2-sensitivity realization (b) Direct form II

Figure 10. Zero-input responses of digital filter $H(z)$ in Eq. (113).

6. Conclusions

This chapter presents analytical approach for synthesis of the minimum L_2-sensitivity realizations for state-space digital filters. The contributions of this chapter are summarized as follows.

Section 3 presents closed form solutions to the L_2-sensitivity minimization problem for second-order state-space digital filters. We have shown that the L_2-sensitivity is expressed by a linear combination of exponential functions, and we can synthesize the minimum L_2-sensitivity realization by only solving a fourth degree polynomial equation, which can be solved analytically.

Section 4 reveals that the L_2-sensitivity minimization problem can be solved analytically for arbitrary filter order if second-order modes are all equal. We derive a general expression of the transfer function of digital filters with all second-order modes equal. We show that the general expression is obtained by a frequency transformation on a first-order prototype FIR digital filter.

Section 5 proves the absence of limit cycles of the minimum L_2-sensitivity realization from the view point of relationship between the controllability and observability Gramians. The minimum L_2-sensitivity realizations were originally known to be low-coefficient sensitivity filter structures. We have succeeded in discovering the novel property of the minimum L_2-sensitivity realizations.

Author details

Shunsuke Yamaki*,
Masahide Abe and Masayuki Kawamata

* Address all correspondence to: yamaki@mk.ecei.tohoku.ac.jp

Department of Electronic Engineering, Graduate School of Engineering, Tohoku University, Sendai, Japan

References

[1] W.-Y. Yan and J. B. Moore. On L_2-sensitivity minimization of linear state-space systems. *IEEE Trans. Circuits Syst. I Fundamental theory and applications*, 39(8):641–648, August 1992.

[2] T. Hinamoto, S. Yokoyama, T. Inoue, W. Zeng, and W.-S. Lu. Analysis and minimization of L_2-sensitivity for linear systems and two-dimensional state-space filters using general controllability and observability gramians. *IEEE Trans. Circuits Syst.*, CAS-49(9):1279–1289, September 2002.

[3] Clifford T. Mullis and Richard A. Roberts. Synthesis of minimum roundoff noise fixed point digital filters. *IEEE Trans. Circuits Syst.*, CAS-23(9):551–562, September 1976.

[4] Clifford T. Mullis and Richard A. Roberts. Roundoff noise in digital filters: Frequency transformations and invariants. *IEEE Trans. Acoust., Speech, Signal Process.*, ASSP-24(6):538–550, December 1976.

[5] Sheng Y. Hwang. Minimum uncorrelated unit noise in state-space digital filtering. *IEEE Trans. Acoust., Speech, Signal Process.*, ASSP-25(4):273–281, August 1977.

[6] M. Kawamata and T. Higuchi. On the absence of limit cycles in a class of state-space digital filters which contains minimum noise realizations. *IEEE Trans. Acoust., Speech, Signal Process.*, ASSP-32(4):928–930, August 1984.

[7] M. Kawamata and T. Higuchi. A unified approach to the optimal synthesis of fixed-point state-space digital filters. *IEEE Trans. Acoust., Speech, Signal Process.*, ASSP-33(4):911–920, August 1985.

[8] V. Tavsanoglu and L. Thiele. Optimal design of state-space digital filters by simultaneous minimization of sensitivity and roundoff noise. *IEEE Trans. Circuits Syst.*, CAS-31(10):884–888, October 1984.

[9] M. Gevers and G. Li. *Parametrizations in Control, Estimation and Filtering Problems*, chapter 5. Springer-Verlag, 1993.

[10] Robert S. H. Istepanian and James F. Whidborne. *Digital Controller Implementation and Fragility*. Springer-Verlag, 2001.

[11] S. Yamaki, M. Abe, and M. Kawamata. On the absence of limit cycles in state-space digital filters with minimum L_2-sensitivity. *IEEE Trans. Circuits Syst. II*, 55(1):46–50, January 2008.

[12] S. Yamaki, M. Abe, and M. Kawamata. A closed form solution to L_2-sensitivity minimization of second-order state-space digital filters. *IEICE Trans. Fundam. Electron., Commun., Comput. Sci.,*, E91-A(5):1268–1273, May 2008.

[13] S. Yamaki, M. Abe, and M. Kawamata. Closed form solutions to L_2-sensitivity minimization of second-order state-space digital filters with real poles. *IEICE Trans. Fundam. Electron., Commun., Comput. Sci.,*, E93-A(5):966–971, May 2010.

[14] S. Yamaki, M. Abe, and M. Kawamata. Derivation of the class of digital filters with all second-order modes equal. *IEEE Trans. Signal Process.*, 59(11):5236–5242, November 2011.

[15] H. Matsukawa and M. Kawamata. Design of variable digital filters based on state-space realizations. *IEICE Trans. Fundam. Electron., Commun., Comput. Sci.,*, E84-A(8):1822–1830, August 2001.

[16] G. Cardano and translated by T. R. Witmer with a foreword by O. Ore. *The Great Art or the Rules of Algebra*. The M. I. T. Press, 1968.

Chapter 9

Two-Rate Based Structures for Computationally Efficient Wide-Band FIR Systems

Håkan Johansson and Oscar Gustafsson

Additional information is available at the end of the chapter

1. Introduction

Many digital signal processing (DSP) systems tend to have a very high computational complexity when they target a large part of the Nyquist band. This corresponds to a wide-band system with one or several so called don't-care bands approaching zero. Examples of such systems include frequency selective filters, fractional-delay filters, and differentiators. This chapter considers finite-length impulse response (FIR) filters due to their attractive implementation features. In particular, they can be implemented with non-recursive structures. In contrast to infinite-length impulse response (IIR) filters, they are therefore always automatically stable and have no bound on the maximal sampling rate, see [1, 2].

For frequency-selective wide-band FIR filters, the frequency-response masking (FRM) technique can be employed for complexity reductions due to its use of sparse (namely periodic) subfilters, see [3–10]. For other functions, the FRM technique cannot be used directly, and one therefore has to seek other methods to reduce the complexity. This chapter discusses such a method which utilizes a two-rate technique, but only for the derivation of efficient single-rate structures. The basic two-rate approach was originally introduced in [11] and has since then been exploited and extended for various contexts as detailed in [12–19] and to be reviewed in this chapter. For single-function systems, it is however necessary to combine the two-rate technique with the FRM approach in order to achieve an overall complexity reduction. For multi-function realizations, complexity savings may be obtained without incorporating the FRM approach but it offers further complexity savings in such cases, as exemplified in [19]. Recent results have shown that the two-rate approach offers dramatic complexity reductions for wide-band systems, especially when combined with the FRM approach.

1.1. Chapter outline

Following this introduction, Section 2 considers the two-rate based structure that is appropriate for so called left-band and right-band systems which have don't-care bands at the low-frequency and high-frequency regions, respectively. Section 3 discusses the extension to so called mid-band systems which have don't-care bands at both the low-frequency and high-frequency regions. In Section 4, multi-function system realizations are considered, whereas Section 5 gives more implementation details. Finally, Section 6 concludes the chapter.

2. Two-rate based structure for left-band and right-band systems

This section will first revisit FIR filters and their computational complexity. After that, the two-rate based structure for left-band and right-band systems will be discussed.

2.1. Complexity of FIR filters

Consider a causal FIR filter with an impulse response $h(n)$, transfer function

$$H(z) = \sum_{n=0}^{N_H} h(n) z^{-n}, \tag{1}$$

and frequency response

$$H(e^{j\omega}) = \sum_{n=0}^{N_H} h(n) e^{-j\omega n}. \tag{2}$$

The order of the system is N_H and the impulse response duration (length) is $N_H + 1$. A direct-form implementation of the filter, corresponding directly to the convolution

$$y(n) = \sum_{k=0}^{N_H} x(n-k) h(k), \tag{3}$$

where $x(n)$ is the input and $y(n)$ the output, requires $N_H + 1$ multiplications and N_H additions to compute each output sample $y(n)$. In the case of a linear-phase frequency response, $h(n)$ is symmetric or anti-symmetric which reduces the number of multiplications to roughly $N_H/2$ [1].

The filter order required is determined by the application and specification. For example, for frequency selective filters, the order is inversely proportional to the transition band

[1] An even-order symmetric (anti-symmetric) linear-phase filter requires $N_H/2 + 1$ ($N_H/2$) multiplications whereas an odd-order linear-phase filter requires $(N_H + 1)/2$ multiplications.

(don't-care band) $\Delta = \omega_s - \omega_c$, where ω_c and ω_s denote the passband and stopband edges, respectively, see [20, 21]. Hence, when the don't-care band decreases towards zero, the order increases rapidly. Then, using a direct-form realization, the computational complexity may become intolerable as it follows the filter order. The same trend exists also for other functions that are not frequency selective filters, like differentiation and integration, as seen in [22].

2.2. Two-rate based structure

To reduce the complexity, we consider here a structure that is derived via a two-rate approach, seen in Fig. 1. This structure is efficient for left-band systems (like a differentiator) targeting the frequency region $\omega \in [0, \omega_c]$, $0 < \omega_c < \pi$. The same structure can also be used for right-band systems targeting the band $\omega \in [\omega_c, \pi]$, $0 < \omega_c < \pi$. The only difference will appear in the design, and we will therefore focus on the left-band case in this chapter, and only comment upon the right-band case in the design section.

For a left-band specification, the basic idea is to first interpolate the input signal $x(n)$ by two through upsampling by two followed by a lowpass filter with transfer function $F(z)$ [2]. Then, a subsequent filter with transfer function $G(z)$ follows that performs the actual function. Finally, downsampling by two takes place to retain the original sampling rate. Using multi-rate theory, see [23], it is readily shown that this scheme corresponds to a linear and time-invariant (LTI) system with a transfer function $H(z)$ that equals the 0th polyphase component of the cascaded filter $F(z)G(z)$, i.e.,

$$H(z) = F_0(z)G_0(z) + z^{-1}F_1(z)G_1(z) \tag{4}$$

where

$$F(z) = F_0(z^2) + z^{-1}F_1(z^2) \tag{5}$$

and

$$G(z) = G_0(z^2) + z^{-1}G_1(z^2). \tag{6}$$

The final realization is thus a single-rate structure. A two-rate technique is only used to derive efficient structures. It is noted here that the order and delay of the overall filter $H(z)$ is $N_H = (N_F + N_G)/2$ and $D_H = (D_F + D_G)/2$, respectively. This can be understood by noting that $F(z)$ and $G(z)$ can be viewed as operating (in principle) at two times the input rate, because the structure is derived by sandwiching $F(z)G(z)$ between upsampling and downsampling by two.

---

[2] The same function can be achieved by sampling the underlying analog signal with a higher sampling rate instead of sampling it slower and then use interpolation in the digital domain. However, this also increases the requirements on the analog-to-digital converters which are power-hungry components and in many cases one of the bottlenecks in overall systems. It is therefore often preferred to perform interpolation in the digital domain.

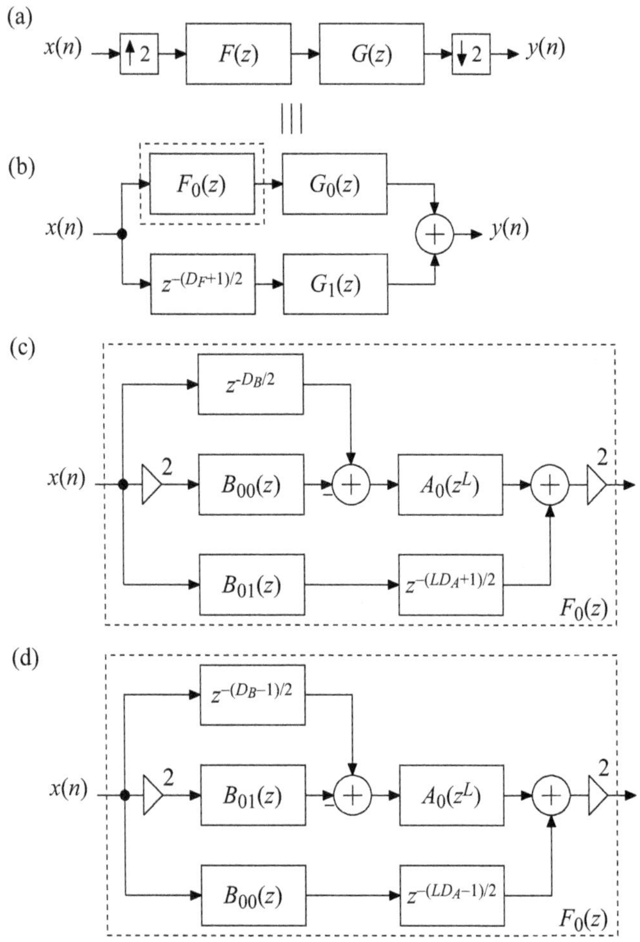

Figure 1. (a) Two-rate approach. (b) Equivalent LTI system when $F(z)$ is an HB filter. (c) and (d) Polyphase component $F_0(z)$ of the HB filter $F(z) = F_0(z^2) + z^{-D_F}$ when realized with the FRM approach for an even-order (c) and odd-order (d) masking filter.

2.2.1. *Filter Types*

It is possible and efficient to let $F(z)$ be a linear-phase half-band (HB) FIR filter [3]. Such a filter has a symmetric impulse response and every second impulse response value is zero, except the center tap which equals unity for an interpolation filter that preserves the signal energy. This corresponds to a pure-delay polyphase component $F_1(z)$, namely $F_1(z) = z^{-(D_F-1)/2}$,

[3] If the delay is of importance, one may need to use a nonlinear-phase (approximately linear-phase) low-delay HB filter $F(z)$ instead. Further, if there are additional requirements in the don't-care band, like attenuation requirements at $\omega = \pi$, a general filter $F(z)$ must be used instead, i.e., a non-HB filter. See [19] for details.

Overall filter $H(z)$	**Half-band filter** $F(z)$	$G(z)$
Type I, even order $2(m+p+1)$	Type I, order $4m+2$	Type I, order $4p+2$
Type II, odd order $2(m+p)+1$	Type I, order $4m+2$	Type I, order $4p$
Type III, even order $2(m+p+1)$	Type I, order $4m+2$	Type III, order $4p+2$
Type IV, odd order $2(m+p)+1$	Type I, order $4m+2$	Type III, order $4p$

Table 1. Linear-phase filter types.

where D_F is the delay of $F(z)$, which is always an odd integer. When $F(z)$ is a linear-phase filter, $G(z)$ is of the same type as that of the overall filter $H(z)$, i.e., a linear-phase filter (nonlinear-phase filter) when $H(z)$ is a linear-phase filter (nonlinear-phase filter). In this section, we focus on linear-phase filters. In the next section, nonlinear-phase applications are considered.

When $F(z)$ is a linear-phase HB filter, it is a symmetric Type I filter with the odd-integer delay D_F. Its delay contribution, $D_F/2$, to the overall delay D_H is therefore an integer plus a half. It is consequently the delay contribution $D_G/2$ of $G(z)$ that determines whether the overall delay is an integer or an integer plus a half. As $D_G/2$ then must be an integer or an integer plus a half to obtain an overall linear-phase filter, D_G must be an integer. Consequently, $G(z)$ is either a Type I or Type III linear-phase FIR filter, i.e., an even-order filter with a symmetric or anti-symmetric impulse response. In other words, the type and order of $G(z)$ determines the type and order of the overall filter, as summarized in Table 1. A formal proof of these facts is given in [17].

The order of $G(z)$ is thus somewhat restricted as it cannot take on all even orders. However, the effective order of $G(z)$ can be reduced by two by setting its first and last impulse response value to zero. In this way, two multiplications and additions may be saved in some cases. For the HB filter $F(z)$, it does not make sense to try to reduce the effective order in this way, as its impulse response is always zero for odd indexes of n (except for the center tap).

2.3. Complexity reduction

Assume that $H(e^{j\omega})$ is to approximate a desired function $D(j\omega)$ in the band $\omega \in [0, \omega_c]$. Due to the principle of interpolation by two in the two-rate based scheme, the effective bandwidth of $G(z)$ is $\omega_c/2$, and thus always less than $\pi/2$. The complexity of $G(z)$ alone will therefore be substantially lower than that of a regular direct-form realization of $H(z)$. (This will be discussed in more detail in the design example considered later in Section 2.4). However, the overall complexity is also determined by the filter $F(z)$. The requirement on this filter is roughly the same as that of the overall filter $H(z)$ and its complexity is therefore relatively high. In other words, a major part of the overall complexity is moved to the filter $F(z)$ and thus to $F_0(z)$ in Fig. 1. Therefore, for a single-function system, there will not be any computational savings using this approach straightforwardly. This is because we can equally well combine the three subfilters into one single conventional filter.

Nevertheless, overall savings can indeed be obtained by utilizing additional complexity saving techniques for the lowpass frequency selective HB filter $F(z)$. Specifically, by realizing $F(z)$ as an FRM filter, see [3–10], we can express the transfer function as

$$F(z) = 2A(z^L)B_0(z) + 2[z^{-LD_A} - A(z^L)]B_1(z) \tag{7}$$

where $A(z^L)$ is a period model filter and $[z^{-LD_A} - A(z^L)]$ is its complement, whereas $B_0(z)$ and $B_1(z)$ are masking filters. Specifically, in the case of a HB filter, as detailed in [5, 7], $A(z)$ is given as

$$A(z) = A_0(z^2) + 0.5z^{-D_A}, \tag{8}$$

with D_A being the delay of $A(z)$, whereas the masking filter are related according to

$$B_1(z) = z^{-D_B} - (-1)^{D_B} B_0(-z), \tag{9}$$

with D_B being the delay of $B_0(z)$. One then finds that $F_0(z)$ becomes, for D_B even:

$$F_0(z) = 2z^{-(LD_A+1)/2} B_{01}(z) + 2[2B_{00}(z) - z^{-D_B/2}] A_0(z^L) \tag{10}$$

and, for D_B odd:

$$F_0(z) = 2z^{-(LD_A-1)/2} B_{00}(z) + 2[2B_{01}(z) - z^{-(D_B-1)/2}] A_0(z^L) \tag{11}$$

where $B_{00}(z)$ and $B_{01}(z)$ are the polyphase components of $B_0(z)$, i.e., $B_0(z) = B_{00}(z^2) + z^{-1}B_{01}(z^2)$. The resulting structures for $F_0(z)$ are depicted in Fig. 1(c) and (d). More details can be found in [7].

As seen, $F_0(z)$ makes use of three subfilters, of which $A_0(z^L)$ is periodic for an integer $L > 1$. A periodic filter is a sparse filter, meaning it has many zero-valued filter coefficients. Specifically, only every Lth impulse response value of $A_0(z^L)$ is non-zero. Consequently, a linear-phase filter $A(z^L)$ of order N_A requires roughly only $N_A/(2L)$ multiplications and N_A/L additions. In this way, substantial overall savings can be obtained as compared to the conventional direct-form structures.

2.4. Design

Filters are typically designed in the minimax (Chebyshev) sense or least-squares sense, or possibly combinations thereof, see [24–26]. The goal of this chapter is to demonstrate that the complexity (number of multiplications and additions) can be reduced when using the two-rate based structures instead of regular structures. This will be done by designing both filter classes to meet the same specification and then comparing the resulting complexities [4]. To this end, the selection of approximation type is irrelevant, as long as one uses the same for both filter classes. In this chapter, we use minimax design, but other designs can of course be used as well after some minor appropriate modifications.

For minimax design, the maximum of the modulus of an error function $E(j\omega)$ is minimized. The error function is typically given as

$$E(j\omega) = W(\omega)[H(e^{j\omega}) - D(j\omega)], \quad \omega \in \Omega. \tag{12}$$

[4] Another type of comparison is to study the approximation error differences between two solutions having the same filter implementation complexity. However, such a comparison is appropriate when using two different design methods applied to the same filter class (structure) which does not apply here.

where $D(j\omega)$ is a desired function to be approximated in the frequency band Ω by the filter frequency response $H(e^{j\omega})$, whereas $W(\omega)$ is a positive weighting function. A conventional FIR filter, with the frequency response in the form of (2), is then designed by solving the following approximation problem.

Approximation problem: Given N_H, find the unknowns $h(n)$ and δ to minimize δ subject to

$$|E(j\omega)| \leq \delta. \tag{13}$$

For a linear-phase filter, we also have the additional symmetry constraints $h(n) = h(N-n)$ or $h(n) = -h(N-n)$.

For a conventional filter, the problem above is a convex optimization problem which has a unique global optimum. It can be found using linear programming, see [27], or the more efficient McClellan-Parks-Rabiner algorithm given in [28]. In practice, one usually has a specification on the desired approximation error δ, say δ_e. The filter will meet this specification if δ after the optimization satisfies $\delta \leq \delta_e$.

For the two-rate based filters, the design becomes more intricate because it contains cascaded and parallel subfilters. This means that the unknowns are not $h(n)$ but instead $f(n)$ and $g(n)$, in general, and $a(n)$, $b_0(n)$, and $g(n)$ when $F(z)$ is realized as an FRM filter. Hence, conventional design methods can no longer be used. Moreover, due to the cascaded subfilters, we are now facing a nonlinear (nonconvex) optimization problem, which means that an overall globally optimum solution cannot be guaranteed. Nevertheless, if carefully designed, even a locally optimum solution for a two-rate based structure can be substantially less complex than the corresponding globally optimum direct-form structure. To ensure a good local optimum, the overall two-rate based filters are designed in three steps as explained below. Although $F(z)$ should here be an FRM HB filter in order to achieve any savings, we will first explain the essential design steps in terms of a regular HB filter for the sake of simplicity. After that, the necessary modifications required for an FRM design will be pointed out.

2.4.1. Basic Three-Step Design Procedure

Given the desired function

$$D(j\omega) = e^{-j\omega(N_G+N_F)/4} D_0(j\omega) \tag{14}$$

and bandwidth $\omega \in [0, \omega_c]$, $\omega_c < \pi$, as well as a targeted approximation error δ_e, perform the following three-step procedure for each combination of filter orders N_G and N_F around estimated required orders $\widehat{N}_G$ and $\widehat{N}_F$:

(1) Design the regular FIR filter $G(z)$, which gives $G_0(z)$ and $G_1(z)$ after polyphase decomposition. It is done by minimizing the maximum of $|E_G(j\omega)|$ in the band $\omega \in [0, \omega_c/2]$ where [5]

$$E_G(j\omega) = G(e^{j\omega}) - e^{-j\omega N_G/2} D_0(j2\omega). \tag{15}$$

[5] For a right-band specification, the band for $G(z)$ is $\omega \in [\omega_c/2, \pi/2]$.

(2) Design a regular lowpass HB FIR filter $F(z)$, which gives $F_0(z)$ and $F_1(z) = z^{-(D_F-1)/2}$, $D_F = N_F/2$, after polyphase decomposition. It is done by minimizing the maximum of $|E_F(j\omega)|$ in the band $\omega \in [\pi - \omega_c/2, \pi]$, where [6],

$$E_F(j\omega) = F(e^{j\omega}). \tag{16}$$

(3) Use $F_0(z)$, $G_0(z)$, and $G_1(z)$ obtained above as the initial solution in a further nonlinear optimization routine that solves the approximation problem stated in (13). If the resulting approximation error δ is smaller than δ_e after the optimization, store the result.

The estimated orders required, $\widehat{N}_G$ and $\widehat{N}_F$, can be found by separately designing $G(z)$ and $F(z)$ to approximate their respective desired functions (as given in Steps 1 and 2, respectively) with the same tolerance as the overall targeted error, i.e., δ_e (or similar as in [17–19]). As the bandwidth of $G(z)$ is always below $\pi/2$, its order is typically below 12 for approximation errors down to some -100 dB, provided a smooth function is targeted, like a differentiator or integrator. Hence, the value of $\widehat{N}_G$ is readily found by designing $G(z, d)$ for all low-order filters, using conventional techniques, and then set $\widehat{N}_G$ to the lowest one for which the approximation error $|E_G(j\omega)|$ is below δ_e. As to the lowpass HB filter $F(z)$, the value $\widehat{N}_F$ can be found via well-known formulas for order estimation, see [20, 21], and a few designs around the estimated value.

Regarding the designs, the problems in Steps 1 and 2 are convex, and thus have unique global optima, provided they are formulated in accordance with the approximation problem stated earlier in this section. These problems can be solved using any regular solver for such problems. As $F(z)$ is a linear-phase filter, it can alternatively be designed using the efficient McClellan-Parks-Rabiner algorithm given in [28]. The problem in Step 3 is nonlinear because of the cascaded subfilters. In the examples of this chapter, we use the general-purpose nonlinear-optimization routine *fminimax* in MATLAB together with the real-rotation theorem, see [29], to solve the problem. The real-rotation theorem states that minimizing $|f|$ is equivalent to minimizing $\Re\{fe^{j\Theta}\}$, $\forall\Theta \in [0, 2\pi]$. The optimization problem is then solved with ω and Θ discretized to dense enough grids. A few hundred and 10–20 points, respectively, are typically sufficient in practice.

2.4.2. Modifications When Using an FRM Filter $F(z)$

When $F(z)$ is an FRM filter, we can use essentially the same design steps as outlined above. However, a difference is that $F(z)$ is now realized in terms of the two subfilters $A(z^P)$ and $B_0(z)$ or, equivalently, $F_0(z)$ is now realized in terms of the three subfilters $A_0(z^L)$, $B_{00}(z)$ and $B_{01}(z)$. This means that three parameters, $\widehat{N}_A$, $\widehat{N}_B$, and $\widehat{L}$, instead of only one parameter, $\widehat{N}_F$, need to be estimated. Given the same approximation error and band edges as before, $F(z)$ as well as $\widehat{N}_A$, $\widehat{N}_B$, and $\widehat{L}$, can be obtained as outlined in [7]. It is noted here that the design of $F(z)$ in Step 2 now corresponds to a nonconvex problem due to cascaded subfilters in the FRM approach. In [7], this is solved via initial linear optimizations and further nonlinear optimization, similar to the approach given above for the two-rate based structure.

[6] For a right-band specification, a highpass filter $F(z)$ is designed instead, in the stopband $\omega \in [0, \omega_c/2]$.

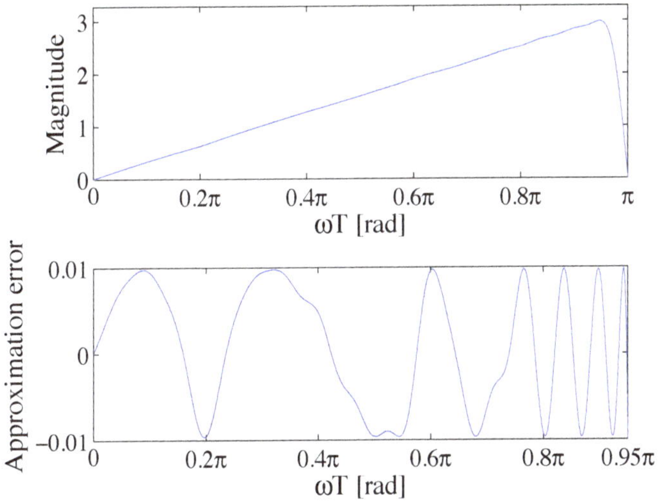

Figure 2. Magnitude response and approximation error of the two-rate and FRM based filter in Example 1. Throughout this chapter, we have used $T = 1$ for simplicity.

2.5. Examples

Consider a first-degree differentiator with the desired function [4]

$$D(j\omega) = e^{-j\omega(N_G+N_F)/4}j\omega \tag{17}$$

in the frequency region $\omega T \in [0, \omega_c]$, $0 < \omega_c < \pi$. This function can be approximated by a Type III linear-phase FIR filter, i.e., by a filter of even order and with an anti-symmetric impulse response, see [4].

Example 1: $\omega_c = 0.95\pi$, and $\delta_e = 0.01$ (−40 dB). Using a conventional differentiator, the specification is met by a 60th-order filter which requires 30 multiplications and 59 additions in an implementation. Using instead the two-rate and FRM based approach, with $L = 5$, we can meet the specification with filter orders 22, 18, and 2, for $A(z)$, $B_0(z)$, and $G(z)$, respectively. The corresponding overall realization requires 17 multiplications and 31 additions. Thus, multiplication and addition savings of 43% and 47%, respectively, are achieved. The savings are however dependent of the bandwidth ω_c as will be illustrated below in Example 2. As always when using linear-phase FRM filters, the price to pay is a somewhat increased delay, and a few more delay elements. In this example, the delay is increased from 30 to 32 samples whereas the number of delay elements is increased from 60 to 64. The increase is thus only 7%. The overall filter frequency response is plotted in Fig. 2.

Example 2: Figure 3 shows the number of multiplications required for the conventional direct-form filter and the two-rate based filter, both approximating first-degree Type III differentiators with approximation errors of $\delta = 0.01, 0.001, 0.0001$ (−40, −60, −80 dB).

As the plots reveal, the complexity savings using the two-rate based filter is increased substantially when the bandwidth approaches π. The break-even point is somewhere around $\omega_c = 0.8\pi$ from which the savings increase approximately linearly with increasing bandwidth. In the region between 0.8π and 0.98π, the savings go from around zero up to some 65%. Similar savings are obtained also for the number of additions as it is proportional to the number of multiplications. Again, a price to pay for the arithmetic complexity reductions is a moderate increase of the delay and number of delay elements, typically between some 5% and 20%.

From the results in [17, 22], the number of multiplications required for a regular Type III differentiator can be estimated as

$$\widehat{M}_{\text{regular}} = \pi \frac{0.810[-\log_{10}(\delta_c)]^{0.919}}{\pi - \omega_c}. \tag{18}$$

For the two-rate based differentiators, we have instead from [17]

$$\widehat{M}_{\text{tworate}} = \pi \frac{0.884[-\log_{10}(\delta_c)]^{0.852}}{\pi - 0.956\omega_c}. \tag{19}$$

Comparing the two expressions, we see that the main difference is the multiplicative constant 0.956 in front of ω_c in the latter expression. This explains why the savings increase with increasing bandwidth, as illustrated in Fig. 3.

2.6. Generalization to $M > 2$

The two-rate based scheme can readily be extended to the one depicted in Fig. 4(a) where the interpolation factor is an arbitrary integer M. Here, the basic principle is thus to first interpolate with M via the interpolation filter $F(z)$. Then the actual function is again approximated by $G(z)$. Finally, downsampling by M occurs. Using multi-rate theory, one finds again that this structure has the LTI system equivalent seen in Fig. 4(b). That is, the overall transfer function is

$$H(z) = F_0(z)G_0(z) + \sum_{m=1}^{M-1} z^{-1}F_m(z)G_{M-m}(z), \tag{20}$$

where $F_m(z)$ and $G_m(z)$ are polyphase components of $F(z)$ and $G(z)$ in the polyphase representations

$$F(z) = \sum_{m=0}^{M-1} z^{-m}F_m(z^M), \quad G(z) = \sum_{m=0}^{M-1} z^{-m}G_m(z^M). \tag{21}$$

Using an Mth-band interpolation filter $F(z)$, also the generalized scheme is appropriate for left-band and right-band systems. It has turned out though that the case with $M = 2$ typically is the most efficient choice which is why that case has been considered in detail in this section. This is because the additional cost of $F(z)$ exceeds the additional savings of $G(z)$ when going from $M = 2$ to $M > 2$. This in turn is due to the fact that the complexity of $G(z)$ is already very low for $M = 2$. A more detailed discussion on this is found in [17].

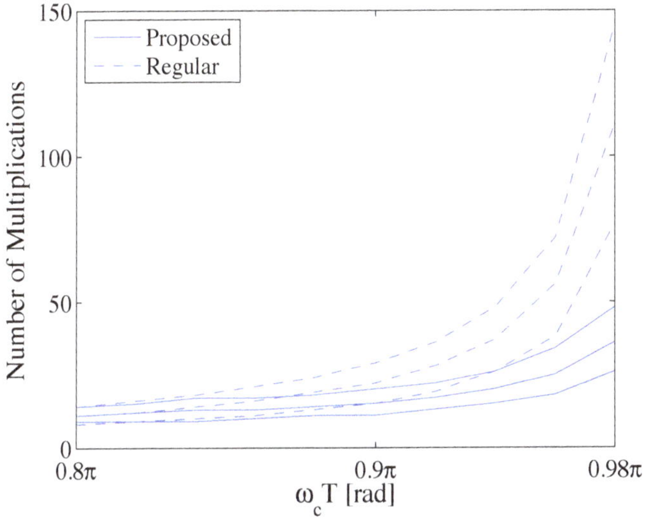

Figure 3. Number of multiplications versus bandwidth $\omega_c T$ for the conventional direct-form filter (dashed line) and proposed two-rate based filter (solid line), for a first-degree Type III differentiator with the approximation errors $\delta_c = 0.01, 0.001, 0.0001$. Throughout this chapter, we have used $T = 1$ for simplicity.

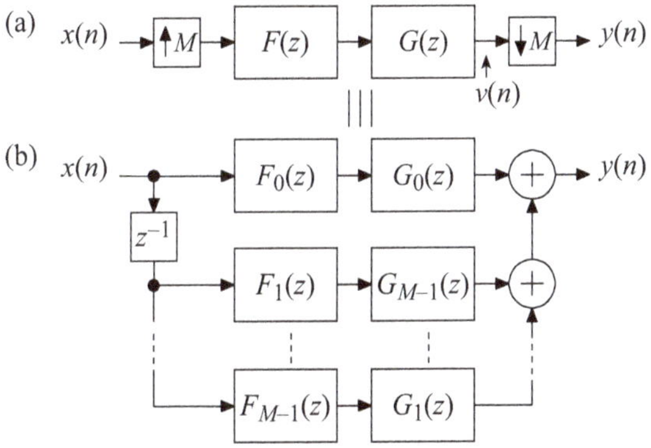

Figure 4. (a) Two-rate approach with arbitrary M. (b) Equivalent single-rate realization.

3. Two-rate based structure for mid-band systems

This section extends the results to mid-band systems which target the region $\omega \in [\omega_{c1}, \omega_{c2}]$, $0 < \omega_{c1} < \omega_{c2} < \pi$. Example applications include fractional-degree differentiators and integrators, see [30–33]. For later discussions, we define the don't-care bands Δ_1 and Δ_2 as

$$\Delta_1 = \omega_{c1}, \quad \Delta_2 = \pi - \omega_{c2}. \tag{22}$$

In principle, we can again make use of the scheme in Fig. 4 with a lowpass filter $F(z)$ but it is not efficient for mid-band systems. This is because the filter $G(z)$ then needs to approximate the desired function in the band between $\omega_{c1}/M = \Delta_1/M$ and $\omega_{c2}/M = (\pi - \Delta_2)/M$. Although this implies that the width of the upper don't-care band of $G(z)$ is increased substantially to roughly $(M-1)\pi/M$ instead of the original $\Delta_2 = \pi - \omega_{c2}$, its lower don't-care band, $\Delta_1/M = \omega_{c1}/M$, becomes M times narrower. This means that the complexity of $G(z)$ may thereby even increase, not decrease. In the left-band case, this is not a problem as there is no don't-care band to the left.

The width of both the lower and the upper don't-care bands of $G(z)$ can be increased by using a bandpass filter $F(z)$ instead of a lowpass filter. This also means that we have to use $M > 2$. Again, it appears that the most efficient case is for the lowest possible M which is here $M = 3$. The reason for this is two-fold. First, odd values of M makes it possible to center the passband of $G(z)$ around $\pi/2$, which maximizes the minimum of its lower and upper don't-care bands. Second, the complexity of $F(z)$ alone reduces with reduced M, in accordance with the discussion in [17] for the left-band case. In addition, the use of $M = 3$ instead of $M > 3$, makes it possible to double the amount of sparsity of $F(z)$, and thus its efficiency, by expressing it as a periodic filter.

Here, $G(z)$ is to approximate $D(j\omega M - j(K-1)\pi)$ in the frequency region $\omega \in [\omega_{c1}^{(G)}, \omega_{c2}^{(G)}]$ where

$$\omega_{c1}^{(G)} = (K-1)\pi/M + \Delta_1/M, \quad \omega_{c2}^{(G)} = K\pi/M - \Delta_2/M, \tag{23}$$

with K being an appropriately chosen odd integer. For $M = 3$, one should use $K = -1$. After the downsampling by M, the above region is mapped to the targeted region $\omega \in [\omega_{c1}, \omega_{c2}]$. Further, $F(z)$ is to approximate M in the same region as that of $G(z)$ and zero in the corresponding image bands created in the upsampling. Hence, $F(z)$ is here a bandpass filter with passband and stopband edges at

$$\omega_{c1}^{(F)} = \omega_{c1}^{(G)}, \quad \omega_{c2}^{(F)} = \omega_{c2}^{(G)} \tag{24}$$

and

$$\omega_{s1}^{(F)} = (K-1)\pi/M - \Delta_1/M, \quad \omega_{s2}^{(F)} = K\pi/M + \Delta_2/M, \tag{25}$$

respectively. Moreover, with $M = 3$ and $K = -1$, $F(z)$ is a symmetric bandpass filter centered on $\pi/2$. Consequently, it can be expressed as

$$F(z) = 3P(z^2) \tag{26}$$

where $P(z)$ is a unity-gain-passband third-band highpass filter. The polyphase decomposition of $P(z)$ is then $P(z) = 1/3 + z^{-1}P_1(z^3) + z^{-2}P_2(z^3)$ which leads to $F(z) = 1 + 3z^{-2}P_1(z^6) + 3z^{-4}P_2(z^6)$ and the polyphase components

$$F_0(z) = 1, \quad F_1(z) = 3z^{-1}P_2(z^2), \quad F_2(z) = 3P_1(z^2). \tag{27}$$

A filter $F(z)$ of the form above requires roughly only one third of the complexity of a general filter of the same order.

	N_H	N_G	N_F	D_H	**DE**	**Mult**	**Add**
Regular	124	-	-	62	124	125	124
Two-rate, regular $F(z) = P(z^2)$	126	10	372	63	126	73	134
Two-rate, FRM $F(z) = P(z^2)$	140	10	412	70	140	41	59

Table 2. Results of Example 3.

3.1. Complexity savings

As opposed to the case of linear-phase overall filters considered in Section 2, we can here achieve complexity savings without using additional FRM techniques. The reason is two-fold. First, as seen above, $F(z) = P(z^2)$ is already sparse. Second, as a mid-band system is often a nonlinear-phase system, the filter coefficient are not symmetric. By using the two-rate based structure, symmetry can partially be utilized as $F(z)$ is a symmetric filter whereas only the low-order $G(z)$ is unsymmetric. As to the sparsity, the degree of sparseness can be increased by realizing $P(z)$ as an FRM third-band filter. Details are given in [18].

3.2. Examples

Example 3: Consider the approximation of a fractional-degree differentiator with the desired function $D(j\omega) = e^{-j\omega(N_G+N_F)/4}(j\omega)^{0.5}$ in the frequency band $\omega \in [0.02\pi, 0.98\pi]$ and for an approximation error of $\delta_e = 0.01$. Figure 5 shows the frequency response and approximation error of the two-rate based design. The filter has been designed using essentially the same three-step procedure described earlier, but after minor appropriate modifications, as detailed in [18]. Table 2 gives the results for the conventional direct-form realization and for the two-rate based realizations, both with a sparse regular bandpass filter and a sparse FRM bandpass filter. The quantity D_H denotes the integer part of the group delay whereas DE denotes the number of delay elements. As seen from the table, substantial savings are achieved using the two-rate based structures, especially when the FRM technique is also utilized. As usual when using the FRM technique, one has to pay a price in a somewhat increased delay. It is also noted that the savings increase/decrease with increased/decreased bandwidth (decreased/increased width of the don't-care bands). This is in line with the basic two-rate based scheme and it was exemplified earlier in Example 2.

4. Multi-function systems

In this section, we will discuss the extension to the realization of multifunction systems. The two-rate based approach is even more efficient for such systems as the same $F(z)$, and thus the same $F_0(z)$, is shared between all functions. We will illustrate this for Farrow-structure based (see [34]) variable fractional-delay (VFD) filters. As an example will reveal, the two-rate based structure offers dramatic complexity reductions in this application, even without using the additional FRM approach. However, incorporating the FRM approach, further complexity savings are obtained.

4.1. Variable fractional-delay filters

Variable fractional-delay filters find applications in many different contexts like interpolation, resampling, delay estimation, and signal reconstruction, see [35–40].

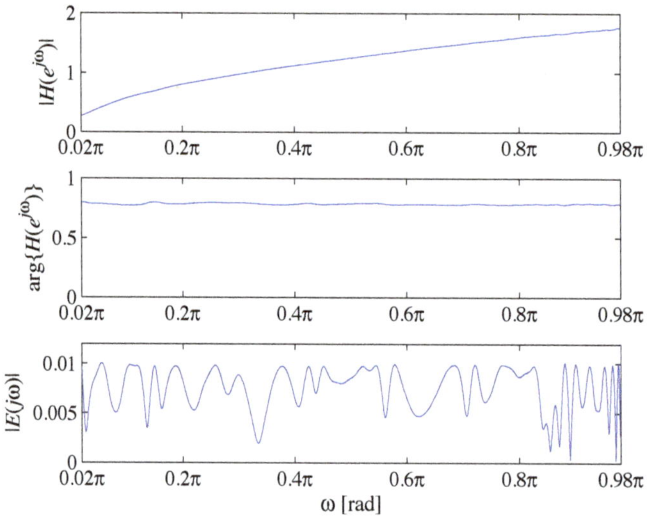

Figure 5. Fractional-degree differentiator responses in Example 3 using the two-rate based structure with an FRM bandpass filter targeting $\omega_{c1} = 0.02\pi$ and $\omega_{c2} = 0.98\pi$.

The VFD filter, with a transfer function $H(z,d)$, is for $z = e^{j\omega}$ to approximate the ideal VFD filter frequency response

$$D(j\omega, d) = e^{-j\omega(D_H+d)} \tag{28}$$

where D_H is a fixed delay which usually is an integer or an integer plus a half. Further, d is the fractional delay. The ideal response should be approximated in the band $\omega \in [0, \omega_c]$, $0 < \omega_c < \pi$, and for all fractional delays $d \in [-0.5, 0.5]$ meaning that a whole sampling period (interval) is covered. In general, the sampling period is T, but we have used $T = 1$ in this chapter for simplicity.

Using the Farrow structure, $H(z,d)$ is expressed in the form

$$H(z,d) = \sum_{k=0}^{L} d^k H_k(z). \tag{29}$$

where $H_k(z)$ are fixed subfilters which, essentially, realize the weighted differentiators $e^{-j\omega D_H} \times (-j\omega)^k/k!$ This follows immediately by truncating the Taylor series expansion of $e^{-j\omega d}$, see [41]. Further, when there are no restrictions on the fixed part of the delay, it is possible and efficient to use linear-phase subfilters $H_k(z)$, thus with symmetric or antisymmetric impulse responses. We then have $D_H = N_H/2$, and the following two different cases. When $G_k(z)$ are of even order N_G, they are of Type I (Type III) for even (odd) values of k. This results in an integer D_H. In the odd-order case, $G_k(z)$ are instead of Type II (Type IV) for even (odd) values of k. In this case, D_H is an integer plus a half. In both

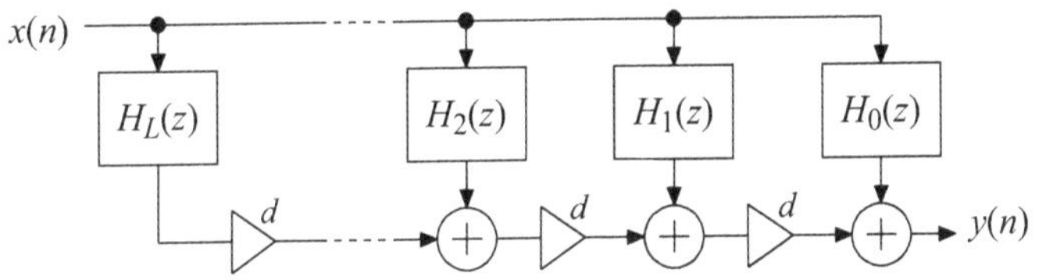

Figure 6. Farrow structure realizing the VFD filter transfer function in (29).

cases, the impulse responses are symmetric (anti-symmetric) for even (odd) values of k, thus $g_k(n) = (-1)^k g_k(N_G - n)$.

Figure 6 shows the regular Farrow structure realizing (29). As seen, the problem amounts to realizing the $L+1$ differentiator functions with ideal responses $(-j\omega T)^k/k!$. In other words, it essentially corresponds to the realization of a multi-function system, although, in this case, the partial outputs are finally combined via the FD multiplications to form only one output. Using now the two-rate based approach introduced in Section 2.2, each $H_k(z)$ is realized as

$$H_k(z) = 2F_0(z)G_{k0}(z) + z^{-(D_F+1)/2}G_{k1}(z) \tag{30}$$

where $F_0(z)$ and $z^{-(D_F-1)/2}$ are again the polyphase components of a linear-phase HB interpolation filter $F(z)$ with a passband gain of two and delay D_F, whereas $G_{k0}(z)$ and $G_{k1}(z)$ are the polyphase components of the subfilters $G_k(z)$. This follows from sandwiching the filter $F(z)G(z,d)$ between the upsampler and downsampler, where $G(z,d)$ approximates an FD filter in the region $[0,\ \omega_c/2]$. That is,

$$G(z,d) = \sum_{k=0}^{L} d^k G_k(z). \tag{31}$$

The overall realization is shown in Fig. 7. It is noted that $F(z)$ again can be realized using the FRM approach in order to further reduce the complexity, as demonstrated in [19]. In this case, $F_0(z)$ is again realized as in Fig. 1(c) or (d).

4.2. Design examples

Example 4: We consider the design of a VFD filter with a bandwidth of $\omega_c = 0.9\pi$. The filter has been designed using essentially the same three-step procedure described earlier. More details are given in [19]. Tables 3 and 4 summarize the results where the number of multiplications and additions covers all fixed subfilters assuming appropriate use of direct-form and transposed direct-form realizations. In addition, L general multipliers and adders are needed for implementing the FD multiply-and-add chain, but this is required in all VFD filter structures. Further, the NRMS and δ_{gd} values given in the tables indicate the normalized root-mean square error and maximum group-delay error as defined by (33) and (36), respectively, in [42], whereas δ_e denotes the maximum of the modulus of the complex error. Further, DE denotes the number of delay elements. It is seen from the table that that the two-rate based structure is considerably more efficient than the regular Farrow structure.

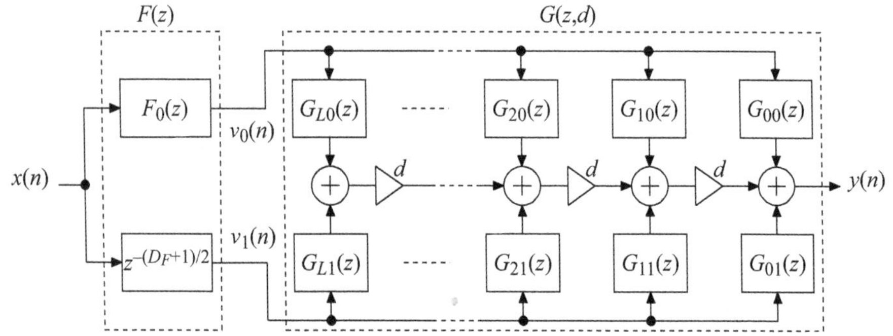

Figure 7. Two-rate based structure realizing the VFD filter transfer function in (29) with $H_k(z)$ as in (30).

Linear-Phase	L	N_H	N_E	N_F	N_A	N_B	P
Reg. Farrow, WLS [44]	7	73	n/a	n/a	n/a	n/a	n/a
Hybrid, WLS[42]	7	117	n/a	n/a	n/a	n/a	n/a
Simplified [43]	9	73	n/a	n/a	n/a	n/a	n/a
Two-rate based	7	75	12	138	n/a	n/a	n/a
Two-rate based with FRM	7	87	12	162	46	26	3

Table 3. Results of Example 4.

Linear-Phase	D_H	**DE**	**Mult**	**Add**	**NRMS (%)**	δ_e **[dB]**	δ_{gd}
Reg. Farrow, WLS [44]	36.5	73	191	374	0.00023	−100.04	0.000446
Hybrid, WLS[42]	58.5	117	148	285	0.00021	−100.21	0.000395
Simplified [43]	36.5	73	91	158	not reported	−102.14*	not reported
Two-rate based	37.5	75	80	149	0.00019	−102.96	0.000318
Two-rate based with FRM	43.5	87	71	129	0.00019	−104.15	0.000263

Table 4. Results of Example 4. *Magnitude and phase delay errors.

It is also more efficient than two alternative approaches whose results are also included in the table, namely for the hybrid structure in [42] and the structure in [43] which meets roughly the same specification. It is also seen that the extended structure that utilizes the FRM technique offers further complexity reductions. The price to pay in this case is however a slight increase of the delay and delay elements, but the figures are still considerably smaller than for the structure in [42]. Compared with the regular Farrow structure in [44] and the one in [43], one has to pay the moderate price of a delay and delay element increase of some 3% using the basic structure in Fig. 7(b) and 19% using the extended structure incorporating the FRM approach.

5. Multiple-constant multiplication techniques for the subfilter implementations

This section will discuss implementation details, design trade-offs, and comparisons when the multiplications in the filters are implemented using multiple constant multiplication (MCM) techniques, which realize a number of constant multiplications using only shifts,

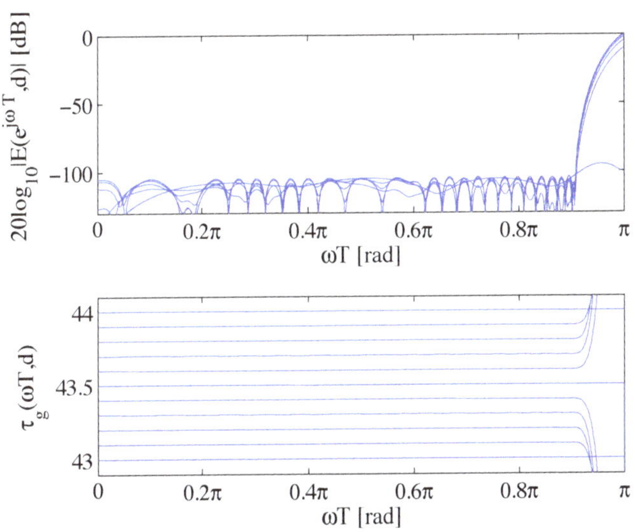

Figure 8. Two-rate and FRM based VFD filter in Example 4. Error and group delay responses for 11 evenly distributed values of d between -0.5 and 0.5.

adders and subtracters. The situation is different here though, than for the most commonly considered transposed direct-form filter realization, as the proposed structures consist of cascaded subfilters. This section will therefore elaborate on these issues and provide design examples. The focus here is on VFD filters using the two-rate based structure without the additional FRM approach.

For dedicated hardware implementations, one can take advantage of MCM techniques to reduce the implementation cost. Multiplications by constant coefficients can be performed using adders, subtracters, and shifts. As adders and subtracters have approximately the same implementation complexity we will refer to both as adders. Efficient realization of constant multiplications is an active research area and much effort has been focused on the case where one input data is multiplied by several constant coefficients. This problem has mainly been motivated by single-rate FIR filters, where for a transposed direct form FIR filter the input is multiplied by several coefficients, see [45–49]. The resulting implementation of several multiplications is denoted multiplier block, as in [45].

Work has also been done for sampling rate change with an integer factor in [50] and rational factor in [51], where it was shown that FIR filters in parallel can be implemented either using one multiplier block or by using a constant matrix multiplication block, as in [52–55], with the first approach requiring more delay elements than the latter. As a Farrow filter also is composed of several FIR filters in parallel we have the same implementation alternatives here, not only the single multiplier block case as reported in [56]. This has been extensively discussed in [57]. In Fig. 9, the approach to implement the subfilters proposed in [56] is shown. This approach typically requires few additions for the multiplier block. However, a separate set of registers is required for each subfilter and the number of structural adders

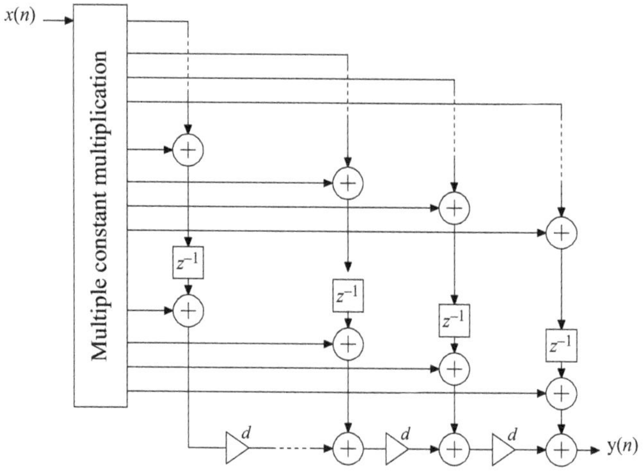

Figure 9. Realization of Farrow filter using transposed direct form subfilters resulting in a multiplier block and several sets of registers.

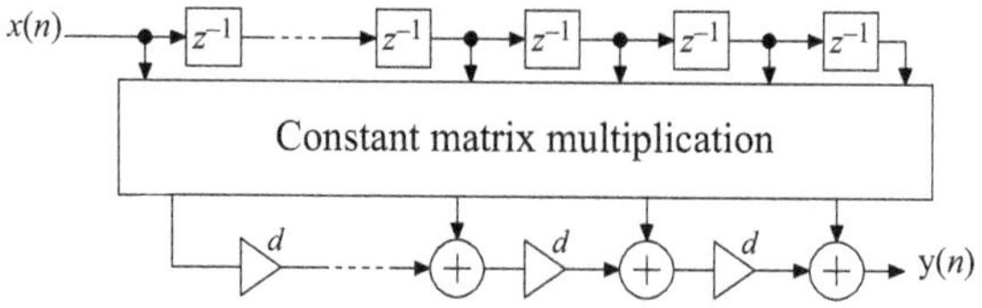

Figure 10. Realization of Farrow filter using direct form subfilters resulting in a constant matrix multiplication and a single set of registers.

is high. Alternatively, in Fig. 10, an approach based on the observation in [50] and further discussed in [57] is shown. Here, only one set of registers is required and the structural adders of the subfilters are merged into the matrix-vector multiplication.

The Farrow filter part of the two-rate based structure in Fig. 7 can be implemented similarly to what is shown in Figs. 9 or 10. For transposed direct form subfilters, as in Fig. 9, the corresponding structure would have two inputs, and, hence, result in a constant matrix multiplication. Using direct form subfilters, as in Fig. 10, requires two sets of registers, one for each input. For the HB filter it is convenient to use a direct form subfilter as the delayed input values are easily obtained from the registers. We note that the input to the lower branch subfilters in Fig. 7 is just a delayed version of the input, which is available from the upper branch subfilter $F_0(z)$. Therefore, it is possible to use the registers of the HB filter as registers for direct form subfilters. The resulting structure is illustrated in Fig. 11.

Naturally, it is also possible to use a transposed direct form HB filter and/or transposed direct form subfilters in the Farrow filter part. From a complexity point of view, a transposed direct form HB filter will have the same number of adders and registers. However, it will not be possible to share registers as shown in Fig. 11.

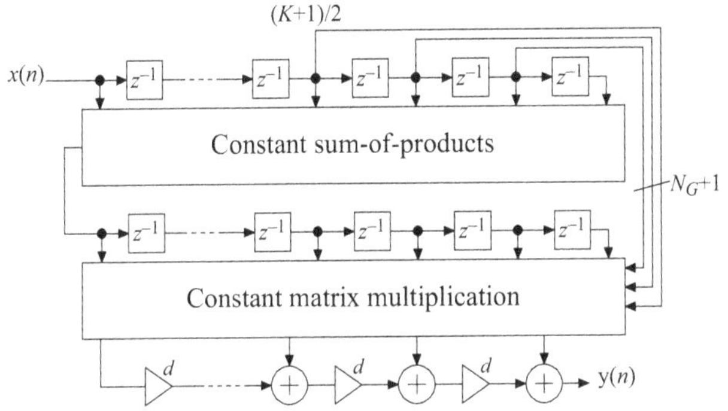

Figure 11. Realization of the filter structure in Fig. 7 using direct form subfilters resulting in a sum-of-products block and a constant matrix multiplication.

5.1. Example and comparisons

Example 5: We consider a specification where the bandwidth is 0.9π and the modulus of the complex error should be below 0.0042. To meet this specification, the Farrow structure in Fig. 6, with subfilters jointly optimized as outlined in detail in [41], requires 45 fixed multipliers, 88 fixed adders, and 5 variable multipliers. The two-rate based structure in Fig. 7, with subfilters jointly optimized as detailed in [14], requires 30 fixed multipliers, 53 fixed adders, and 5 variable multipliers. Thus, in terms of number of multiplications and additions, the two-rate based structure is superior.

To refine the comparison when MCM techniques are applied we must quantize the filter coefficients. For a relative comparison, one can use simple rounding. We found that the original Farrow structure requires 11 bits to fulfil the requirements whereas the structure in Fig. 7 requires 13 fractional bits. The slightly larger number of bits for the two-rate approach is explained by the fact that a cascaded filter must meet the requirements which leads to a somewhat more stringent requirement on the subfilters, at least when simple rounding is used.

For the regular Farrow structure in Fig. 6, together with the realization in Fig. 9 a total of 33 adders are required for the multiplier block using the RAG-n algorithm in [45]. This is an optimal result since there are 33 different (odd) coefficients as discussed in [58], and, hence, there is no need to apply the slightly more efficient algorithms in [47–49]. Furthermore, 80 structural adders and 118 registers are required for the FIR subfilters. Further, five general multipliers and four additional adders are required (for both the regular Farrow and two-rate based filters). Alternatively, using the proposed structure in Fig. 10 the constant matrix multiplication can be realized using 96 adders using the algorithm in [53]. In addition, 26 structural adders are required. One observation is that separating the symmetric and anti-symmetric subfilters may reduce the complexity as some algorithms work better for fewer columns. If this is utilized, the number of adders can be reduced to 107 by applying the algorithm in [52] to the resulting two matrices and adding and subtracting the results. The number of registers is now decreased to 30, whereas the number of general multipliers and additional adders is constant.

For the two-rate based structure in Fig. 7, the HB filter requires 43 structural adders and 29 registers. The sum of products are realized by computing the corresponding multiplier

Filter structure	Mult	Add	Registers
Farrow with transposed direct-form subfilters	5	117	118
Farrow with direct-form subfilters	5	111	30
Proposed two-rate based structure	5	85	35

Table 5. Results of Example 5.

block with RAG-n and transposing it. For the constant matrix multiplication, 38 adders are required. This number is not reduced by separating the symmetric and anti-symmetric subfilters. A total of six additional registers are required as well as the five general multipliers and four additional adders.

The results are summarized in Table 5. It is seen that the two-rate based structure still has the lowest complexity for most implementation technologies as five registers will typically be less complex to implement than 26 adders. Furthermore, whereas the use of transposed direct form subfilters for the Farrow filter, as proposed in [50, 57], reduces the number of adders related to the multiplication, it is for the two-rate approach still more efficient to use direct form subfilters.

6. Conclusion

This chapter has reviewed recent two-rate based structures and their design for obtaining efficient wide-band FIR systems. Left-band, right-band, and mid-band systems, as well as single-function and multi-function systems, were covered. Several design examples were given, for differentiators and VFD filters (a special case of multi-function systems), revealing dramatic complexity savings for wide-band specifications. More details can be found in [12–19].

Author details

Håkan Johansson and Oscar Gustafsson

Division of Electronics Systems, Department of Electrical Engineering, Linköping University, Sweden

References

[1] M. Renfors and Y. Neuvo. The maximum sampling rate of digital filters under hardware speed constraints. *IEEE Trans. Circuits Syst.*, 28(3):196–202, Mar. 1981.

[2] A. Fettweis. On assessing robustness of recursive digital filters. *Eur. Trans. Telecomm. Relat. Technol.*, 1(2):103–109, Mar.–Apr. 1990.

[3] Y. C. Lim. Frequency-response masking approach for the synthesis of sharp linear phase digital filters. *IEEE Trans. Circuits, Syst.*, CAS-33(4):357–364, Apr. 1986.

[4] T. Saramäki. Finite impulse response filter design. In S.K. Mitra and J.F. Kaiser, editors, *Handbook for Digital Signal Processing*, chapter 4, pages 155–277. New York: Wiley, 1993.

[5] T. Saramäki, Y. C. Lim, and R. Yang. The synthesis of half-band filter using frequency-response masking technique. *IEEE Trans. Circuits Syst. II*, 42(1):58–60, Jan. 1995.

[6] P. S. R. Diniz, L. C. R. de Barcellos, and S. L. Netto. Design of high-resolution cosine-modulated transmultiplexers with sharp transition band. *IEEE Trans. Signal Processing*, 52(5):1278–1288, May 2004.

[7] H. Johansson. Two classes of frequency-response masking linear-phase FIR filters for interpolation and decimation. *Circuits, Syst., Signal Processing*, 25(2):175–200, Apr. 2006.

[8] R. Bregovic, Y. C. Lim, and T. Saramäki. Frequency-response masking-based design of nearly perfect-reconstruction two-channel FIR filterbanks with rational sampling factors. *IEEE Trans. Circuits Syst I: Regular Papers*, 55(7):2002–2012, July 2008.

[9] Y. Wei and Y. Lian. Frequency-response masking filters based on serial masking schemes. *Circuits, Syst., Signal Processing*, 29(1):7–24, Feb. 2010.

[10] J. Yli-Kaakinen and T. Saramäki. An efficient algorithm for the optimization of FIR filters synthesized using the multistage frequency-response masking approach. *Circuits, Syst., Signal Processing*, 30(1):157–183, Feb. 2011.

[11] N. P. Murphy, A. Krukowski, and I. Kale. Implementation of a wide-band integer and fractional delay element. *Electron. Lett.*, 30(20):1658–1659, 1994.

[12] G. Jovanovic-Dolecek and J. Diaz-Carmona. One structure for wide-bandwidth and high-resolution fractional delay filter. In *Proc. IEEE Int. Conf. Electr. Circuits Syst.*, 2002.

[13] E. Hermanowicz. On designing a wideband fractional delay filter using the Farrow approach. In *Proc. XII European Signal Processing Conf.*, Vienna, Austria, Sept. 6–10 2004.

[14] E. Hermanowicz and H. Johansson. On designing minimax adjustable wideband fractional delay FIR filters using two-rate approach. In *Proc. European Conf. Circuit Theory Design*, Cork, Ireland, Aug. 29–Sept. 1 2005.

[15] H. Johansson, O. Gustafsson, K. Johansson, and L. Wanhammar. Adjustable fractional-delay FIR filters using the Farrow structure and multirate techniques. In *Proc. IEEE Asia Pacific Conf. Circuits Syst.*, 2006.

[16] J. Diaz-Carmona and G. Jovanovic-Dolecek. Frequency-based optimization design for fractional delay FIR filters with software-defined radio applications. *Int. J. Digital Multimedia Broadcasting*, 53(6):1–6, June 2010.

[17] Z. U. Sheikh and H. Johansson. A class of wide-band linear-phase FIR differentiators using a two-rate approach and the frequency-response masking technique. *IEEE Trans. Circuits Syst. I: Regular papers*, 58(8):1827–1839, Aug. 2011.

[18] Z. U. Sheikh and H. Johansson. Efficient wide-band FIR LTI systems derived via multi-rate techniques and sparse bandpass filters. *IEEE Trans. Signal Processing, to appear*, 60(7):3859–3863, July 2012.

[19] H. Johansson and E. Hermanowicz. Two-rate based low-complexity variable fractional-delay FIR filter structures. *IEEE Trans. Circuits Syst. I: Regular papers, to appear*, 2012.

[20] J. F. Kaiser. Nonrecursive digital filter design using I_0-sinh window function. In *Proc. Int. Symp. Circuits Syst.*, volume 3, pages 20–23, Apr. 1974.

[21] K. Ichige, M Iwaki, and R. Ishii. Accurate estimation of minimum filter length for optimum FIR digital filters. *IEEE Trans. Circuits Syst. II*, 47(10):1008–1016, Oct. 2000.

[22] Z. U. Sheikh, A. Eghbali, and H. Johansson. Linear-phase FIR digital differentiator order estimation. In *Proc. European Conf. Circuit Theory Design*, Linköping, Sweden, Aug. 29–31 2011.

[23] P. P. Vaidyanathan. *Multirate Systems and Filter Banks*. Prentice Hall, 1993.

[24] L. Rabiner, J. McClellan, and T. Parks. FIR digital filter design technique using weighted-Chebyshev approximation. *IEEE Proc.*, 63(4):595–610, Apr. 1975.

[25] G. Mollova. Compact formulas for least-squares design of digital differentiators. *Electron. Lett.*, 35(20):1695–1697, Sept. 1999.

[26] J. J. Shyu, S. C. Pei, and Y. D. Huang. Least-squares design of variable maximally linear FIR differentiators. *IEEE Trans. Signal Processing*, 57(11):4568–4573, Nov. 2009.

[27] S. G. Nash and A. Sofer. *Linear and Nonlinear Programming*. McGraw-Hill, 1996.

[28] J. H. McClellan, T. W. Parks, and L.R. Rabiner. A computer program for designing optimum FIR linear phase digital filters. *IEEE Trans. Audio Electroacoust.*, AU-21:506–526, Dec. 1973.

[29] T. W. Parks and C. S. Burrus. *Digital Filter Design*. John Wiley and Sons, 1987.

[30] Y. Q. Chen and K. L. Moore. Discretization schemes for fractional-order differentiators and integrators. *IEEE Trans. Circuits Syst. I.*, 49:363–367, Mar. 2002.

[31] R. S. Barbosa, J. A. T. Machado, and M. F. Silva. Time-domain design of fractional differintegrators using least-squares. *Signal Processing*, 86(10):2567–2581, Oct. 2006.

[32] C. C. Tseng and S.-L. Lee. Design of fractional order digital differentiator using radial basis function. *IEEE Trans. Circuits Syst. I: Regular Papers*, 57(7):1708–1718, July 2010.

[33] B. T. Krishna. Studies on fractional order differentiators and integrators: A survey. *Signal Processing*, 91(3):386–426, Mar. 2011.

[34] C. W. Farrow. A continuously variable delay element. In *Proc. IEEE Int. Symp., Circuits, Syst.*, volume 3, pages 2641–2645, Espoo, Finland, June 7–9 1988.

[35] F. M. Gardner. Interpolation in digital modems–Part I: Fundamentals. *IEEE Trans. Comm.*, 41(3):502–508, Mar. 1993.

[36] T. I Laakso, V. Välimäki, M. Karjalainen, and U. K. Laine. Splitting the unit delay–Tools for fractional delay filter design. *Signal Processing Mag.*, 13(1):30–60, Jan. 1996.

[37] S. R. Dooley and A. K. Nandi. On explicit time delay estimation using the Farrow structure. *Signal Processing*, 72:53–57, Jan. 1999.

[38] H. Johansson and P. Löwenborg. Reconstruction of nonuniformly sampled bandlimited signals by means of digital fractional delay filters. *IEEE Trans. Signal Processing*, 50(11):2757–2767, Nov. 2002.

[39] M. Olsson, H. Johansson, and P. Löwenborg. Time-delay estimation using Farrow-based fractional-delay FIR filters: filter approximation vs. estimation errors. In *Proc. XIV European Signal Processing Conf.*, Florence, Italy, Sept. 4–8 2006.

[40] S. Tertinek and C. Vogel. Reconstruction of nonuniformly sampled bandlimited signals using a differentiator-multiplier cascade. *IEEE Trans. Circuits Syst. I: Regular papers*, 55(8):2273–2286, Sept. 2008.

[41] H. Johansson and P. Löwenborg. On the design of adjustable fractional delay FIR filters. *IEEE Trans. Circuits Syst. II*, 50(4):164–169, Apr. 2003.

[42] T.-B. Deng. Hybrid structures for low-complexity variable fractional delay filters. *IEEE Trans. Circuits Syst. I: Regular Papaers*, 57(4):897–910, Apr. 2010.

[43] J. Yli-Kaakinen and T. Saramäki. A simplified structure for FIR filters with an adjustable fractional delay. In *Proc. IEEE Int. Symp. Circuits Syst.*, New Orleans, USA, May 27–30 2007.

[44] T.-B. Deng. Symmetric structures for odd-order maximally flat and weighted-least-squares variable fractional-delay filters. *IEEE Trans. Circuits Syst. I: Regular Papers*, 54(12):2718–2732, Dec. 2007.

[45] A. G. Dempster and M. D. Macleod. Use of minimum-adder multiplier blocks in FIR digital filters. *IEEE Trans. Circuits Syst. II*, 42(9):569–577, Sept. 1995.

[46] R. I. Hartley. Subexpression sharing in filters using canonic signed digit multipliers. *IEEE Trans. Circuits Syst. II*, 43(10):677–688, Oct. 1996.

[47] Y. Voronenko and M. Püschel. Multiplierless multiple constant multiplication. *ACM Trans. Algorithms*, 2006.

[48] O. Gustafsson. A difference based adder graph heuristic for multiple constant multiplication problems. In *Proc. IEEE Int. Symp. Circuits Syst.*, pages 1097–1100, New Orleans, USA, May 27–30, 2007.

[49] L. Aksoy, E.O. Günes, and P. Flores. Search algorithms for the multiple constant multiplications problem: Exact and approximate. 34(5):151–162, August 2010.

[50] O. Gustafsson and A. G. Dempster. On the use of multiple constant multiplication in polyphase fir filters and filter banks. In *Proc. Nordic Signal Processing Symp.*, pages 53–56, Espoo, Finland, June 9–11, 2004.

[51] O. Gustafsson and H. Johansson. Efficient implementation of FIR filter based rational sampling rate converters using constant matrix multiplication. In *Proc. Fortieth Asilomar Conf. Signals, Systems and Computers ACSSC '06*, pages 888–891, 2006.

[52] A. G. Dempster, O. Gustafsson, and J. O. Coleman. Towards an algorithm for matrix multiplier blocks. In *Proc. European Conf. Circuit Theory Design*, Krakow, Poland, Sept. 1–4, 2003.

[53] M. D. Macleod and A. G. Dempster. A common subexpression elimination algorithm for low-cost multiplierless implementation of matrix multipliers. *Electronics Lett.*, 40(11):651–652, May 2004.

[54] N. Boullis and A. Tisserand. Some optimizations of hardware multiplication by constant matrices. *IEEE Trans. Computers*, 54(10):1271–1282, Oct. 2005.

[55] L. Aksoy, E. Costa, P. Flores, and J. Monteiro. Optimization algorithms for the multiplierless realization of linear transforms. *ACM Trans. Design Automation Electronic Syst.*, vol. 17, no. 1, article 3, Jan. 2012.

[56] A. G. Dempster and N. P. Murphy. Efficient interpolators and filter banks using multiplier blocks. *IEEE Trans. Signal Processing*, 48(1):257–261, Jan. 2000.

[57] M. Abbas, O. Gustafsson, and H. Johansson. On the fixed-point implementation of fractional-delay filters based on the farrow structure. *IEEE Trans. Circuits Syst I: Regular Papers*, 2012. to appear.

[58] O. Gustafsson. Lower bounds for constant multiplication problems. *IEEE Trans. Circuits Syst. II*, 54(11):974–978, Nov. 2007.

Chapter 10

Particle Swarm Optimization of Highly Selective Digital Filters over the Finite-Precision Multiplier Coefficient Space

Seyyed Ali Hashemi and Behrouz Nowrouzian

Additional information is available at the end of the chapter

1. Introduction

Digital filters find wide variety of applications in modern digital signal processing systems [1, 2]. As a result of the recent progress in such systems, there is an ever growing demand for sharp transition band digital filters. These narrow transition bandwidth digital filters are usually designed by using the frequency response masking (FRM) approach [3]. The computational efficiency of the FRM technique makes it suitable for different applications, e.g. in audio signal processing and data compression [4].

Practical design of digital filters is based on optimization for satisfying the given design specifications together with the hardware architecture. However, the optimization may be carried out in terms of fixed configurations but variable multiplier coefficient values. On the other hand, the problem may concern the optimization of the hardware architecture without taking the multiplier coefficient values into consideration.

In order to optimize the given design specifications, the multiplier coefficient values can be determined in infinite precision by using hitherto optimization techniques. However, in an actual hardware implementation of the digital filters, the infinite precision multipliers should be quantized to their finite precision counterparts, but these finite precision multiplier coefficients may no longer satisfy the given design specifications. Consequently, from a hardware implementation point of view, there is a need for finite precision optimization techniques, capable of finding the optimized digital filter rapidly while keeping the computational complexity at a desired level. In principle, there exist two different techniques for the optimization of digital filters, namely, gradient-based and heuristic optimization approaches.

Gradient-based optimization techniques have been studied widely. In [5], an integer programming technique was developed for the optimization of digital filters over a discrete multiplier coefficient space. In [6], a Remez exchange algorithm was used for the optimization of FRM finite impulse response (FIR) digital filters and it was shown that this algorithm may provide a speed advantage over the linear programming approach. However, both these techniques suffer from sub-optimality problems. In [7], unconstrained weighted least-squares criterion was used to develop another technique for the optimization of digital filters. Convex optimization approaches such as semi-definite programming [8] and second-order cone programming [9] have also been applied to the optimization of digital filters. However, if a large number of constraints are present, these optimization techniques may become computationally inefficient in terms of time consumption and speed.

Heuristic optimization algorithms have emerged as promising candidates for the design and discrete optimization of digital filters, particularly due to the fact that they are capable of automatically finding near-optimum solutions while keeping the computational complexity of the algorithm at moderate levels. Simulated annealing (SA) and genetic algorithms (GAs) were widely used in the design and optimization of digital filters [10–12]. Particle swarm optimization (PSO) and seeker optimization algorithm (SOA) are two newly developed algorithms suitable for the optimization of various digital filters due to their few number of implementation parameters and high speed of convergence [13, 14]. It was shown that SOA has advantages over PSO in terms of the speed of convergence and global search ability [15]. Tabu search (TS) [16], ant colony optimization (ACO) [17], immune algorithm (IA) [18] and differential evolution (DE) [19, 20] are alternative candidates for the optimization of digital filters. All the foregoing techniques allow a robust search of the solution space through a parallel search in all directions without any recourse to gradient information. However, the aforementioned techniques were developed for infinite precision optimization of digital filters which require the user to perform a quantization step for a hardware implementation.

In [21–23], a technique was developed for finite-precision design and optimization of FRM digital filters using GAs. finite-precision optimization of FRM FIR digital filters using PSO was studied in [24, 25] and finite-precision optimization of infinite impulse response based (IIR-based) FRM digital filters was studied in [26, 27]. PSO was originally proposed by Kennedy and Eberhart in 1995 as a new intelligent optimization algorithm which simulates the migration and aggregation of a flock of birds seeking food [28]. It adopts a strategy based on particle swarm and parallel global random search, that may exhibit superior performance to other intelligent algorithms in computational speed and memory. In PSO, a potential candidate solution is represented as a particle in a multidimensional search space, where each dimension represents a distinct optimization variable. The particles in the multidimensional search space are characterized by corresponding fitness values. They make movements in the search space towards regions characterized by high fitness values.

The conventional FRM digital filters incorporate FIR interpolation digital subfilters. These digital subfilters are usually of high orders, rendering the resulting overall FRM digital filters as not economical, since the resulting digital filters occupy large chip areas and consume high amounts of power in their VLSI hardware implementations. In general, the multiplication operation is the most cost-sensitive part in such an implementation. Therefore, there is every incentive to reduce the number of multiplication operations in the digital filter realization. This problem may be circumvented by employing IIR interpolation digital subfilters [29, 30].

There is a vast body of literature available for the design and optimization of digital IIR filters [31–33]. However, all the aforementioned designs are based on the exact transfer function coefficients which leads to an uneconomical hardware realization of such filters. In order to realize the constituent IIR interpolation digital subfilters on a hardware platform, the bilinear-lossless-discrete-integrator (bilinear-LDI) digital filter design approach is employed [34]. These digital subfilters are realized as a sum/difference of a pair of bilinear-LDI digital allpass networks. The salient features of the bilinear-LDI digital filters are that they lend themselves to fast two-cycle parallel digital signal processing speeds, while being minimal in the number of digital multiplication operations (and, practically, minimal in number of digital addition and unit-delay operations).

The starting point in the design of FRM digital filters is to find the multiplier coefficients constituent in the FRM digital filter in infinite precision by using the hitherto gradient-based optimization techniques (e.g. Parks-McClellan approach [35] for FIR digital filters) followed by a quantization step. The quantization can be performed by constraining the multiplier coefficients values to conform to certain number systems such as the signed power-of-two (SPT) system. SPT is a computationally efficient number system which can further reduce the hardware complexity of the FRM IIR digital filters. In this number system, each multiplier coefficient is represented with only a few non-zero bits within its wordlength, permitting the decomposition of the multiplication operation into a finite series of shift and add operations. Digital filters incorporating SPT multiplier coefficient representation are commonly referred to as *multiplierless* digital filters [36]. However, the SPT representation of a given number is not unique, resulting in redundancy in the multiplier coefficient representation. This redundancy can adversely affect the corresponding computational complexity due to recourse to compare operations repetitively.

The canonical signed digit (CSD) number system is a special case of the SPT number system which circumvents the above redundancy problem by limiting the number of non-zero bits in the representation of the multiplier coefficients. It is usually used in combination with subexpression sharing and elimination, which in turn results in substantial reduction in the cost of the VLSI hardware implementation of the digital filters [37]. In CSD number system, no two (or more) non-zero bits can appear consecutively in the representation of the multiplier coefficients, reducing the maximum number of non-zero bits by a factor of two in terms of shift and add operations [38].

After multiplier coefficient quantization, the resulting FRM digital filter may no longer satisfy the given target design specifications. Therefore, the next step in the design of FRM digital filters is to perform a further optimization to make the finite precision FRM digital filter to conform to the design specifications. This can be achieved by resorting to a finite-precision optimization technique such as PSO.

A direct application of the conventional PSO algorithm to the optimization of the above FRM digital filters gives rise to three separate problems:

- The first problem arises because in the course of optimization, the multiplier coefficient update operations lead to values that may no longer conform to the desired CSD wordlength, etc. (due to random nature of velocity and position of particles). This problem is resolved by generating indexed look-up tables (LUTs) of permissible CSD multiplier coefficient values, and by employing the indices of LUTs to represent FRM digital filter multiplier coefficient values.
- The second problem stems from the fact that in case of FRM IIR digital filters, the resulting FRM IIR digital filters may no longer be bounded-input-bounded-output (BIBO) stable. This problem can be resolved by generation and successive augmentation of template LUTs until the BIBO stability constraints remain satisfied [23].
- Finally, the third problem arises because even in case of having indexed LUTs, the particles may go over the boundaries of LUTs in course of optimization (due to the inherent limited search space). This can be resolved by introducing *barren layers*. A barren layer is a region, with a certain width

and certain entries, which is added to the problem space such that the particles tend to shy away from such a region. The width of the barren layers is calculated based on a worst case scenario that may happen in the particles movements in the search space. However, the entries of barren layers are different for different problems and depend on the topology of the search space and the fitness function used in the problem.

This chapter discusses in detail the design, realization and discrete PSO of FRM IIR digital filters. FRM IIR digital filters are designed by FIR masking digital subfilters together with IIR interpolation digital subfilters. The FIR filter design is straightforward and can be performed by using hitherto techniques. The IIR digital subfilter design topology consists of a parallel combination of a pair of allpass networks such that its magnitude-frequency response matches that of an odd order elliptic minimum Q-factor (EMQF) transfer function. This design is realized using the bilinear-LDI approach, with multiplier coefficient values represented as finite-precision CSD numbers.

The above FRM digital filters are optimized over the discrete multiplier coefficient space, resulting in FRM digital filters which are capable of direct implementation in digital hardware platform without any need for further optimization. A new PSO algorithm is developed to tackle three different problems. In this PSO algorithm, a set of indexed LUTs of permissible CSD multiplier coefficient values is generated to ensure that in the course of optimization, the multiplier coefficient update operations constituent in the underlying PSO algorithm lead to values that are guaranteed to conform to the desired CSD wordlength, etc. In addition, a general set of constraints is derived in terms of multiplier coefficients to guarantee that the IIR bilinear-LDI interpolation digital subfilters automatically remain BIBO stable throughout the course of PSO algorithm. Moreover, by introducing barren layers, the particles are ensured to automatically remain inside the boundaries of LUTs in course of optimization.

2. The conventional PSO algorithm

Let us consider an optimization problem consisting of N design variables, and let us refer to each solution as a particle. Let us further consider a swarm of K particles in the N-dimensional search space. The position of the k-th particle in the search space can be assigned a N-dimensional position vector $X_k = \{x_{k1}, x_{k2}, \ldots, x_{kN}\}$. In this way, the element x_{kj} (for $j = 1, 2, \ldots, N$) represents the j-th coordinate of the particle X_k.

The PSO optimization fitness function maps each particle X_k in the search space to a fitness value. In addition, the particle X_k is assigned a N-dimensional velocity vector $V_k = \{v_{k1}, v_{k2}, \ldots, v_{kN}\}$. The PSO optimization search is directed towards promising regions by taking into account the velocity vector V_k together with the best previous position of the k-th particle $X_{best_k} = \{x_{best_{k1}}, x_{best_{k2}}, \ldots, x_{best_{kN}}\}$, and the best global position of the swarm $G_{best} = \{g_{best_1}, g_{best_2}, \ldots, g_{best_N}\}$ (i.e. the location of the particle with the best fitness value).

The conventional PSO is initialized by spreading the particles X_k through the search space in a random fashion. Then, the particles make movements through the search space towards regions characterized by high fitness values with corresponding velocities V_k. The movement of each particle is governed by the best previous location of the same particle X_{best_k}, and by the global best location G_{best}. The velocity of particle movement is determined from the previous best location of the particle, the global best location, and the previous velocity.

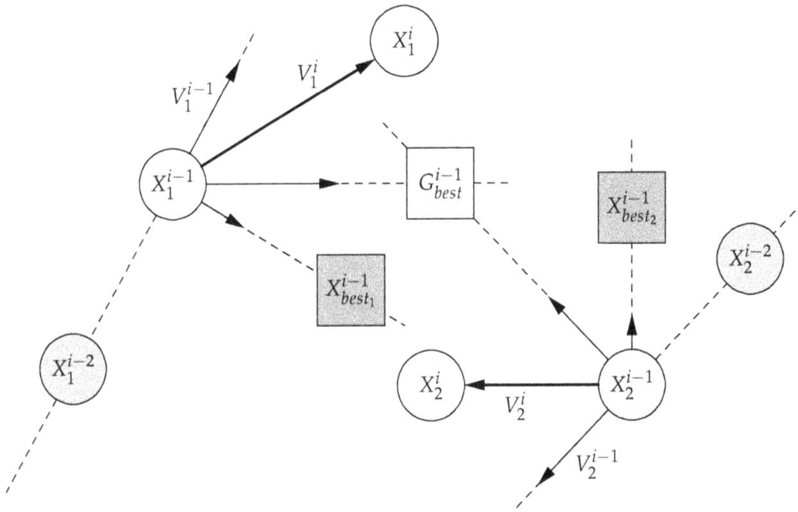

Figure 1. Movement of Particles in PSO Algorithm

The velocity and position of each particle in the i-th iteration throughout the course of PSO are updated in accordance with the equations:

$$v_{kj}^i = wv_{kj}^{i-1} + c_1 r_1 (x_{best_{kj}}^{i-1} - x_{kj}^{i-1}) + c_2 r_2 (g_{best_j}^{i-1} - x_{kj}^{i-1}) \tag{1}$$

$$\text{if} \quad v_{kj}^i < v_{min} \quad ; \quad v_{kj}^i = v_{min}$$

$$\text{if} \quad v_{kj}^i > v_{max} \quad ; \quad v_{kj}^i = v_{max}$$

$$x_{kj}^i = x_{kj}^{i-1} + v_{kj}^i \tag{2}$$

The parameter w represents an inertia weight; c_1 and c_2 are the correction (learning) factors, and r_1 and r_2 are random numbers in the interval $[0, 1]$. The velocity is limited between v_{min} and v_{max} to avoid very large particle movements in the search space, where $v_{min} < 0$ and $v_{max} > 0$. Fig. 1 illustrates how the particles move in a two-dimensional search space ($N = 2$). In this figure, two particles are present in the swarm, i.e. $K = 2$.

The first term in the right hand side of movement update Eqn. (1), weighted by w, signifies the dependence of the current particle velocity on its value in the previous iteration. The second term, weighted by c_1, signifies an attractor to pull the particle towards its previous best position. The third term, weighted by c_2 controls the movement of the particle towards the global best position.

In addition to the update Eqns. (1) and (2), one can limit the coordinates in a particle between two user defined values $x_{j_{min}}$ and $x_{j_{max}}$ in order to limit the search space. However, This operation increases the complexity and consumes time.

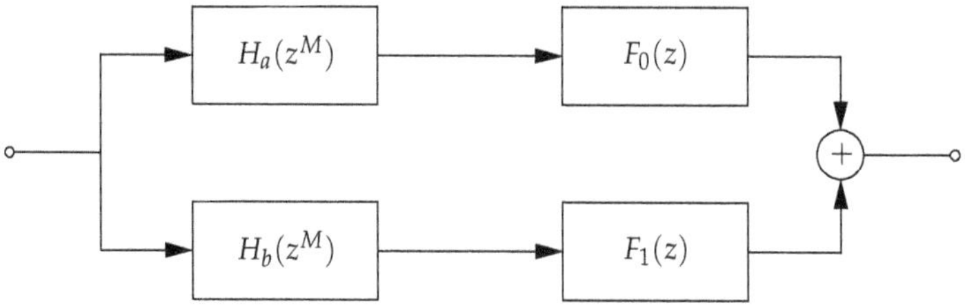

Figure 2. FRM Digital Filter Block Diagram

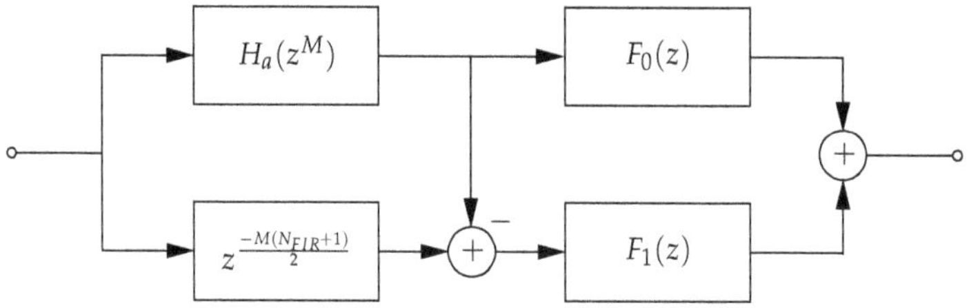

Figure 3. Block Diagram Representation of Frequency-Response Masking

3. The conventional FRM design approach

3.1. Design of lowpass FRM digital filters

The block diagram in Fig. 2 shows a conventional FRM digital filter, where $H_a(z)$ represents a FIR interpolation lowpass digital subfilter, and where $H_b(z)$ represents a power complementary counterpart of $H_a(z)$ in accordance with

$$|H_a(e^{j\omega})|^2 + |H_b(e^{j\omega})|^2 = 1 \tag{3}$$

Here, z represents the discrete-time complex frequency, and ω represents the corresponding (normalized) real frequency variable. Moreover, $F_0(z)$ and $F_1(z)$ represent FIR masking digital subfilters, while $H_a(z^M)$ and $H_b(z^M)$ represent M-fold interpolated versions of $H_a(z)$ and $H_b(z)$, respectively. In case of FIR digital interpolation subfilters, for a linear-phase filter $H_a(z)$ of order N_{FIR}, the relationship between $H_b(z)$ and $H_a(z)$ is as follows:

$$H_b(z) = z^{(N_{FIR}+1)/2} - H_a(z) \tag{4}$$

and hence $H_b(z)$ can be implemented by subtracting the output of $H_a(z)$ from the delayed version of the input, as shown in Fig. 3.

The FRM digital filter in Fig. 2 has an overall transfer function

$$H(z) = H_a(z^M)F_0(z) + H_b(z^M)F_1(z) \tag{5}$$

	Filter	Passband Edge	Stopband Edge
Case I	$H(z)$	$\frac{2I_L\pi + \omega_p}{M}$	$\frac{2I_L\pi + \omega_a}{M}$
	$F_0(z)$	$\frac{2I_L\pi + \omega_a}{M}$	$\frac{2(I_L+1)\pi - \omega_a}{M}$
	$F_1(z)$	$\frac{2I_L\pi - \omega_p}{M}$	$\frac{2I_L\pi + \omega_p}{M}$
Case II	$H(z)$	$\frac{2I_L\pi - \omega_a}{M}$	$\frac{2I_L\pi - \omega_p}{M}$
	$F_0(z)$	$\frac{2(I_L-1)\pi + \omega_a}{M}$	$\frac{2I_L\pi - \omega_a}{M}$
	$F_1(z)$	$\frac{2I_L\pi - \omega_p}{M}$	$\frac{2I_L\pi + \omega_p}{M}$

Table 1. Edge Frequencies of the Overall FRM FIR filter and Masking Subfilters

The masking digital subfilters $F_0(z)$ and $F_1(z)$ are employed to suppress the unwanted image bands produced by the interpolated digital subfilters $H_a(z^M)$ and $H_b(z^M)$. The masking filters are made to have equal order (by zero padding) in order to ensure that their phase characteristics are similar. The corresponding interpolated digital subfilters $H_a(z^M)$ and $H_b(z^M)$ can realize transition bands which are a factor of M sharper than those of $H_a(z)$ and $H_b(z)$, without increasing the number of required non-zero digital multipliers. The magnitude frequency-response of the various subfilters incorporated by the FRM digital filter design approach are shown in Fig. 4.

Here, Case I design is when the transition band of $H(z)$ is extracted from that of $H_a(z^M)$ and Case II design is when the transition band of $H(z)$ is extracted from that of $H_b(z^M)$. The edge frequencies of the overall digital FRM filter and its constituent subfilters are given in Table 1, where I_L represents the number of image lobes to be masked given by:

$$I_L = \begin{cases} \left\lfloor \frac{M\omega_p}{2\pi} \right\rfloor & \text{Case I} \\ \left\lceil \frac{M\omega_a}{2\pi} \right\rceil & \text{Case II} \end{cases} \tag{6}$$

where $\lfloor\ \rfloor$ denotes the largest integer from the lower side, and $\lceil\ \rceil$ signifies the smallest integer from the upper side.

3.2. Design of bandpass FRM digital filters

In general, it is possible to extend the conventional FRM approach for the design of bandpass or bandstop FRM digital filters. However, the resulting FRM digital filters are constrained to have identical lower and upper transition bandwidths. In [39], this restriction was relaxed by realizing the bandstop FRM FIR digital filter as a parallel combination of a corresponding pair of lowpass and highpass FIR

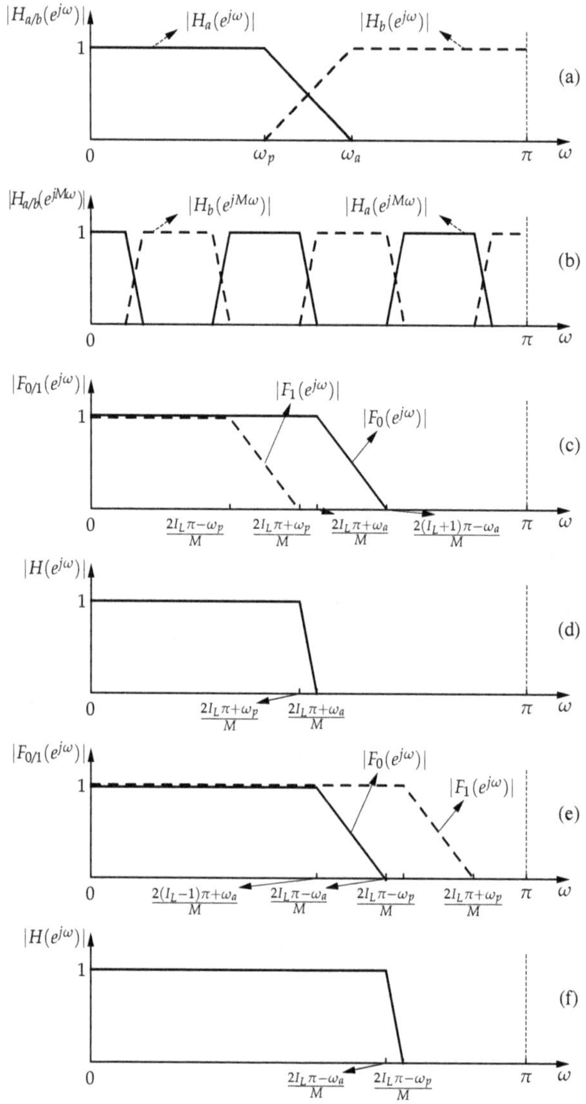

Figure 4. Magnitude Frequency-Response of FRM Digital Filter. (a) Magnitude Frequency-Response of the Bandedge-Shaping Digital Subfilters $H_a(z)$ and $H_b(z)$. (b) Magnitude Frequency-Response of the M-Interpolated Complementary Digital Subfilters $H_a(z^M)$ and $H_b(z^M)$. (c) Magnitude Frequency-Response of the Masking Digital Subfilters $F_0(z)$ and $F_1(z)$ for Case I. (d) Magnitude Frequency-Response of the Overall FRM Digital Filter $H(z)$ for Case I. (e) Magnitude Frequency-Response of the Masking Digital Subfilters $F_0(z)$ and $F_1(z)$ for Case II. (f) Magnitude Frequency-Response of the Overall FRM Digital Filter $H(z)$ for Case II [3].

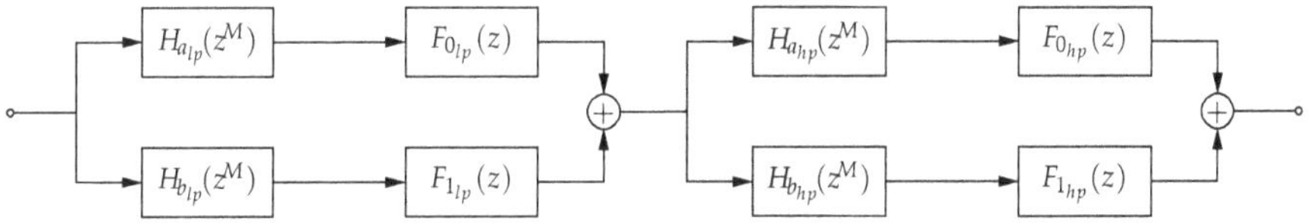

Figure 5. Bandpass FRM Digital Filter Block Diagram

digital filters. The latter lowpass and highpass FRM digital filters were obtained using a variation of the conventional FRM approach.

Let the desired bandpass FRM digital filter $H(z)$ have a lower transition bandwidth which is not identical to its upper transition bandwidth. $H(z)$ can be realized as a cascade combination of a pair of lowpass and highpass FRM digital filters, so that

$$H(z) = H_{lp}(z)H_{hp}(z) \tag{7}$$

where $H_{lp}(z)$ represents a lowpass and $H_{hp}(z)$ represents a highpass FRM digital filter. In this way, $H_{lp}(z)$ and $H_{hp}(z)$ can be obtained with the help of Eqn. (5) as

$$H_{lp}(z) = H_{a_{lp}}(z^M)F_{0_{lp}}(z) + H_{b_{lp}}(z^M)F_{1_{lp}}(z) \tag{8}$$

$$H_{hp}(z) = H_{a_{hp}}(z^M)F_{0_{hp}}(z) + H_{b_{hp}}(z^M)F_{1_{hp}}(z) \tag{9}$$

The lower transition bandwidth is governed by the constituent transition bandwidth of the highpass FRM digital filter, while the upper transition bandwidth is governed by the constituent transition bandwidth of the lowpass FRM digital filter. The realization for bandpass FRM digital filter are as shown in Fig. 5.

4. Design of FRM digital filters incorporating IIR interpolation digital subfilters

In the case of FRM IIR digital filters, $H_a(z)$ and $H_b(z)$ (in section 3) act as IIR interpolation digital subfilters. The masking filters $F_0(z)$ and $F_1(z)$ are not changed (i.e. they are still equal order FIR digital filters). Therefore, Eqn. (5) is still valid for the FRM IIR digital filter.

The IIR interpolation digital subfilter $H_a(z)$ is chosen to have an odd order N_{IIR}. Odd-ordered elliptic transfer functions can be represented as a sum of or difference between two allpass transfer functions [40]. Therefore, $H_a(z)$ can be realized as the addition of two allpass digital networks $G_0(z)$ and $G_1(z)$ as follows:

$$H_a(z) = \frac{G_0(z) + G_1(z)}{2} \tag{10}$$

where $G_0(z)$ is odd-ordered and $G_1(z)$ is even-ordered. The interesting fact is that the difference between $G_0(z)$ and $G_1(z)$ results in a filter that is power complementary to $H_a(z)$, and can subsequently be used as the power complementary interpolation digital subfilter $H_b(z)$ as in the following:

$$H_b(z) = \frac{G_0(z) - G_1(z)}{2} \tag{11}$$

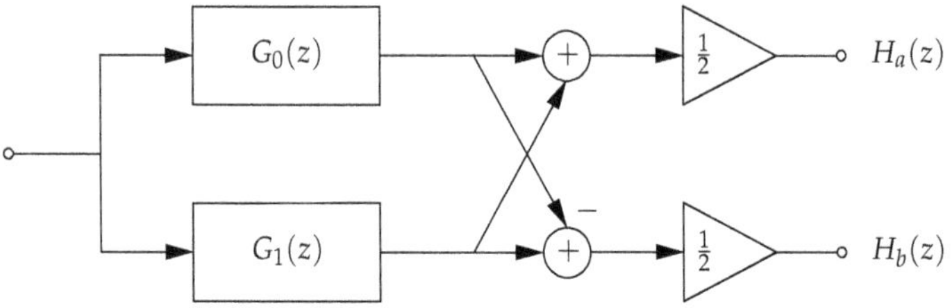

Figure 6. Block Diagram of Interpolation and Complementary Filters as a Parallel Combination of Two Allpass Networks

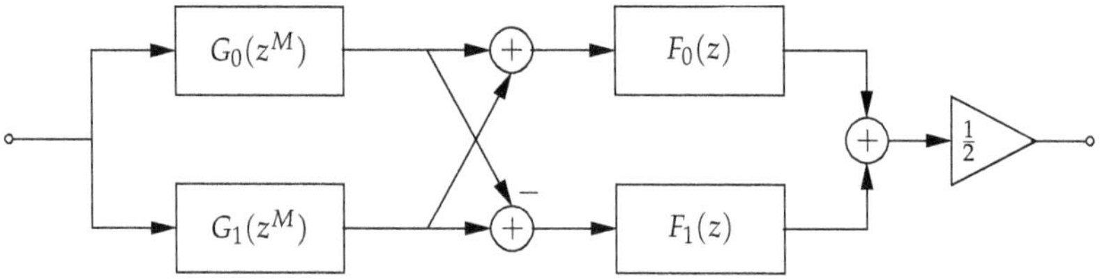

Figure 7. FRM Digital Filter Realization in Terms of Allpass Digital Networks $G_0(z)$ and $G_1(z)$

It can be easily verified that $H_a(z)$ and $H_b(z)$ are power complementary digital filters [29], i.e. they satisfy Eqn. (3). In addition, it is well known that this structure halves the number of multiplier coefficients required for the implementation of FRM digital filters and therefore is the most economical realization since it requires a total of only N_{IIR} multiplier coefficients to realize both $H_a(z)$ and $H_b(z)$. The overall transfer function of $H(z)$ given by Eqn. (5) can be expressed as:

$$H(z) = \frac{G_0(z^M) + G_1(z^M)}{2} F_0(z) + \frac{G_0(z^M) - G_1(z^M)}{2} F_1(z) \tag{12}$$

The block diagram in Fig. 6 shows the IIR interpolation digital subfilters $H_a(z)$ and $H_b(z)$ realized as a parallel combination of two allpass networks. It should be noted that if $H_a(z)$ is a lowpass filter, $H_b(z)$, which is the power complementary of $H_a(z)$, is a highpass filter. Fig. 7 shows an overall FRM IIR digital filter realization.

One may rearrange the structure in Fig. 7 by using Eqns. (10-11). This can be performed by defining two digital subfilters as follows:

$$A(z) = \frac{F_0(z) + F_1(z)}{2} \tag{13}$$

$$B(z) = \frac{F_0(z) - F_1(z)}{2} \tag{14}$$

Then $H(z)$ in Eqn. (12) simplifies to:

$$H(z) = G_0(z^M)A(z) + G_1(z^M)B(z) \tag{15}$$

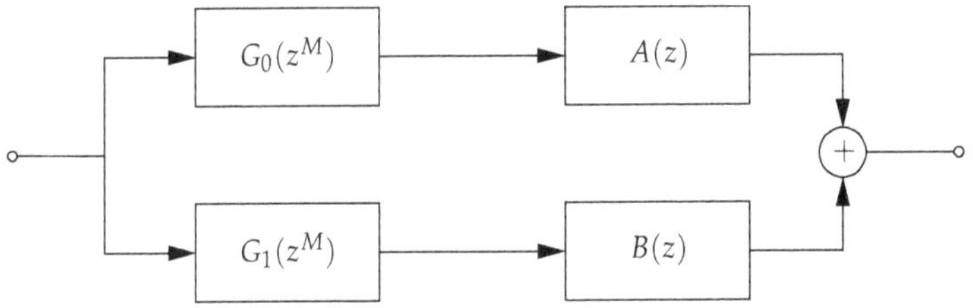

Figure 8. Alternative Structure of the Overall FRM IIR Digital Filter

Fig. 8 shows the block diagram representing Eqn. (15).

The advantage of realizing the FRM IIR digital filter as shown in Fig. 8 is that two adders shown in Fig. 7 are removed and they are no longer required. This subsequently simplifies the hardware implementation of the overall FRM IIR digital filter. However, it should be noted that the FIR masking digital subfilters $F_0(z)$ and $F_1(z)$ are made to be equal order using zero padding, and this results in the masking filters being moderately sparse. This is not the case when $A(z)$ and $B(z)$ are used instead. Therefore, the gain in hardware that could be achieved by using the realization in Fig. 8 is offset by a greater number of non-zero multiplier coefficients required in the realization of FRM IIR digital filters.

5. Realization of IIR interpolation digital subfilters using Elliptic Filters with Minimum Q-factor (EMQF)

Bilinear-LDI transformation falls into the category of digital filter realization techniques that transform an analog reference filter to its digital counterpart. Therefore, in order to determine the multiplier coefficient values of the IIR interpolation digital subfilters $H_a(z)$ and $H_b(z)$ constituent in the FRM IIR digital filter, a suitable analog reference filter $H_a(s)$ and its power complementary analog filter $H_b(s)$ have to be determined, where s is the analog frequency domain variable. Once $H_a(s)$ and $H_b(s)$ have been determined, the interpolation digital subfilters $H_a(z)$ and $H_b(z)$ are derived by using bilinear-LDI technique (see Section 6).

EMQF filters have several advantages for the design of FRM IIR digital filters. The squared ripple in the passband region of $H_a(z)$ and the squared ripple in the stopband region of $H_b(z)$ are equal as indicated by Eqn. (3). On the other hand, the squared ripple in the stopband region of $H_a(z)$ and the squared ripple in the passband region of $H_b(z)$ are equal. In addition, depending on whether the design specifications require a Case I or Case II FRM technique, either $H_a(z)$ or $H_b(z)$ could determine the maximum passband and stopband ripple of the overall FRM IIR digital filter $H(z)$. Consequently, the interpolation filter $H_a(z)$ is chosen to have equal passband and stopband squared tolerances. In this way, the resulting $H_b(z)$ also displays equal passband and stopband squared tolerances. These characteristics can be generalized for the analog reference subfilters $H_a(s)$ and $H_b(s)$. Therefore, there is a need for an analog reference filter $H_a(s)$ that together with its power complement $H_b(s)$ can exactly satisfy the passband and stopband relations in the FRM IIR filter. EMQF filters can successfully comply with the specifications present in the FRM IIR filter design. In addition, an EMQF transfer function can be easily designed by using bilinear-LDI transformation technique or any other structure consisting of two digital allpass networks in parallel. Furthermore, filters having EMQF transfer functions are minimally sensitive to component variations.

Despite all the advantages of EMQF filters, they suffer from not being able to independently specify passband and stopband ripples [41],[42] of the filter. Additionally, EMQF filters have exceedingly low passband attenuation.

All the poles of an EMQF transfer function reside on a circle in the s domain rendering them to have equal magnitudes. Given a squared passband and stopband tolerance of δ_p and δ_a, respectively, for an EMQF filter, the passband ripple Δ_p and minimum stopband attenuation Δ_a can be obtained as follows [43]:

$$\Delta_p = -10\log(1-\delta_p) \tag{16}$$

$$\Delta_a = -10\log(\delta_a) \tag{17}$$

The required passband and stopband edge frequencies for the analog reference filter $H_a(s)$ can be determined using design specifications along with Table 1. Frequency wrapping from digital to analog domain, and vice versa, has to be taken into account in accordance with:

$$\Omega_A = \frac{2}{T}\tan(\frac{\omega_d T}{2}) \tag{18}$$

where Ω_A is the analog frequency variable, where ω_d is the digital frequency variable, and where T is the sampling period.

Once the transfer function of the analog reference filter $H_a(s)$ is determined, it is represented as a sum of two allpass analog filters $G_0(s)$ and $G_1(s)$. In addition, $H_b(s)$, which is the power complementary of $H_a(s)$ is represented as the difference of $G_0(s)$ and $G_1(s)$. The poles of $G_0(s)$ and $G_1(s)$ are determined by cyclically distributing the poles of the reference filter $H_a(s)$ [43]. In the next section, belinear-LDI design technique is used to transform the two allpass networks $G_0(s)$ and $G_1(s)$ into digital domain.

6. Implementation of EMQF interpolation subfilters using bilinear-LDI design approach

In this section, the design procedure in [34, 44] is briefly explained to design and implement digital filters $G_0(z)$ and $G_1(z)$ using the the bilinear-LDI approach. This approach transforms analog reference filters $G_0(s)$ and $G_1(s)$ to obtain their digital filter counterparts $G_0(z)$ and $G_1(z)$.

The bilinear frequency transformation maps the analog frequency variable s to its digital domain counterpart z in accordance with:

$$s = \frac{2}{T}\frac{z-1}{z+1} \tag{19}$$

where T represents the sampling period, for mapping the transfer function of a prototype reference filter from the analog domain to the digital domain. The bilinear transform maps the left half of the complex s-plane to the interior of the unit circle in the z-plane. Therefore, BIBO stable filters in the s domain are converted to filters in the z domain which preserve that stability. Similarly, if the analog reference filter is minimum-phase, the previous characteristic of bilinear transform guarantees that the resulting digital filter is also minimum-phase. It also preserves the sensitivity properties of the analog reference filter. However, bilinear transform may result in a digital filter that has delay-free loops in its implementation. Unfortunately, delay-free loops prevent the implementation of a digital filter to be realizable in hardware platform.

The LDI frequency transformation ensures the absence of delay-free loops in the digital implementation and is given by

$$s = \frac{1}{T}\left(z^{\frac{1}{2}} - z^{-\frac{1}{2}}\right) \tag{20}$$

The LDI frequency transformation maps the hardware implementation of the analog reference filter to digital domain. While the LDI frequency transformation guarantees that there are no delay-free loops in the implementation of the digital filter, it does this to the cost of resulting in a digital filter having poor magnitude-frequency responses. Moreover, it is incapable of preserving the BIBO stability properties of the analog reference filter.

The bilinear-LDI approach is a combination of the two above mentioned realization techniques. In bilinear-LDI transform, a precompensation is performed to the reference analog filter. Then, the conventional LDI design technique is applied to a network resulting from the precompensated analog prototype filter. The precompensation is such that the application of the LDI design technique results in a filter that exactly matches the bilinear frequency transform of the uncompensated analog prototype filter.

The resulting bilinear-LDI digital filters have several desirable features from a hardware realization point of view. They are minimal in the number of digital multiplication operations. Although they are not minimal in the number of digital adders and unit-delays, the additional adders and the additional unit delay lead to certain advantages when the concept of generalized delay unit is used for the realization of the network [34]. Moreover, The bilinear-LDI digital filters lend themselves to fast two-cycle parallel digital signal processing speeds and they exhibit exceptionally low passband sensitivity to their multiplier coefficient values, resulting in small coefficient wordlengths.

As discussed in Section 5, the analog reference filter $H_a(s)$ is decomposed into two allpass analog networks $G_0(s)$ and $G_1(s)$. The digital allpass networks $G_0(z)$ and $G_1(z)$ are obtained from $G_0(s)$ and $G_1(s)$ using the bilinear-LDI design approach.

It should be pointed out that $G_0(s)$ is an odd-ordered allpass function. Therefore, it has a pole on the real axis in the s domain. On the other hand, $G_1(s)$ ends up having an even-ordered allpass function. It is well known that an allpass transfer function can be written in the general form [34]:

$$G(s) = \frac{P(-s)}{P(s)} \tag{21}$$

where $P(s)$ is a Hurwitz polynomial of order, say, $\tilde{n}$. Moreover, $P(s)$ can be expressed as:

$$P(s) = \mathrm{Ev}P(s) + \mathrm{Od}P(s) \tag{22}$$

where $\mathrm{Ev}P(s)$ denotes the even and $\mathrm{Od}P(s)$ denotes the odd part of $P(s)$.

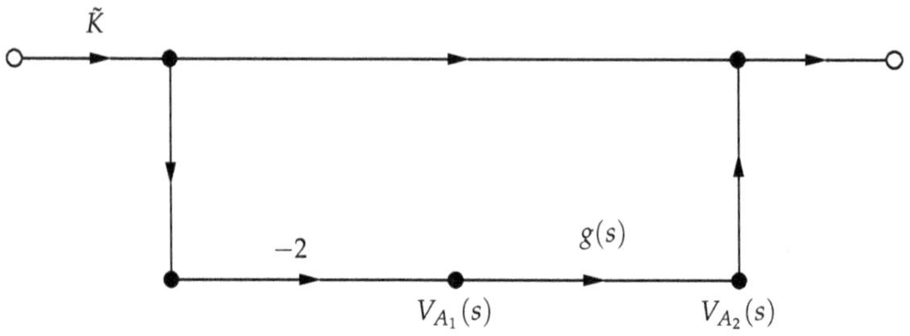

Figure 9. Signal Flow Graph of $G(s)$

By simple manipulation of Eqns. (21) and (22) one can get

$$G(s) = \tilde{K}\frac{1 - Z(s)}{1 + Z(s)} \tag{23}$$

Here, $\tilde{K} = 1$ or -1, and $Z(s)$ is a realizable reactive impedance given by

$$Z(s) = \begin{cases} \dfrac{\text{Od}P(s)}{\text{Ev}P(s)} & \text{for even } \tilde{n} \\ \dfrac{\text{Ev}P(s)}{\text{Od}P(s)} & \text{for odd } \tilde{n} \end{cases} \tag{24}$$

where $\tilde{n}$ is the order of $G(s)$ (odd when realizing $G_0(s)$ and even when realizing $G_1(s)$). The impedance $Z(s)$ has a zero at $s = 0$ for even $\tilde{n}$ and a pole at $s = 0$ for odd $\tilde{n}$, while having a zero at $s = \infty$ both for even $\tilde{n}$ and for odd $\tilde{n}$.

The bilinear-LDI digital realization of $G(s)$ is achieved by using the following steps:

- The transfer function $G(s)$ is decomposed in the form

$$G(s) = \tilde{K}[1 - 2g(s)] \tag{25}$$

  where

$$g(s) = \frac{Z(s)}{1 + Z(s)} \tag{26}$$

  Here, $G(s)$ can be realized as the transfer function of the signal-flow graph in Fig. 9.

  Furthermore, $g(s)$ represents a lowpass or highpass analog filter that can be realized as the transfer function of the voltage divider network in Fig. 10.

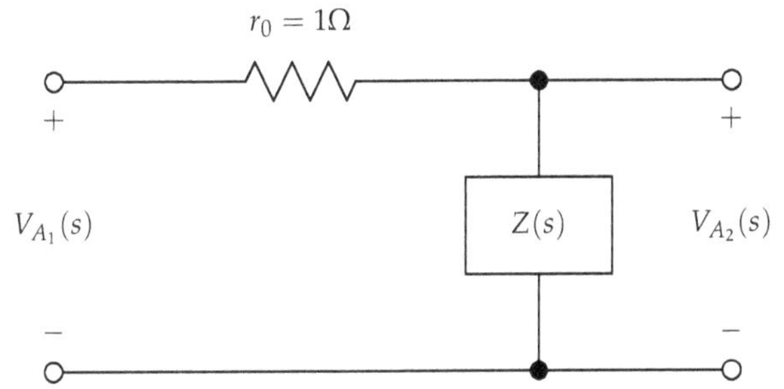

Figure 10. Voltage Divider Circuit for $g(s)$

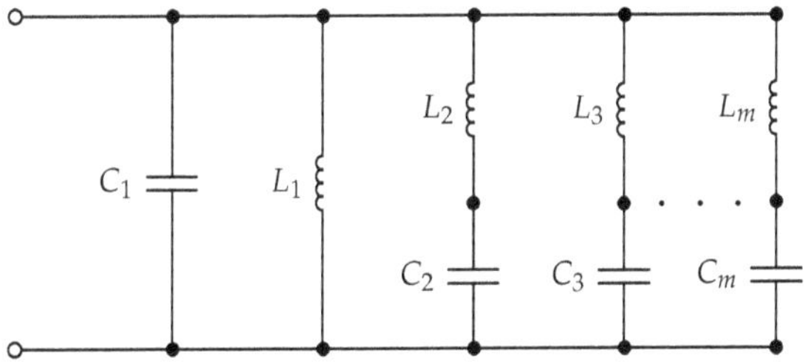

Figure 11. Realization of Impedance $Z(s)$

Finally, $Z(s)$ represents realizable reactances (consisting of capacitors and inductors only) and can be decomposed into its Foster II canonical form, as in Fig. 11, in accordance with

$$Z(s) = \frac{1}{Y(s)} \tag{27}$$

$$Y(s) = sC_1 + \frac{1}{sL_1} + \sum_{i=2}^{m} \frac{sC_i}{s^2 C_i L_i + 1} \tag{28}$$

where $m = \tilde{n}/2$ for even $\tilde{n}$ and $m = (\tilde{n}+1)/2$ for odd $\tilde{n}$, and where C_i represent capacitances and L_i represent inductances (for $i = 1, 2, \ldots, m$), and inductor L_1 is only present for even $\tilde{n}$.

- The impedance $Z(s)$ in Fig. 11 is substituted into Fig. 10 and the precompensation is applied to the resulting network. This amounts to a modification of circuit elements in accordance with:

$$V'_{A_1}(s) = \frac{V_{A_1}(s)}{1 - sT/2} \tag{29}$$

The resistance in r_0 in Fig. 10 is modified to:

$$r'_0 = z^{\frac{1}{2}} r_0 \tag{30}$$

and

$$L_1^{'} = L_1 \tag{31}$$

$$C_1^{'} = C_1 + \frac{T}{2} + \frac{T^2}{4L_1} + \sum_{i=2}^{m} \frac{C_i \frac{T^2}{4L_i}}{C_i + \frac{T^2}{4L_i}} \tag{32}$$

$$L_i^{'} = L_i \left[\frac{C_i + \frac{T^2}{4L_i}}{C_i} \right]^2 \tag{33}$$

$$C_i^{'} = \frac{C_i^2}{C_i + \frac{T^2}{4L_i}} \tag{34}$$

with $r_0 = 1\Omega$ and for $i = 2, 3, ..., m$.

- Since the voltage/current signal-flow graph of the precompensated network [34] consists of analog integrators only and it has no analog differentiators, it can be used for bilinear-LDI realization method. Therefore, the analog integrators in the signal-flow graph of the precompensated network are replaced by LDI digital integrators, and by impedance-scaling, the resulting network is scaled by $z^{-\frac{1}{2}}$ to eliminate any half-delay elements. The resulting digital network is displayed in Fig. 12. The multiplier coefficients in Fig. 12 are as follows:

$$m_{L_i} = \frac{T}{L_i^{'}} \tag{35}$$

$$m_{C_i} = \frac{T}{C_i^{'}} \tag{36}$$

for $i = 1, 2, ..., m$.

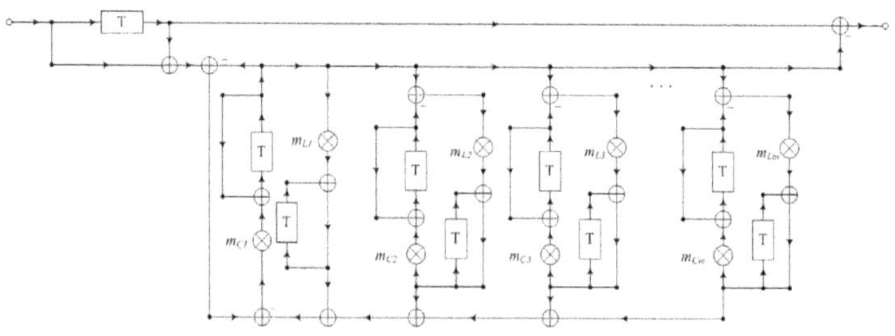

Figure 12. Realization of the Bilinear-LDI Digital Allpass Network $G(z)$ [34]

7. Constraints for guaranteed BIBO stability

In order for the FRM digital filter consisting of CSD multiplier coefficients $\hat{m}_{FRM}$ to be BIBO stable, it is both necessary and sufficient for the bilinear-LDI IIR interpolation digital subfilters $H_a(z)$ and $H_b(z)$ to be BIBO stable. Likewise, in order for the interpolation digital subfilters $H_a(z)$ and $H_b(z)$ to be BIBO stable, it is both necessary and sufficient for the bilinear-LDI allpass digital networks $G_0(z)$ and $G_1(z)$ to be BIBO stable. In this way, it is required that the bilinear-LDI digital allpass networks $G_0(z)$ and $G_1(z)$ remain BIBO stable throughout the course of the PSO algorithm.

In the course of PSO algorithm, the infinite-precision multiplier coefficients m_{L_i} and m_{C_i} can only take quantized values $\hat{m}_{L_i}$ and $\hat{m}_{C_i}$ that belong to $CSD(L,l,f)$. In order for the bilinear-LDI digital allpass networks $G_0(z)$ and $G_1(z)$ to remain BIBO stable, it is required that the values of the corresponding quantized reactive elements $\hat{L}_i$ and $\hat{C}_i$ remain positive [45] in the course of optimization. This is due to the properties of the bilinear frequency transformation from analog to digital domain. In order to find the conditions for BIBO stability and in accordance with Eqns. (35) and (36), one has:

$$\hat{L}_i' = \frac{T}{\hat{m}_{L_i}} \tag{37}$$

$$\hat{C}_i' = \frac{T}{\hat{m}_{C_i}} \tag{38}$$

Moreover, in accordance with Eqns. (31-34), one has:

$$\hat{L}_1' = \hat{L}_1 \tag{39}$$

$$\hat{C}_1' = \hat{C}_1 + \frac{T}{2} + \frac{T^2}{4\hat{L}_1} + \sum_{i=2}^{m} \frac{\hat{C}_i \frac{T^2}{4\hat{L}_i}}{\hat{C}_i + \frac{T^2}{4\hat{L}_i}} \tag{40}$$

$$\hat{L}_i' = \hat{L}_i \left[\frac{\hat{C}_i + \frac{T^2}{4\hat{L}_i}}{\hat{C}_i} \right]^2 \tag{41}$$

$$\hat{C}_i' = \frac{\hat{C}_i^2}{\hat{C}_i + \frac{T^2}{4\hat{L}_i}} \tag{42}$$

where $\hat{L}_1 = \infty$ for odd-ordered allpass network $G_0(z)$.

By substituting Eqns. (37) and (38) into Eqns. (39-42), and by solving the resulting equations for the reactive elements $\hat{L}_i$ and $\hat{C}_i$, one can obtain

$$\hat{L}_1 = \frac{T}{\hat{m}_{L_1}} \tag{43}$$

$$\hat{C}_1 = \frac{T\left\{\frac{4}{\hat{m}_{C_1}} - \hat{m}_{L_1} - 4\left(\sum_{i=2}^{m} \frac{1}{\frac{4}{\hat{m}_{L_i}} - \hat{m}_{C_i}}\right) - 2\right\}}{4} \tag{44}$$

$$\hat{L}_i = \frac{T(\hat{m}_{L_i}\hat{m}_{C_i} - 4)^2}{16\hat{m}_{L_i}} \tag{45}$$

$$\hat{C}_i = \frac{-4T}{\hat{m}_{C_i}(\hat{m}_{L_i}\hat{m}_{C_i} - 4)} \tag{46}$$

From Eqns. (43-46), $\hat{L}_i > 0$ and $\hat{C}_i > 0$ provide that

$$\hat{m}_{L_1} > 0 \tag{47}$$

$$\hat{m}_{L_i} > 0 \tag{48}$$

$$\hat{m}_{C_i} < \frac{4}{\hat{m}_{L_i}} \tag{49}$$

$$\hat{m}_{C_1} < \frac{4}{\left\{\hat{m}_{L_1} + 4\left(\sum_{i=2}^{m} \frac{1}{\frac{4}{\hat{m}_{L_i}} - \hat{m}_{C_i}}\right) + 2\right\}} \tag{50}$$

Then, in order to make the CSD FRM digital filter BIBO stable, it is necessary and sufficient to choose the values of the multiplier coefficients $\hat{m}_{FRM} \in CSD(L,l,f)$ such that the inequality constraints (47-50) are satisfied. The equations and corresponding condition required for BIBO stability are summarized in Table 2.

In order to make the CSD lowpass digital IIR FRM filter BIBO stable, it is necessary and sufficient to choose the values of the multiplier coefficients $\hat{m}_{L_i}$, $\hat{m}_{C_i} \in CSD(L,l,f)$ such that the inequality constraints of Table 2 are satisfied.

It should be pointed out that constraint (49) is most stringent when $\hat{m}_{L_i}$ is at its largest possible value. Similarly, constraint (50) is most stringent when $\hat{m}_{L_1}$, $\hat{m}_{L_i}$ and $\hat{m}_{C_i}$'are all at their largest possible values (while $\hat{m}_{L_i}$ and $\hat{m}_{C_i}$ still adhere to constraint $\hat{m}_{C_i} < 4\,(\hat{m}_{L_i})^{-1}$).

8. Proposed PSO of FRM IIR digital filters

The proposed particle swarm optimization of BIBO stable FRM IIR digital filters is carried out over the CSD multiplier coefficient space $CSD(L_{0\text{ or }1}, l_{0\text{ or }1}, f_{0\text{ or }1})$, where $L_{0\text{ or }1}$ represents the multiplier coefficient wordlength, where $l_{0\text{ or }1}$ represents the maximum number of non-zero digits, and where $f_{0\text{ or }1}$ represents the number of fractional part digits (for FIR or IIR digital subfilters, respectively).

Element	Equation	Inequality Constraints
$\hat{L}_1$	$\frac{T}{\hat{m}_{L_1}}$	$\hat{m}_{L_1} > 0$
$\hat{C}_1$	$\frac{1}{4}T\left\{\frac{4}{\hat{m}_{C_1}} - \hat{m}_{L_1} - 4\left(\sum_{i=2}^{m}\frac{1}{\frac{4}{\hat{m}_{L_i}} - \hat{m}_{C_i}}\right) - 2\right\}$	$\hat{m}_{C_1} < 4\left\{\hat{m}_{L_1} + 4\left(\sum_{i=2}^{m}\frac{1}{\frac{4}{\hat{m}_{L_i}} - \hat{m}_{C_i}}\right) + 2\right\}^{-1}$
$\hat{L}_i$	$\frac{T(\hat{m}_{L_i}\hat{m}_{C_i} - 4)^2}{16\hat{m}_{L_i}}$	$\hat{m}_{L_i} > 0$
$\hat{C}_i$	$\frac{-4T}{\hat{m}_{C_i}(\hat{m}_{L_i}\hat{m}_{C_i} - 4)}$	$\hat{m}_{C_i} < 4\left(\hat{m}_{L_i}\right)^{-1}$

Table 2. Relations for Elements of Back-Transformed Reactance

The starting point of any stochastic algorithm plays an important role in the convergence behavior of the optimization algorithm [46]. Therefore, it is important to generate the initial swarm in proper positions in the search space rather than complete random generation of the initial population. In order to achieve this, the following technique is employed:

8.1. Initiation of PSO

To start the PSO algorithm from a good position in the search space the infinite precision multiplier coefficient values of the seed particle are generated by using classical techniques as discussed in previous sections. These infinite precision multiplier coefficient values are turned into their finite precision counterparts by simply rounding them to their closest CSD values. This seed particle is used as the center of the swarm and a cloud of particles are generated randomly around the seed particle. It should be noted that the distance of the randomly generated particles should not be far from the seed particle. In this way, the initial swarm contains particles which have high chances of being near the optimal solution. The multiplier coefficient values of the swarm are taken from a set of CSD LUTs which are constructed as follows:

8.2. FRM IIR digital filter template LUTs

It is necessary and sufficient to choose the values of the multiplier coefficients, such that the inequality constraints (47-50) are satisfied. In order to achieve this, the LUTs are constructed as follows:

- One LUT is constructed for all multiplier coefficient values $\hat{m}_{FIR} \in CSD(L_0, l_0, f_0)$ for the masking digital subfilters $F_0(z)$ and $F_1(z)$. The values of L_0, l_0 and f_0 are determined empirically based on the amplitude frequency-response of the masking digital subfilters $F_0(z)$ and $F_1(z)$.
- A LUT is constructed for all multiplier coefficient values $\hat{m}_{IIR} \in CSD(L_1, l_1, f_1)$ for the digital allpass networks $G_0(z)$ and $G_1(z)$. Once again, the values of L_1, l_1 and f_1 are determined empirically. Also, it is expedient to assume that $\hat{m}_{IIR}$ have only positive values.
- The above CSD LUT is used to form one size-reduced LUT per the multiplier coefficient for digital allpass networks $G_0(z)$ and $G_1(z)$, where each size-reduced LUT initially includes CSD values bounded from below by the smallest representable value belonging to $CSD(L_1, l_1, f_1)$, and

from above by the corresponding value of the finite-wordlength coefficients for the seed particle. The size-reduced LUTs are augmented before PSO process commences. The purpose of this augmentation is to ensure that the exploration space include as many of those CSD multiplier coefficients $\hat{m}_{L_1}$, $\hat{m}_{C_1}$, $\hat{m}_{L_i}$ and $\hat{m}_{C_i}$ which still satisfy the BIBO stability constraints (47-50).

The above constructed LUTs are used as template LUTs. There are two problems concerning the PSO of FRM IIR digital filters over the CSD multiplier coefficient space. To overcome these problems, the template LUTs must be further processed. These two problems and the way to solve them are discussed in the following.

8.3. PSO indirect search method

In PSO, the required new particle position is obtained from the previous position of the particle through the addition of a random (normalized) velocity value. However, by directly applying the conventional PSO to the above optimization over the CSD multiplier coefficients, one may obtain new particle positions whose coordinate values are no longer in $CSD(L_{0\text{ or }1}, l_{0\text{ or }1}, f_{0\text{ or }1})$. In order to overcome this problem, the optimization search is carried out indirectly via the indices to the LUT CSD values (as opposed to LUT CSD values themselves). In this way, the CSD coordinate values for each particle position are obtained by integer indices to the CSD LUTs. The key point in the indirect search rests with ensuring that the index set is closed, i.e. by ensuring that each index points to a valid CSD value in the LUT, and that the resulting particle in the course of PSO adheres to the prespecified CSD number format.

If the velocity values are replaced by their closest integer values, the update equations become modified to

$$\hat{v}^i_{kj} = [w\hat{v}^{i-1}_{kj} + c_1 r_1(\hat{x}^{i-1}_{best_{kj}} - \hat{x}^{i-1}_{kj}) + c_2 r_2(\hat{g}^{i-1}_{best_j} - \hat{x}^{i-1}_{kj})]^1 \tag{51}$$

$$\text{if} \quad \hat{v}^i_{kj} < \hat{v}_{min} \quad ; \quad \hat{v}^i_{kj} = \hat{v}_{min}$$

$$\text{if} \quad \hat{v}^i_{kj} > \hat{v}_{max} \quad ; \quad \hat{v}^i_{kj} = \hat{v}_{max}$$

$$\hat{x}^i_{kj} = \hat{x}^{i-1}_{kj} + \hat{v}^i_{kj} \tag{52}$$

Here, $\hat{x}_{kj}$, $\hat{v}_{kj}$, $\hat{x}_{best_{kj}}$, $\hat{g}_{best_j}$, $\hat{v}_{min}$ and $\hat{v}_{max}$ are all integer values where $\hat{v}_{min} < 0$ and $\hat{v}_{max} > 0$. In addition, w is limited in the interval $[0, 0.5)$ (as discussed shortly).

8.4. Barren layers

Due to their finite length, the template LUTs inevitably lead to a bounded optimization search space. In order to ensure that the particles do not cross over to the outside of the search space in the course of PSO, the search space is constructed as a combination of two regions, namely the interior and barren layers. The barren layer is constructed to yield relatively low fitness values, and is represented as header and footer in the template LUT. There are two problems concerning the construction of the barren layers:

[1] $[R]$ denotes rounding R to its closest integer, where R is assumed to be a real value.

8.4.1. barren layer entries

The first problem in the construction of barren layers concerns how to make the fitness values in the barren layer relatively low. This problem can be resolved by filling the header part by unrealistically large, and the footer part by unrealistically small CSD multiplier coefficient values.

8.4.2. barren layer width

The second problem, on the other hand, concerns how to determine the width of the barren layer such that the particles do not cross over to the outside of the search space even under the worst case scenario. These two problems relate to the number of entries and the CSD values of the entries in header and footer parts of the template LUTs. To overcome this problem, let us consider the j-th variable in the k-th particle is in the boundaries of one of the template LUTs in iteration $i-1$. The worst case scenario occurs when x_{kj}^{i-1} moves toward the barren layer with the peak permissible velocities (v_{max} for the header, and v_{min} for the footer). If in the i-th iteration x_{kj}^{i} is in the footer:

$$\hat{x}_{best_{kj}}^{i} > \hat{x}_{kj}^{i} \tag{53}$$

$$\hat{g}_{best_j}^{i} > \hat{x}_{kj}^{i} \tag{54}$$

and if it is in the header:

$$\hat{x}_{best_{kj}}^{i} < \hat{x}_{kj}^{i} \tag{55}$$

$$\hat{g}_{best_j}^{i} < \hat{x}_{kj}^{i} \tag{56}$$

Eqns. (53-56) show that the velocity of the particle in iteration $i+1$ tends to move the particle in a direction opposite to the direction of the barren regions. Here, the worst case happens when $r_1 = r_2 = 0$. In this way, the number of entries L_f in the footer part, and the number of entries L_h in the header part is determined in accordance with

$$\begin{aligned} L_f &= |\hat{v}_{min}| + [w|\hat{v}_{min}|] + [w[w|\hat{v}_{min}|]] + \ldots \\ &\leq |\hat{v}_{min}| + \frac{|\hat{v}_{min}|}{2} + \frac{|\hat{v}_{min}|}{4} + \ldots \\ &= 2|\hat{v}_{min}| \end{aligned} \tag{57}$$

$$\begin{aligned} L_h &= \hat{v}_{max} + [w\hat{v}_{max}] + [w[w\hat{v}_{max}]] + \ldots \\ &\leq \hat{v}_{max} + \frac{\hat{v}_{max}}{2} + \frac{\hat{v}_{max}}{4} + \ldots \\ &= 2\hat{v}_{max} \end{aligned} \tag{58}$$

Let us recall that since $0 \leq w < 0.5$,

$$\text{if} \quad v : \text{positive integer} \quad \Rightarrow \quad [wv] \leq \frac{v}{2} \tag{59}$$

In addition, after some iterations $\hat{v}_{kj}^{i+1} = 0$. Otherwise, if $w \geq 0.5$, $\hat{v}_{kj}^{i+1}$ can never become zero, and the width of the barren layer will be infinity.

The augmented LUTs remains fixed in the course of PSO, restricting automatic particle movement inside the limited search space. Modifying the index values inside each particle by adding the current indices to the length of the footer barren region, L_f, PSO algorithm is ready to start the optimization of FRM digital filters.

9. Design methodology

The design methodology for the proposed PSO of BIBO stable bilinear-LDI based FRM IIR digital filters over the CSD multiplier coefficient space can be summarized as follows:

1. *Designing the interpolation digital subfilter*: the first step in determining the interpolation subfilter specifications is to fix the interpolation factor M from a pre-specified range. This is done in a way that the order of the FIR masking filters is kept minimal. Using the passband edge frequency ω_p and stopband edge frequency ω_a and the expressions for boundary frequencies given in Table 1, one can determine the filter case and calculate the approximate passband edge $\tilde{\theta}$ and stopband edge $\tilde{\phi}$ of the digital interpolation lowpass subfilter $H(e^{j\omega})$, for every value of the user specified range of interpolation factors M. The order of the FIR masking filters depends on the minimum distance between consecutive image replicas of either the interpolated subfilter $H_a(e^{jM\omega})$ or its complement $H_b(e^{jM\omega})$. Then, displacement λ_M and distance $\tilde{D}_M$ for each interpolation factor M are given as:

$$\lambda_M = \max[|(\frac{\pi}{2} - \tilde{\theta})|, |(\frac{\pi}{2} - \tilde{\phi})|] \tag{60}$$

$$\tilde{D}_M = \frac{\pi}{M} - \frac{2\lambda}{M} \tag{61}$$

To minimize the length of FIR-masking filters, the value of M that results in the largest value of $\tilde{D}_M$ is chosen. This determines the optimal interpolation factor M as well as the approximate passband edge $\tilde{\theta}$ and stopband edge $\tilde{\phi}$ of the digital interpolation subfilter $H(e^{j\omega})$. EMQF filters have the property of equal square magnitude ripple size in the passband and stopband. Therefore, of the two ripple specifications, whichever gives the smallest tolerance in the squared magnitude response determines both the passband ripple R_p and stopband attenuation R_a of the interpolation digital subfilter $H_a(e^{j\omega})$. The interpolation digital subfilter order N_{IIR} is then determined using R_p, R_a, $\tilde{\theta}$ and $\tilde{\phi}$. N_{IIR} must be rounded to the nearest larger odd integer so that it can be implemented by a parallel combination of two allpass networks. With the order N_{IIR}, and passband and stopband ripples R_p and R_a fixed, the ratio of the analog passband edge θ_A and stopband edge ϕ_A is a constant k given by [47]

$$D = \frac{10^{0.1R_a} - 1}{10^{0.1R_p} - 1}$$

$$q = 10^{\frac{-\log(16D)}{N_{IIR}}}$$

$$q = q_0 + 2q_0^5 + 15q_0^9 + 150q_0^{13}$$

$$k_p = \left[\frac{1-2q_0}{1+2q_0}\right]^2$$

$$k = \sqrt{1-k_p^2}$$

In order to satisfy the passband edge specification, the digital passband edge $\omega_p = \tilde{\theta}$ for Case I filters. The digital stopband edge ω_a is then determined using the analog ratio k. (Here, frequency warping from digital to analog domain, and vice versa, given by Eqn. (18) needs to be taken into account.) Similarly, $\omega_a = \tilde{\phi}$ for Case II filters, and ω_p can be determined by using ratio k. Also, using given ripple specifications along with the boundary frequencies described in Table 1, one can determine the transfer function of the FIR masking filters $F_0(e^{j\omega})$ and $F_1(e^{j\omega})$.

2. *Generation of seed FRM digital filter particle*: The seed FRM digital filter particle is formed as follows:
   - A particle with B_1 coordinates is formed in which each coordinate serves as an index of the corresponding CSD LUT for each multiplier coefficient constituent in the interpolation digital subfilters. For FRM IIR digital filters, the multiplier coefficients correspond to the bilinear-LDI allpass digital networks $G_0(z)$ and $G_1(z)$.
   - A particle with B_2 coordinates is formed in which each coordinate serves as an index of the corresponding CSD LUT for each multiplier coefficient in the FIR masking digital subfilters $F_0(z)$ and $F_1(z)$.
3. *Generation of Initial Swarm*: An initial swarm of K particles is formed by generating a random cloud around the seed particle as discussed in section 8.1.
4. *Fitness Evaluation*: The fitness function for CSD FRM IIR digital filters is defined in accordance with

$$fitness_{magnitude} = -20log[max(\varepsilon_p, \varepsilon_a)] \tag{62}$$

$$fitness_{group-delay} = \varsigma_p \tag{63}$$

$$fitness = fintess_{magnitude} - fitness_{group-delay} \tag{64}$$

where

$$\varepsilon_p = \underbrace{max}_{\omega\in\Delta\omega_p} [W_p|H(e^{j\omega}) - 1|] \tag{65}$$

$$\varepsilon_a = \underbrace{max}_{\omega\in\Delta\omega_a} [W_a|H(e^{j\omega})|] \tag{66}$$

$$\varsigma_p = \underbrace{max}_{\omega\in\Delta\omega_p} [W_{gd}|\tau(\omega) - \mu_\tau|] \tag{67}$$

with $\Delta\omega_p$ representing the passband frequency region(s), with $\Delta\omega_a$ representing the stopband frequency region(s), and with $\tau(\omega)$ representing the group-delay frequency response of the FRM IIR digital filter. Here, W_p, W_a, and W_{gd} represent weighting factors for the passband and stopband magnitude responses, and for the group-delay response, respectively. Moreover, μ_τ represents the average group-delay over the passband region.

In [48], a convenient way to represent digital networks in terms of matrix representation is presented. This technique can be used to find the magnitude and group delay frequency response of the digital network in Fig. 12. Let us consider the input to the digital network in Fig. 12 to be x_D and the output of it to be y_D. In addition, let the output of the i-th time delay in Fig. 12 to be x_i and the input to the i-th time delay to be y_i. The transfer function matrix of the network, **T**, can be found as

$$\mathbf{y} = \mathbf{T}\mathbf{x} \tag{68}$$

where $\mathbf{y} = [y_D, y_1, y_2, \ldots, y_{2m+1}]^t$ [2]and $\mathbf{x} = [x_D, x_1, x_2, \ldots, x_{2m+1}]^t$, and **T** is a $(2m+2) \times (2m+2)$ matrix with the entries obtained as Eqn. (69).

$$\mathbf{T} = \begin{bmatrix} 0 & 1 & -1 & 0 & 0 & 0 & \cdots & 0 & 0 \\ 1 & 0 & 0 & 0 & 0 & 0 & \cdots & 0 & 0 \\ m_{C_1} & m_{C_1} & 1-m_{C_1}(1+\sum_{i=1}^{m} m_{L_i}) & -m_{C_1} & m_{C_1}m_{L_2} & -m_{C_1} & \cdots & m_{C_1}m_{L_m} & -m_{C_1} \\ 0 & 0 & m_{L_1} & 1 & 0 & 0 & \cdots & 0 & 0 \\ 0 & 0 & m_{C_2}m_{L_2} & 0 & 1-m_{C_2}m_{L_2} & m_{C_2} & \cdots & 0 & 0 \\ 0 & 0 & m_{L_2} & 0 & -m_{L_2} & 1 & \cdots & 0 & 0 \\ \vdots & \vdots & \vdots & \vdots & \vdots & \vdots & \ddots & & \\ 0 & 0 & m_{C_m}m_{L_m} & 0 & 0 & 0 & & 1-m_{C_m}m_{L_m} & m_{C_m} \\ 0 & 0 & m_{L_m} & 0 & 0 & 0 & & -m_{L_m} & 1 \end{bmatrix} \tag{69}$$

Since $x_i = z^{-1}y_i$, the transfer function $G(z) = \frac{y_D}{x_D}$ can be found as

$$G(z) = z^{-1}\mathbf{e}[\mathbf{I} - z^{-1}\mathbf{D}]^{-1}\mathbf{c} \tag{70}$$

where **e** is a row vector and **c** is a column vector of length $2m+1$, and where **I** is the identity matrix and **D** is a $(2m+1) \times (2m+1)$ matrix in accordance with

$$\mathbf{T} = \begin{bmatrix} 0 & \mathbf{e} \\ \mathbf{c} & \mathbf{D} \end{bmatrix} \tag{71}$$

[2] $\mathbf{X}^t$ denotes the transpose of the matrix **X**.

The matrix **T** is also useful in finding the group delay of $H(z)$. The group-delay of $H(e^{j\omega})$ is given by

$$\tau(\omega) = -\text{Im}\left\{\frac{1}{H(e^{j\omega})}\frac{dH(e^{j\omega})}{d\omega}\right\} \tag{72}$$

With the help of Eqn. (12), the expression $\frac{dH(e^{j\omega})}{d\omega}$ can be written as

$$\begin{aligned}\frac{dH(e^{j\omega})}{d\omega} = \frac{1}{2}\Big[&\frac{dG_0(e^{j\omega})}{d\omega}(F_0(e^{j\omega}) + F_1(e^{j\omega}))+\\ &\frac{d(F_0(e^{j\omega}) + F_1(e^{j\omega}))}{d\omega}G_0(e^{j\omega})+\\ &\frac{dG_1(e^{j\omega})}{d\omega}(F_0(e^{j\omega}) - F_1(e^{j\omega}))+\\ &\frac{d(F_0(e^{j\omega}) - F_1(e^{j\omega}))}{d\omega}G_1(e^{j\omega})\Big]\end{aligned} \tag{73}$$

The derivative of FIR filters can be easily found from their transfer function. In order to find the derivative of the digital allpass networks $G_0(z)$ and $G_1(z)$, the following expression can be used

$$\frac{dG(e^{j\omega})}{d\omega} = -je^{-j\omega}\sum_{i=1}^{2m+1} G_{xi}(e^{j\omega})G_{iy}(e^{j\omega}) \tag{74}$$

where $G_{xi}(z)$ is the transfer function between x_D and y_i, and where $G_{iy}(z)$ is the transfer function between x_i and y_D. The transfer functions $G_{xi}(z)$ and $G_{iy}(z)$ can be found from the transfer function matrix **T** as follows

$$G_{xi}(z) = a_{xi} + z^{-1}\mathbf{e}_{xi}[\mathbf{I} - z^{-1}\mathbf{D}]^{-1}\mathbf{c} \tag{75}$$

$$G_{iy}(z) = a_{iy} + z^{-1}\mathbf{e}[\mathbf{I} - z^{-1}\mathbf{D}]^{-1}\mathbf{c}_{iy} \tag{76}$$

where a_{xi} and a_{iy} are scalars, $\mathbf{e}_{xi}$ is a row vector and $\mathbf{c}_{iy}$ is a column vector of length $2m+1$, in accordance with $[a_{xi} \quad \mathbf{e}_{xi}]$ is the i-th row of the matrix **T**, and $[a_{iy} \quad \mathbf{c}_{iy}^t]^t$ is the i-th column of the matrix **T**. Having the expressions for $H(e^{j\omega})$ and $\frac{dH(e^{j\omega})}{d\omega}$, the group delay can be obtained in accordance with Eqn. (72).

The passband and stopband weighting factors W_p and W_a are easily determined from user specifications. The group-delay weighting factor is set as

$$W_{gd} = \frac{\zeta \times fitness_{magnitude}}{fitness_{group-delay}} \tag{77}$$

where ζ is a fixed constant such that $0 < \zeta < 1$, and where $fitness_{magnitude}$ and $fitness_{group-delay}$ are obtained by examining the seed FRM digital filter particle. The weighting factor for the group-delay increases as $\zeta \to 1$.

Maximum Passband Ripple A_p	0.1[dB]
Minimum Stopband Loss A_a	40[dB]
Lower Stopband-Edge Normalized Frequency ω_{a_1}	0.31π[Rad]
Lower Passband-Edge Normalized Frequency ω_{p_1}	0.33π[Rad]
Upper Passband-Edge Normalized Frequency ω_{p_2}	0.60π[Rad]
Upper Stopband-Edge Normalized Frequency ω_{a_2}	0.61π[Rad]
Normalized Sampling Period T	1[s]
Lowpass Filter Interpolation Factor M_{lp}	6
Highpass Filter Interpolation Factor M_{hp}	5

Table 3. Design Specifications for Bandpass FRM IIR Digital Filter

K	w	c_1	c_2	ϑ_{min}	ϑ_{max}	L_f	L_h
700	0.4	2	2	−5	5	10	10

Table 4. PSO Design Parameters for Bandpass FRM IIR Digital Filter

L_0	l_0	f_0	L_1	l_1	f_1
11	3	10	12	3	7

Table 5. CSD Parameters for Bandpass FRM IIR Digital Filter

10. Application examples

10.1. Bandpass FRM IIR digital filter design example

Consider the design of a bandpass FRM IIR digital filter satisfying the magnitude response design specifications given in Table 3 over the CSD multiplier coefficient space.

The parameters for the PSO of bandpass FRM IIR digital filter is shown in Table 4 and the CSD parameters are presented in Table 5.

Given the design specification in Table 3, The order of the digital allpass networks $G_{0_{lp}}(z)$, $G_{1_{lp}}(z)$, $G_{0_{hp}}(z)$ and $G_{1_{hp}}(z)$ are found to be 3, 4, 3 and 4, respectively. In addition, the digital masking subfilters $F_{0_{lp}}(z)$, $F_{1_{lp}}(z)$, $F_{0_{hp}}(z)$ and $F_{1_{hp}}(z)$ have the same length as the previous example, i.e. 24, 42, 25 and 35 respectively, resulting in $N = 140$. In this example a set of fifteen CSD LUTs are required, fourteen LUTs for the multiplier coefficients $m_{C_{0,1}}$, $m_{C_{0,2}}$, $m_{C_{0,3}}$, $m_{L_{0,2}}$, $m_{L_{0,3}}$, $m_{C_{1,1}}$, $m_{L_{1,1}}$, $m_{C_{1,2}}$ and $m_{L_{1,2}}$ constituent in the digital allpass networks $G_{0_{lp}}(z)$, $G_{1_{lp}}(z)$, $G_{0_{hp}}(z)$ and $G_{1_{hp}}(z)$, and one template LUT for all the multiplier coefficients constituent in the masking digital subfilters $F_{0_{lp}}(z)$, $F_{1_{lp}}(z)$, $F_{0_{hp}}(z)$ and $F_{1_{hp}}(z)$.

Finally, by using Parks McClellan approach, the subfilters $F_{0_{lp}}(z)$, $F_{1_{lp}}(z)$, $F_{0_{hp}}(z)$ and $F_{1_{hp}}(z)$ can be designed. Also, by using the EMQF technique, the digital allpass networks $G_{0_{lp}}(z)$, $G_{1_{lp}}(z)$, $G_{0_{hp}}(z)$ and $G_{1_{hp}}(z)$ can be designed. Consequently, the magnitude and group delay frequency responses of the overall infinite-precision bandpass FRM IIR digital filter $H(z)$ is obtained as shown in Figs. 13 and 14.

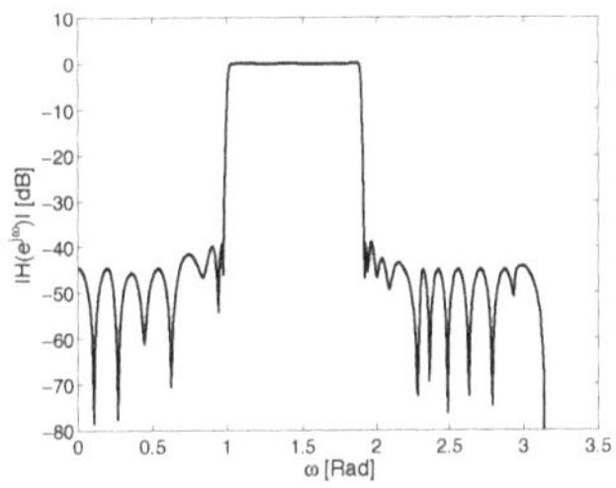

Figure 13. Magnitude Frequency-Response of the Overall Infinite-Precision Bandpass FRM IIR Digital Filter $H(e^{j\omega})$

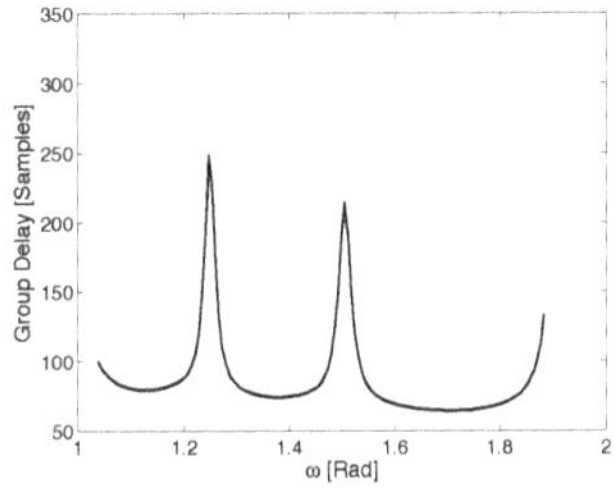

Figure 14. Group Delay Frequency-Response of the Overall Infinite-Precision Bandpass FRM IIR Digital Filter $H(e^{j\omega})$

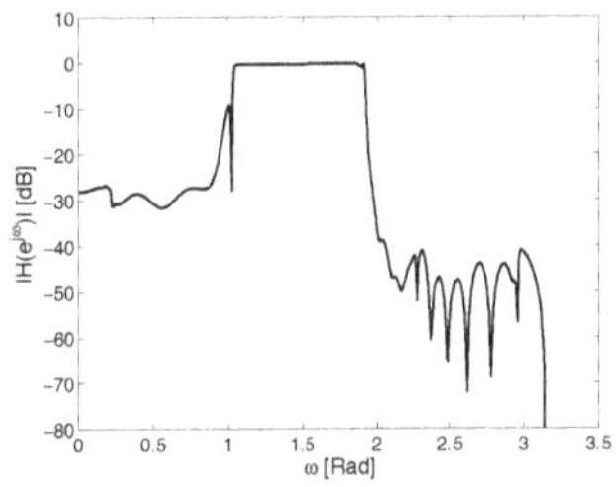

Figure 15. Magnitude Frequency-Response of the Overall CSD Bandpass FRM IIR Digital Filter $H(e^{j\omega})$ Before PSO

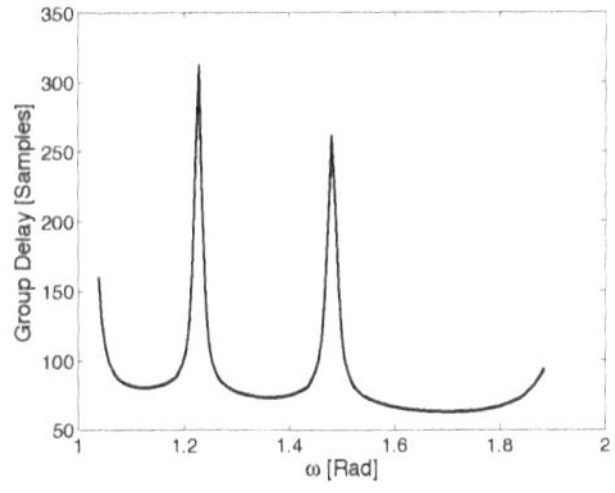

Figure 16. Group Delay Frequency-Response of the Overall CSD Bandpass FRM IIR Digital Filter $H(e^{j\omega})$ Before PSO

Multiplier	CSD Representation	Decimal Value
$m_{C_{0,1}}$	$00001.000\bar{1}00\bar{1}$	0.9297
$m_{C_{0,2}}$	$00010.000\bar{1}0\bar{1}0$	1.9219
$m_{L_{0,2}}$	00000.1000010	0.5156
$m_{C_{1,1}}$	$00001.00\bar{1}00\bar{1}0$	0.8594
$m_{C_{1,2}}$	$10000.\bar{1}0\bar{1}0000$	15.375
$m_{L_{1,1}}$	$00001.000\bar{1}0\bar{1}0$	0.9219
$m_{L_{1,2}}$	$00000.0010\bar{1}0\bar{1}$	0.0859

Table 6. Digital Multiplier Values for the Lowpass Section of the Bandpass FRM IIR Digital Filter

Based on the infinite-precision bandpass FRM IIR digital filter, the corresponding CSD FRM IIR initial digital filter is obtained to have a magnitude and group delay frequency responses as shown in Figs. 15 and 16.

By applying the proposed PSO to the initial FRM IIR digital filter and after about 160 iterations, the PSO converges to the optimal bandpass FRM IIR digital filter having a magnitude frequency response as shown in Fig. 17. In addition, Fig. 18 gives us a closer look to the magnitude frequency response of the passband region of the bandpass FRM IIR digital filter. Fig. 19 illustrates the group delay frequency response of the optimized bandpass FRM IIR digital filter. The values of the multiplier coefficients for the lowpass and highpass sections of the bandpass FRM IIR digital filter are obtained as summarized in Tables 6 and 7.

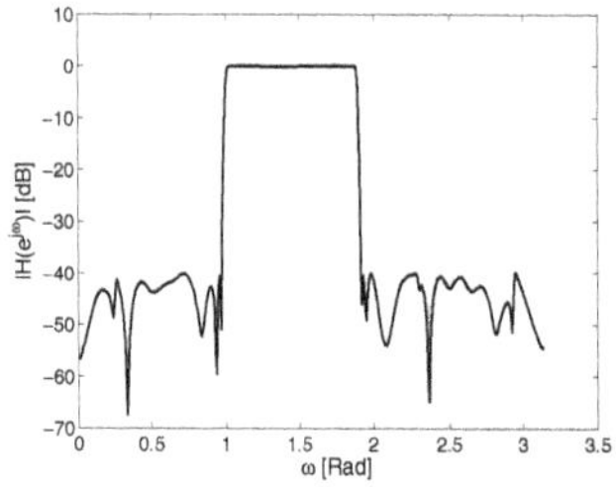

Figure 17. Magnitude Frequency-Response of the Overall CSD Bandpass FRM IIR Digital Filter $H(e^{j\omega})$ After PSO

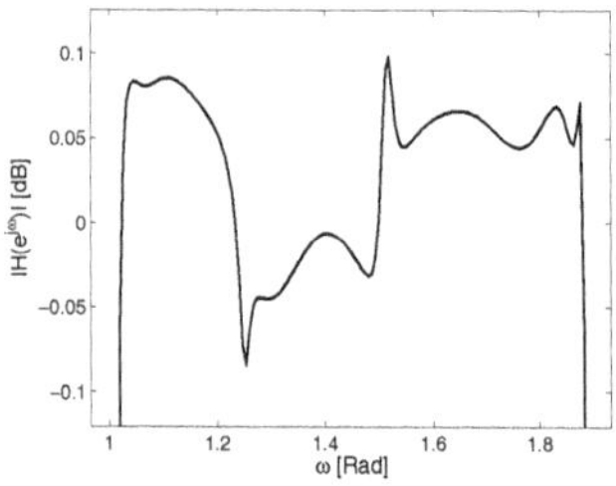

Figure 18. Magnitude Frequency-Response of the Passband Region of the Overall CSD Bandpass FRM IIR Digital Filter $H(e^{j\omega})$ After PSO

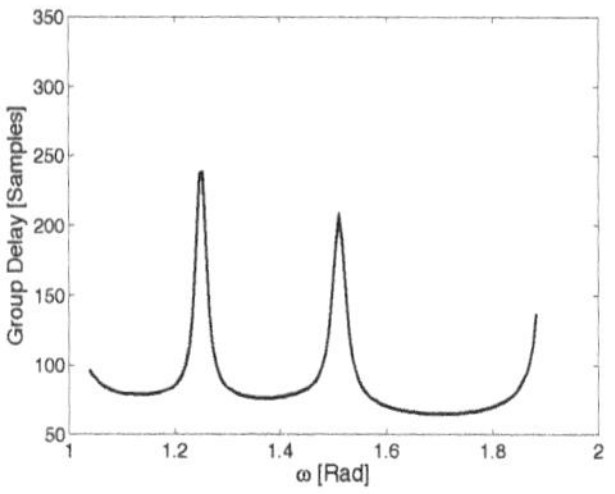

Figure 19. Group Delay Frequency-Response of the Overall CSD Bandpass FRM IIR Digital Filter $H(e^{j\omega})$ After PSO

Multiplier	CSD Representation	Decimal Value
$m_{C_{0,1}}$	$00001.00\bar{1}0\bar{1}00$	0.8438
$m_{C_{0,2}}$	$00010.000100\bar{1}$	2.0547
$m_{L_{0,2}}$	$00000.100000\bar{1}$	0.4922
$m_{C_{1,1}}$	$00001.0\bar{1}00010$	0.7656
$m_{C_{1,2}}$	10000.0100001	16.2578
$m_{L_{1,1}}$	$00001.0000\bar{1}01$	0.9766
$m_{L_{1,2}}$	$00000.0010\bar{1}0\bar{1}$	0.0859

Table 7. Digital Multiplier Values for the Highpass Section of the Bandpass FRM IIR Digital Filter

Frequency-Response Characteristic	Before PSO	After PSO
Maximum Passband Ripple A_p	0.8982[dB]	0.0978[dB]
Minimum Stopband Loss A_a	9.1715[dB]	40.0172[dB]
Maximum Group Delay	312[Samples]	239[Samples]

Table 8. Frequency-Response Analysis of the CSD Bandpass FRM IIR Digital Filter Before and After PSO

Table 8 represents the comparison of the CSD bandpass FRM IIR digital filters before and after PSO.

Author details

Seyyed Ali Hashemi and Behrouz Nowrouzian

Department of Electrical and Computer Engineering, University of Alberta, Edmonton, Alberta, Canada

References

[1] D. R. Wilson, D. R. Corrall, and R. F. Mathias, "The Design and Application of Digital Filters," *IEEE Transactions on Industrial Electronics and Control Instrumentation*, vol. IECI-20, pp. 68–74, 1973.

[2] P. P. Vaidyanathan, "Multirate digital filters, filter banks, polyphase networks, and applications: a tutorial," *Proceedings of the IEEE*, vol. 78, pp. 56–93, 1990.

[3] Y. C. Lim, "Frequency-Response Masking Approach for the Synthesis of Sharp Linear Phase Digital Filters," *IEEE Transactions on Circuits and Systems*, vol. 33, no. 4, pp. 357–364, 1986.

[4] ——, "A Digital Filter Bank for Digital Audio Systems," *IEEE Transactions on Circuits and Systems*, vol. 33-8, p. 848 âĂŞ 849, Aug. 1986.

[5] Y. C. Lim, S. R. Parker, and A. G. Constantinides, "Finite Word Length FIR Filter Design Using Integer Programming over a Discrete Coefficient Space," *IEEE Transactions on Acoustics, Speech and Signal Processing*, vol. 30, pp. 661–664, 1982.

[6] T. Saramaki and Y. C. Lim, "Use of the Remez Algorithm for Designing FIR Filters Utilizing the Frequency-Response Masking Approach," in *1999 IEEE International Symposium on Circuits and Systems. ISCAS '99*, vol. 3, 1999, pp. 449–455.

[7] Y. J. Yu and Y. C. Lim, "FRM Based FIR Filter Design - the WLS Approach," in *2002 IEEE International Symposium on Circuits and Systems. ISCAS 2002*, vol. 3, 2002, pp. III–221 – III–224.

[8] W.-S. Lu and T. Hinamoto, "Optimal Design of Frequency-Response-Masking Filters Using Semidefinite Programming," *IEEE Transactions on Circuits and Systems I: Fundamental Theory and Applications*, vol. 50, pp. 557–568, 2003.

[9] ——, "Optimal Design of FIR Frequency-Response-Masking Filters Using Second-Order Cone Programming," in *Proceedings of 2003 IEEE International Symposium on Circuits and Systems. ISCAS '03*, vol. 3, 2003, pp. III–878 – III–881.

[10] L. Cen and Y. Lian, "Hybrid Genetic Algorithm for the Design of Modified Frequency-Response Masking Filters in a Discrete Space," *Circuits, Systems, and Signal Processing*, vol. 25, pp. 153–174, April 2006.

[11] S. Chen, R. H. Istepanian, and B. L. Luk, "Digital IIR Filter Design Using Adaptive Simulated Annealing," *Digital Signal Processing*, vol. 11, no. 3, pp. 241–251, July 2001.

[12] K. sang Tang, K. fung Man, S. Kwong, and Z. feng Liu, "Design and Optimization of IIR Filter Structure Using Hierarchical Genetic Algorithms," *IEEE Transactions on Industrial Electronics*, vol. 45, no. 3, pp. 481–487, 1998.

[13] C. Dai, W. Chen, and Y. Zhu, "Seeker Optimization Algorithm for Digital IIR Filter Design," *IEEE Transactions on Industrial Electronics*, vol. 57, no. 5, pp. 1710–1718, May 2010.

[14] M. Najjarzadeh and A. Ayatollahi, "FIR Digital Filters Design: Particle Swarm Optimization Utilizing LMS and Minimax Strategies," in *IEEE International Symposium on Signal Processing and Information Technology, 2008. ISSPIT 2008.*, 2008, pp. 129–132.

[15] C. Dai, W. Chen, Y. Zhu, and X. Zhang, "Seeker Optimization Algorithm for Optimal Reactive Power Dispatch," *IEEE Transactions on Power Systems*, vol. 24, no. 3, pp. 1218–1231, August 2009.

[16] A. Kalinli and N. Karaboga, "A New Method for Adaptive IIR Filter Design Based on Tabu Search Algorithm," *AEU - International Journal of Electronics and Communications*, vol. 59, no. 3, pp. 111–117, May 2005.

[17] N. Karaboga, A. Kalinli, and D. Karaboga, "Designing Digital IIR Filters Using Ant Colony Optimisation Algorithm," *Engineering Applications of Artificial Intelligence*, vol. 17, no. 3, pp. 301–309, April 2004.

[18] A. Kalinli and N. Karaboga, "Artificial Immune Algorithm for IIR Filter Design," *Engineering Applications of Artificial Intelligence*, vol. 18, no. 5, pp. 919–929, Dec 2005.

[19] R. Storn, "Designing Nonstandard Filters with Differential Evolution," *IEEE Signal Processing Magazine*, vol. 22, no. 1, pp. 103–106, Jan 2005.

[20] N. Karaboga, "Digital IIR Filter Design Using Differential Evolution Algorithm," *EURASIP Journal on Applied Signal Processing*, vol. 2005, no. 8, pp. 1269–1276, Jan 2005.

[21] P. Mercier, S. M. Kilambi, and B. Nowrouzian, "Optimization of FRM FIR Digital Filters Over CSD and CDBNS Multiplier Coefficient Spaces Employing a Novel Genetic Algorithm," *Journal of Computers*, vol. 2, no. 7, pp. 20–31, Sept. 2007.

[22] S. Bokhari and B. Nowrouzian, "DCGA Optimization of Lowpass FRM IIR Digital Filters Over CSD Multiplier Coefficient Space," in *52nd IEEE International Midwest Symposium on Circuits and Systems*, August 2009, pp. 573–576.

[23] S. Bokhari, B. Nowrouzian, and S. A. Hashemi, "A novel technique for DCGA optimization of guaranteed BIBO stable IIR-based FRM digital filters over the CSD multiplier coefficient space," in *proceedings of 2010 IEEE International Symposium on Circuits and Systems (ISCAS)*, 2010, pp. 2710–2713.

[24] S. A. Hashemi and B. Nowrouzian, "Particle swarm optimization of FRM FIR digital filters over the CSD multiplier coefficient space," in *proceedings of 53rd IEEE International Midwest Symposium on Circuits and Systems (MWSCAS)*, 2010, pp. 1246–1249.

[25] ——, "A novel discrete particle swarm optimization for FRM FIR digital filters," *Journal of Computers*, vol. 7, no. 7, July 2012.

[26] — —, "Discrete particle swarm optimization of magnitude response of iir-based frm digital filters," in *proceedings of 17th IEEE International Conference on Electronics, Circuits, and Systems, 2010. ICECS 2010.*, December 2010.

[27] — —, "A novel finite-wordlength particle swarm optimization technique for frm iir digital filters," in *proceedings of 2011 IEEE International Symposium on Circuits and Systems (ISCAS)*, May 2011, pp. 2745–2748.

[28] J. Kennedy and R. Eberhart, "Particle Swarm Optimization," in *Proceedings of IEEE International Conference on Neural Networks*, vol. 4, 1995, pp. 1942–1948.

[29] M. D. Lutovac and L. D. Milić, "IIR Filters Based on Frequency-Response Masking Approach," in *Telecommunications in Modern Satellite, Cable and Broadcasting Service, TELSIKS 2001*, Sept. 2001, pp. 163–170.

[30] H. Johansson and L. Wanhammar, "High-speed Recursive Filtering Using the Frequency-Response Masking Approach," in *Proceedings of the IEEE Int. Symposium on Circuits and Systems*, 1997, pp. 2208–2211.

[31] J. Sun, W. Fang, and W. Xu, "A Quantum-Behaved Particle Swarm Optimization With Diversity-Guided Mutation for the Design of Two-Dimensional IIR Digital Filters," *IEEE Transactions on Circuits and Systems II: Express Briefs*, vol. 57, no. 2, pp. 141–145, 2010.

[32] A. Slowik and M. Bialko, "Design and Optimization of IIR Digital Filters with Non-Standard Characteristics Using Particle Swarm Optimization Algorithm," in *14th IEEE International Conference on Electronics, Circuits and Systems, ICECS 2007*, 2007, pp. 162–165.

[33] B. Luitel and G. K. Venayagamoorthy, "Particle Swarm Optimization with Quantum Infusion for the design of digital filters," in *IEEE Swarm Intelligence Symposium, SIS 2008*, 2008, pp. 1–8.

[34] B. Nowrouzian and L. S. Lee, "Minimal Multiplier Realisation of Bilinear-LDI Digital Allpass Networks," in *IEE Proceedings on Devices and Systems, G Circuits*, vol. 136, Jun. 1989, pp. 114–117.

[35] T. Parks and J. McClellan, "Chebyshev Approximation for Nonrecursive Digital Filters with Linear Phase," *IEEE Transactions on Circuit Theory*, vol. CT-19, pp. 189–194, 1972.

[36] Y. C. Lim, R. Yang, D. Li, and J. Song, "Signed Power-of-Two Term Allocation Scheme for the Design of Digital Filters," *IEEE Transactions on Circuits and Systems II: Analog and Digital Signal Processing*, vol. 46, pp. 577–584, 1999.

[37] R. I. Hartley, "Subexpression Sharing in Filters Using Canonic Signed Digit Multipliers," *IEEE Transactions on Circuits and Systems II: Analog and Digital Signal Processing*, vol. 43, pp. 677–688, 1996.

[38] A. T. G. Fuller, B. Nowrouzian, and F. Ashrafzadeh, "Optimization of FIR Digital Filters over the Canonical Signed-Digit Coefficient Space Using Genetic Algorithms," in *1998 Midwest Symposium on Circuits and Systems*, 1998, pp. 456–459.

[39] R. Yang, Y. C. Lim, and S. R. Parker, "Design of sharp linear-phase FIR bandstop filters using the frequency-response-masking technique," *Circuits, Systems, and Signal Processing*, vol. 17, no. 1, pp. 1–27, Jan. 1998.

[40] A. Willson and H. Orchard, "Insights into Digital Filters Made as the Sum of Two Allpass Functions," *IEEE Trans. On Circuits And Syst.*, vol. 42, pp. 129–137, Mar. 1995.

[41] D. Rabrenovic and M. Lutovac, "Elliptic filters with minimal Q-factors," in *IEE Electronics Letters Online*, vol. 30, no. 3, Feb. 1994, pp. 206–207.

[42] L. D. Milić and M. D. Lutovac, "Design of Multiplierless Elliptic IIR Filters with a Small Quantization Error," *IEEE Transactions on Signal Processing*, vol. 47, no. 2, pp. 469–479, Feb. 1999.

[43] M. D. Lutovac and L. D. Milić, "Design of Computationally Efficient Elliptic IIR Filters with a Reduced Number of Shift-and-Add Operations in Multipliers," *IEEE Transactions on Signal Processing*, vol. 45, no. 7, pp. 2422–2430, Oct. 1997.

[44] B. Nowrouzian, "A Novel Approach to the Exact Design of LDI Symmetrical Digital and Switched-Capacitor Filters," in *Proceedings of 33rd Midwest Symposium on Circuits and Systems*, vol. 2, Aug. 1990, pp. 967–972.

[45] V. Valkenburg, *Introduction to Modern Network Synthesis*. John Wiley and Sons, Inc., 1965.

[46] F. van den Bergh and A. P. Engelbrecht, "A Cooperative Approach to Particle Swarm Optimization," *IEEE Transactions on Evolutionary Computation*, vol. 8, no. 3, pp. 225–239, 2004.

[47] A. Antoniou, *Digital Filters: Analysis, Design, and Applications*. McGraw Hill, Inc., 1993.

[48] R. E. Crochiere, "Digital Network Theory and its Application to the Analysis and Design of Digital Filters," Ph.D. dissertation, M.I.T, Dep. of Elec. Eng., M.I.T, Cambridge, MA, May 1974.

Chapter 11

Analytical Design of Two-Dimensional Filters and Applications in Biomedical Image Processing

Radu Matei and Daniela Matei

Additional information is available at the end of the chapter

1. Introduction

The field of two-dimensional filters and their design methods has known a large development due to its importance in image processing (Lim, 1990; Lu & Antoniou, 1992). There are methods based on numerical optimization and also analytical methods relying on 1D prototypes. A commonly-used design technique for 2D filters is to start from a specified 1D prototype filter and transform its transfer function using various frequency mappings in order to obtain a 2D filter with a desired frequency response. These are essentially spectral transformations from s to z plane, followed by z to (z_1, z_2) mappings, approached in early papers (Chakrabarti & Mitra, 1977; Hirano & Aggarwal, 1978; Harn & Shenoi, 1986; Nie & Unbehauen, 1989). Generally these transformations conserve stability, so from 1D prototypes various stable recursive 2D filters can be obtained. The most common types are directional, fan-shaped, diamond-shaped and circular filters. Diamond filters are commonly used as anti-aliasing filters in the conversion between signals sampled on the rectangular sampling grid and the quincunx sampling grid. Various design methods for diamond-shaped filters were studied in (Tosic, 1997; Lim & Low, 1997; Low & Lim, 1998; Ito, 2010; Matei, 2010).

There are several classes of filters with orientation-selective frequency response, useful in tasks like edge detection, motion analysis, texture segmentation etc. Some relevant papers on directional filters and their applications are (Danielsson, 1980; Paplinski, 1998; Austvoll, 2000). An important class of orientation-selective filters are steerable filters, synthesized as a linear combination of a set of basis filters (Freeman & Adelson, 1991) and steerable wedge filters (Simoncelli & Farid, 1996). A directional filter bank (DFB) for image decomposition in the frequency domain was proposed in (Bamberger, 1992). In (Qunshan & Swamy, 1994) various 2D recursive filters are approached. Fan-shaped, also known as wedge-shaped filters find interesting applications. Design methods for IIR and FIR fan filters are presented in some early papers (Kayran & King, 1983; Ansari, 1987). An efficient design method for recursive fan filters is presented in (Zhu & Zhenya, 1990). An implementation of recursive fan filters using all-pass sections is given in (Zhu & Nakamura, 1996). In (Mollova, 1997), an analytical least-squares technique for FIR filters, in particular fan-type, is proposed. Design methods for efficient 2D FIR filters were treated in papers like (Zhu et al., 1999; Zhu et al., 2006). Zero-phase filters were studied as well (Psarakis, 1990). Different types of 2D filters derived from 1D prototypes through spectral transformations were treated in (Matei, 2011a).

We propose in this chapter some new design procedures for particular classes of 2D filters; the described methods are mainly analytical but also include numerical approximations. Various types of 2D filters will be approached, both recursive (IIR) and non-recursive (FIR). The design methods will focus however on recursive filters, since they are the most efficient.

The proposed design methods start from either digital or analog 1D prototypes with a desired characteristic. In this chapter we will mainly use analog prototypes, since the design turns out to be simpler and the 2D filters result of lower complexity. This analog prototype filter is described by a transfer function in the complex variable s, which can be factorized as a product of elementary functions of first or second order. The prototype transfer function results from an usual approximation (Butterworth, Chebyshev, elliptic) and the shape of the frequency response corresponds to the desired characteristic of the 2D filter.

The next design stage consists in finding the specific complex frequency transformation from the axis s to the complex plane (z_1, z_2), of the general form $F:\mathbb{C}\to\mathbb{C}^2$, $s\to F(z_1, z_2)$. This mapping will be determined for each type of 2D filter separately, starting from the geometrical specification of its shape in the frequency plane. Once found this particular mapping, the 2D filter function results directly by applying this transformation to each factor function of the prototype. Thus, the 2D filter transfer function $H(z_1, z_2)$ results directly factorized, which is a major advantage in its implementation. The proposed design method applies the bilinear transform as an intermediate step in determining the 1D to 2D frequency mapping. All the proposed design techniques are mainly analytical but also involve numerical optimization, in particular rational approximations (e.g. Chebyshev-Padé). Some of the designed 2D filters result with complex coefficients. This should not be a serious shortcoming, since such IIR filters are also used (Nikolova et al., 2011).

In this chapter we will approach two main classes of 2D filters. The first one comprises three types of orientation-selective filters, as follows: square-shaped (diamond-type) IIR filters, with arbitrary orientation in the frequency plane; fan-type IIR filters with specified orientation and aperture angles; and very selective IIR multi-directional filters (in particular two-directional and three-directional), which are useful in detecting and extracting simultaneously lines with different orientations from an image.

The other class discussed here refers to FIR filters. From this category we will approach zero-phase filters with circular frequency response. Zero-phase filters, with real transfer functions, are often used in image processing since they do not introduce any phase distortions. All these types of 2D filters are analyzed in detail in the following sections.

Stability of the two-dimensional recursive filters is also an important issue and is much more complicated than for 1D filters. For 2D filters, in general, it is quite difficult to take stability constraints into account during approximation stage (O'Connor & Huang, 1978). Therefore, various techniques were developed to separate stability from approximation. If the designed filter becomes unstable, some stabilization procedures are needed (Jury et al., 1977). Various stability conditions for 2D filters have been found (Mastorakis, 2000).

The medical image processing field has known a rapid development due to imaging value in assisting and assessing clinical diagnosis (Semmlow, 2004; Berry, 2007; Dougherty, 2011). In particular, the currently used vascular imaging technique is x-ray angiography, mainly in diagnosing cardio-vascular pathologies. A frequent application of cardiac imaging is the localization of narrowed or blocked coronary arteries. Fluorescein angiography is the best technique to view the retinal circulation and is useful for diagnosing retinal or optic nerve condition and assessing disorders like diabetic retinopathy, macular degeneration, retinal vein occlusions etc. There are many papers approaching various methods and techniques aiming at improving angiogram images. In papers like (Frangi et al., 1998) the multiscale analysis is used, with the purpose of vessel enhancement and detection. Usual approaches include Hessian-based filtering, based on the multiscale local structure of an image and directional features of vessels (Truc et al., 2007). In cardio-vascular imaging, an essential pre-processing task is the enhancement of coronary arterial tree, commonly using gradient or other local operators. In (Khan et al., 2004) a decimation-free directional filter bank is used. An adaptive vessel detection scheme is proposed in (Wu et al., 2006) based on Gabor filter response. Filtering is an elementary operation in low level computer vision and a pre-processing stage in many biomedical image processing applications. Some edge-preserving filtering techniques for biomedical image smoothing have been proposed (Rydell et al., 2008; Wong et al., 2004). At the end of this chapter some simulation results are given for biomedical image filtering using some of the proposed 2D filters, namely the directional narrow fan-filter with specified orientation and the zero-phase circular filter.

2. Analog and digital 1D prototype filters used in 2D filter design

This section presents the types of analog and digital 1D recursive prototype filters which will be further used to derive the desired 2D filter characteristics. An analog IIR prototype filter of order N has a transfer function in variable s of the general form:

$$H_P(s) = \frac{P(s)}{Q(s)} = \sum_{i=0}^{M} p_i \cdot s^i \Big/ \sum_{j=0}^{N} q_j \cdot s^j \tag{1}$$

This general transfer function can be factorized into simpler rational functions of first and second order. Such a second-order rational function (biquad) can be written:

$$H_b(s) = k\left(s^2 + b_1 s + b_0\right)/\left(s^2 + a_1 s + a_0\right) \quad (2)$$

where generally the second-order polynomials at numerator and denominator have complex-conjugated roots, and k is a constant. For typical approximations – Chebyshev or elliptic – usually $b_1=0$, therefore the numerator has imaginary zeros. For odd-order filters, the denominator contains at least a first-order factor $(s+\alpha)$. An elliptic approximation with very low ripple can be used for an almost maximally-flat low-order filter. Next we consider two such low-pass (LP) prototypes with imposed specifications. The first is an elliptic LP analog filter of order $N=6$, cutoff frequency $\omega_c=0.4\pi$, peak-to-peak ripple $R_p=0.04$dB, stop-band attenuation $R_s=38$dB. Its transfer function can be factorized into three biquad functions like (2): $H_P(s)=k \cdot H_{b1}(s) \cdot H_{b2}(s) \cdot H_{b3}(s)$ where $k=2.375$ and:

$$H_{b1}(s) = (s^2 + 39.195)/(s^2 + 0.2221s + 2.8797) \quad (3)$$

$$H_{b2}(s) = (s^2 + 6.5057)/(s^2 + 0.9172s + 2.4291) \quad (4)$$

$$H_{b3}(s) = (s^2 + 4.2217)/(s^2 + 2.0448s + 1.5454) \quad (5)$$

The frequency response magnitude of this LP filter for $\omega \in [-\pi, \pi]$ is shown in Fig. 1(a).

The second prototype is an elliptic LP analog filter with parameters: $N=4$, $\omega_c=0.4\pi$, $R_p=0.05$ db,$R_s=36$db. Its transfer function is written as a product of two biquad functions like (2): $H_P(s)=k \cdot H_{b1}(s) \cdot H_{b2}(s)$, where $k \cong 0.01$ and

$$H_{b1}(s) = (s^2 + 33.385)/(s^2 + 0.5894s + 2.2398) \quad (6)$$

$$H_{b2}(s) = (s^2 + 6.4226)/(s^2 + 1.9691s + 1.5266) \quad (7)$$

The frequency response magnitude of this LP filter for $\omega \in [-\pi, \pi]$ is shown in Fig.1 (b). The simplest analog LP filter has a transfer function $H_j(s)=\alpha/(s+\alpha)$, where the value α gives the selectivity (Fig.1(c)). If the filter characteristic is shifted to a given frequency $\omega_{01} \in [-\pi, \pi]$, the transfer function becomes:

$$H_{jS}(s) = \alpha/(s + \alpha + j \cdot \omega_{01}) \quad (8)$$

In Fig.1 (d) the shifted filter response magnitude for ω_{01}=0.416π is shown. Another useful analog prototype is the selective second-order (resonant) filter with central frequency ω_0:

$$H_r(s) = \alpha s \big/ \left(s^2 + \alpha s + \omega_0^2\right) \tag{9}$$

The transfer function magnitude for such a filter with α=0.1 and ω_0=1.3 is shown in Fig. 1 (e). This will be further used as a prototype for two-directional filters.

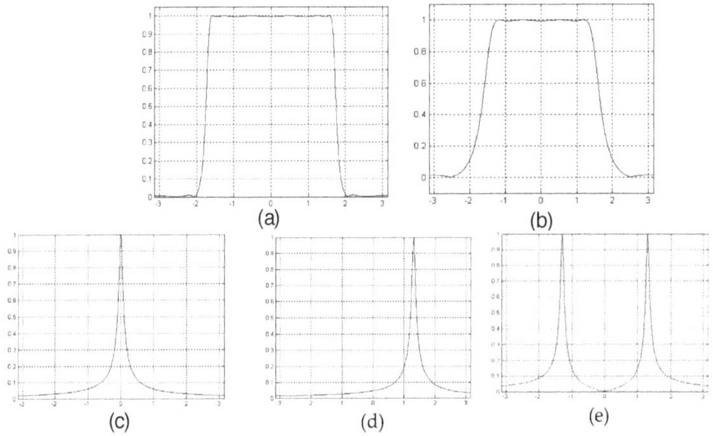

Figure 1. Frequency response magnitudes of: (a) LP elliptic prototype of order 6; (b) LP elliptic prototype of order 4; very selective first-order filter with central frequencies ω_0=0 (c) and ω_0=0.416π (d); (e) selective band-pass filter with ω_0=1.3

A useful zero-phase prototype can be obtained from the general function (1) by preserving only the magnitude characteristics of the 1D filter; this prototype will be further used to design 2D zero-phase FIR filters of different types, specifically circular filters, with real-valued transfer functions. In order to obtain a zero-phase filter, we consider the magnitude characteristics of $H_P(j\omega)$, defined by the absolute value $|H_P(j\omega)| = |P(j\omega)| / |Q(j\omega)|$. We look for a series expansion of the magnitude $|H_P(j\omega)|$ that has to be an approximation as accurate as possible on the frequency domain $[-\pi, \pi]$. The most convenient for our purpose is the Chebyshev series expansion, because it yields an efficient approximation of a given function, which is uniform along the desired interval. The Chebyshev series in powers of the frequency variable ω for a given function on a specified interval can be easily found using a symbolic computation software like MAPLE. However, we will finally need a trigonometric expansion of $|H_P(j\omega)|$, namely in cos($n\omega$), rather than a polynomial expansion in powers of ω. Therefore, prior to Chebyshev series calculation, we apply the change of variable:

$$\omega = \arccos(x) \Leftrightarrow x = \cos(\omega) \tag{10}$$

and so we get the polynomial expansion in variable x:

$$\left|H_P(\arccos(x))\right| \cong \sum_{n=0}^{N} a_n \cdot x^n = a_0 + a_0 x + a_2 x^2 + a_3 x^3 + \ldots + a_N x^N \tag{11}$$

where the number of terms N is chosen large enough to ensure the desired precision. The next step is to substitute back $x = \cos\omega$ in the polynomial expression (11), therefore we obtain the factorized function in $\cos\omega$, with $n + 2m = N$:

$$\left|H_P(\omega)\right| \cong \sum_{n=0}^{N} a_n \cdot \cos^n(\omega) = k \cdot \prod_{i=1}^{n} (\cos\omega + a_i) \cdot \prod_{j=1}^{m} (\cos^2 \omega + a_{1j} \cos\omega + a_{2j}) \tag{12}$$

Next let us consider a recursive digital filter of order N with the transfer function:

$$H_P(z) = \frac{P(z)}{Q(z)} = \sum_{i=0}^{M} p_i \cdot z^i \Big/ \sum_{j=0}^{N} q_j \cdot z^j \tag{13}$$

This general transfer function with $M = N$ can be factorized into first and second order rational functions. For an odd order filter, $H_P(z)$ has at least one first-order factor:

$$H_1(z) = \left(b_1 z + b_0\right) / \left(z + a_0\right) \tag{14}$$

The transfer function also contains second-order (biquad) functions, where in general the numerator and denominator polynomials have complex-conjugated roots:

$$H_2(z) = \left(b_2 z^2 + b_1 z + b_0\right) / \left(z^2 + a_1 z + a_0\right) \tag{15}$$

We will further use the term *template*, common in the field of cellular neural networks, for the coefficient matrices of the numerator and denominator of a 2D transfer function $H(z_1, z_2)$.

3. Diamond-type recursive filters

In this section a design method is proposed for 2D square-shaped (diamond-type) IIR filters. The design relies on an analog 1D maximally-flat low-pass prototype filter. To this filter a frequency transformation is applied, which yields a 2D filter with the desired square shape in the frequency plane. The proposed method combines the analytical approach with numerical approximations.

3.1. Specification of diamond-type filters in the frequency plane

The standard diamond filter has the shape in the frequency plane as shown in Fig.2 (a). It is a square with a side length of $\pi\sqrt{2}$, while its axis is tilted by an angle of $\varphi = \pi/4$ radians about the two frequency axes. Next we will consider the orientation angle φ about the ω_2- (vertical) axis. In this chapter a more general case is approached, i.e. a 2D diamond-type filter with a square shape in the frequency plane, but with arbitrary axis orientation angle, as shown in Fig. 2(e). Next we refer to them as diamond-type filters, since they are more general than the diamond filter from Fig. 2 (a).

The diamond-type filter in Fig.2 (e) is derived as the intersection of two oriented low-pass filters whose axes are perpendicular to each other, for which the shape in the frequency plane is given in Fig.2 (c), (d). Correspondingly, the diamond-type filter transfer function $H_D(z_1, z_2)$ results as a product of two partial transfer functions:

$$H_D(z_1, z_2) = H_1(z_1, z_2) \cdot H_2(z_1, z_2) \quad (16)$$

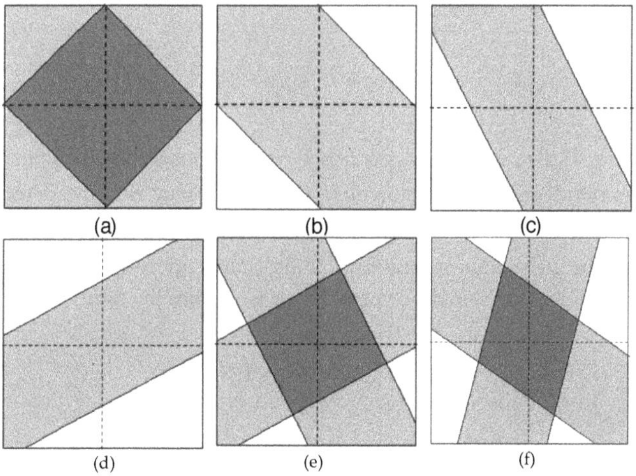

Figure 2. (a) diamond filter; (b) wide-band oriented filter; (c), (d) wide-band oriented filters with orientations forming an angle $\varphi = \pi/2$; (e) square-shaped filter resulted as product of the above oriented filters; (f) rhomboidal filter

The frequency characteristic of $H_2(z_1, z_2)$ is ideally identical to the frequency characteristic of $H_1(z_1, z_2)$ rotated by an angle of $\varphi = \pi/2$. Since this rotation of axes implies the frequency variable change: $\omega_1 \to \omega_2$, $\omega_2 \to -\omega_1$, the transfer function $H_2(z_1, z_2)$ can be derived from $H_1(z_1, z_2)$ as $H_2(z_1, z_2)=H_1(z_2, z_1^{-1})$. A more general filter belonging to this class is a rhomboidal filter, as shown in Fig.2 (f). In this case the two oriented LP filters may have different bandwidths and their axes are no longer perpendicular to each other.

3.2. Design method for diamond-type filters

The issue of this section is to find the transfer function $H_{2D}(z_1, z_2)$ of the desired 2D filter using a complex frequency transformation $s \to F(z_1, z_2)$. From a prototype $H_P(s)=H_P(j\omega)$ (which varies on one axis only), a 2D oriented filter is obtained by rotating the axes of the plane (ω_1, ω_2) by an angle φ. The rotation is defined by the following linear transformation, where ω_1, ω_2 are the original frequency variables and $\bar{\omega}_1$, $\bar{\omega}_2$ the rotated ones:

$$\begin{bmatrix} \omega_1 \\ \omega_2 \end{bmatrix} = \begin{bmatrix} \cos\varphi & \sin\varphi \\ -\sin\varphi & \cos\varphi \end{bmatrix} \cdot \begin{bmatrix} \bar{\omega}_1 \\ \bar{\omega}_2 \end{bmatrix} \tag{17}$$

The spatial orientation is specified by an angle φ with respect to ω_1-axis, defined by the 1D to 2D frequency mapping $\omega \to \omega_1\cos\varphi + \omega_2\sin\varphi$. By substitution, we obtain the oriented filter transfer function $H_\varphi(\omega_1, \omega_2)=H_P(\omega_1\cos\varphi + \omega_2\sin\varphi)$. In the complex plane (s_1, s_2) the above frequency transformation becomes:

$$s \to s_1\cos\varphi + s_2\sin\varphi \tag{18}$$

The oriented filter $H_\varphi(\omega_1, \omega_2)$ has the frequency response magnitude section along the line $\omega_1\cos\varphi + \omega_2\sin\varphi=0$, identical with prototype $H_P(\omega)$, and constant along the perpendicular line (filter longitudinal axis) $\omega_1\sin\varphi-\omega_2\cos\varphi=0$. The usual method to obtain a discrete filter from an analog prototype is the bilinear transform. If the sample interval takes the value $T=1$, the bilinear transform for s_1 and s_2 in the complex plane (s_1, s_2) has the form:

$$s_1 = 2(z_1-1)/(z_1+1) \quad s_2 = 2(z_2-1)/(z_2+1) \tag{19}$$

This method is straightforward, still the resulted 2D filter will present linearity distortions in its shape, which increase towards the limits of the frequency plane as compared to the ideal frequency response. This is mainly due to the so-called frequency warping effect of the bilinear transform, expressed by the continuous to discrete frequency mapping:

$$\omega = (2/T)\cdot \operatorname{arctg}\left(\omega_a T/2\right) \tag{20}$$

where ω is a frequency of the discrete filter and ω_a is the corresponding frequency of the analog filter. This error can be corrected by applying a pre-warping. Taking $T = 1$ in (20), we substitute the mappings:

$$\omega_1 \to 2 \cdot \mathrm{arctg}(\omega_1/2) \qquad \omega_2 \to 2 \cdot \mathrm{arctg}(\omega_2/2) \tag{21}$$

In order to include the nonlinear mappings (21) into the frequency transformation, a rational approximation is needed. One of the most efficient is Chebyshev-Padé, which gives uniform approximation over a specified range. We get the accurate approximation for $\omega \in [-\pi, \pi]$:

$$\mathrm{arctg}(\omega/2) \cong 0.4751 \cdot \omega \big/ \left(1 + 0.05 \cdot \omega^2\right) \tag{22}$$

Substituting the nonlinear mappings (21) with approximate expression (22) into (18) we get the 1D to 2D mapping which includes the pre-warping along both frequency axes:

$$s \to F_\varphi(s_1, s_2) = 0.95 \cdot \left[\frac{s_1 \cos\varphi}{1 - 0.05 \cdot s_1^2} + \frac{s_2 \sin\varphi}{1 - 0.05 \cdot s_2^2}\right] \tag{23}$$

Applying the bilinear transform (19) along the two axes we obtain the mapping $s \to F_\varphi(z_1, z_2)$ in matrix form, where $\mathbf{z_1} = [1 \;\; z_1 \;\; z_1^2]$ and $\mathbf{z_2} = [1 \;\; z_2 \;\; z_2^2]$:

$$s \to F_\varphi(z_1, z_2) = k \cdot M_\varphi(z_1, z_2) / N_\varphi(z_1, z_2) = k \cdot (\mathbf{z_1} \times \mathbf{M}_\varphi \times \mathbf{z_2}^T) / (\mathbf{z_1} \times \mathbf{N}_\varphi \times \mathbf{z_2}^T) \tag{24}$$

Here $k = 1.5233$ and the matrices $\boldsymbol{M}_\varphi$ and $\mathbf{N}_\varphi$ of size 3×3 are given by:

$$\mathbf{M}_\varphi = \cos\varphi \cdot \begin{bmatrix} -1 & -3 & -1 \\ 0 & 0 & 0 \\ 1 & 3 & 1 \end{bmatrix} + \sin\varphi \cdot \begin{bmatrix} -1 & 0 & 1 \\ -3 & 0 & 3 \\ -1 & 0 & 1 \end{bmatrix} \quad \mathbf{N}_\varphi = \begin{bmatrix} 1 & 3 & 1 \\ 3 & 9 & 3 \\ 1 & 3 & 1 \end{bmatrix} \tag{25}$$

Substituting the mapping (24) into the expression (2) of the biquad transfer function $H_b(s)$ with $b_1 = 0$, we get the 2D transfer function $H_B(z_1, z_2)$ in the matrix form:

$$H_B(z_1, z_2) = \left(\mathbf{z_1} \times \mathbf{B_1} \times \mathbf{z_2}^T\right) \big/ \left(\mathbf{z_1} \times \mathbf{A_1} \times \mathbf{z_2}^T\right) \tag{26}$$

where $\mathbf{z_1}$ and $\mathbf{z_2}$ are the vectors:

$$\mathbf{z}_1 = \begin{bmatrix} 1 & z_1 & z_1^2 & z_1^3 & z_1^4 \end{bmatrix} \quad \mathbf{z}_2 = \begin{bmatrix} 1 & z_2 & z_2^2 & z_2^3 & z_2^4 \end{bmatrix} \tag{27}$$

and the 5×5 templates B_1, A_1 are given by the expressions:

$$\mathbf{B}_1 = k^2 \cdot \mathbf{M}_\varphi * \mathbf{M}_\varphi + b_0 \cdot \mathbf{N}_\varphi * \mathbf{N}_\varphi\ ; \quad \mathbf{A}_1 = k^2 \cdot \mathbf{M}_\varphi * \mathbf{M}_\varphi + a_1 \cdot k \cdot \mathbf{M}_\varphi * \mathbf{N}_\varphi + a_0 \cdot \mathbf{N}_\varphi * \mathbf{N}_\varphi \tag{28}$$

For instance, corresponding to the third biquad function $H_{b3}(s)$ given by (5), the following 5×5 templates result according to the expressions (28):

$$B_1 = \begin{bmatrix} 0.2464 & 0.9407 & 1.1418 & 0.4027 & 0.0671 \\ 0.9407 & 3.2233 & 3.8917 & 1.6092 & 0.4027 \\ 1.1418 & 3.8917 & 6.1484 & 3.8917 & 1.1418 \\ 0.4027 & 1.6092 & 3.8917 & 3.2233 & 0.9407 \\ 0.0671 & 0.4027 & 1.1418 & 0.9407 & 0.2464 \end{bmatrix}$$

$$A_1 = \begin{bmatrix} 0.0947 & 0.1941 & 0.0732 & -0.0163 & 0.0245 \\ 0.1941 & -0.2738 & -0.7181 & 0.0774 & 0.3112 \\ 0.0732 & -0.7181 & 1.0000 & 2.8851 & 1.2743 \\ -0.0163 & 0.0774 & 2.8851 & 3.6570 & 1.1768 \\ 0.0245 & 0.3112 & 1.2743 & 1.1768 & 0.3131 \end{bmatrix}$$

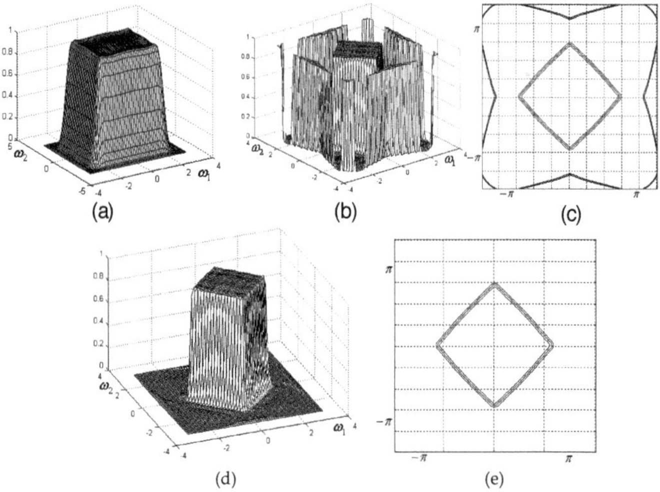

Figure 3. (a) LP correction filter characteristic; frequency response magnitudes and contour plots of: (b), (c) uncorrected diamond-type filter; (d), (e) corrected diamond-type filter

The characteristics of a diamond-type filter with orientation angle $\varphi=\pi/4$ and based on the prototype filter of order 6 given by the factors (3)-(5) is shown in Fig.3 (b), (c). As can be noticed, the filter characteristic corrected by pre-warping has a good linearity, however it still twists towards the margins of the frequency plane. These marginal linearity distortions can be corrected using an additional LP filter. For instance, we can choose as prototype an 1D elliptic digital filter of order $N=3$, pass-band ripple $R_p=0.1$dB, stop-band attenuation $R_s=40$ dB and cutoff frequency $\omega_c=0.6$, which has the coefficients given by the vectors:

$$\mathbf{B}_{C1}=[0.3513 \quad 1.01 \quad 1.01 \quad 0.3513] \qquad \mathbf{A}_{C1}=[1 \quad 0.9644 \quad 0.6701 \quad 0.088] \tag{29}$$

The 2D low-pass filter is separable and results by applying successively the 1D filter along the two frequency axes; the 4×4 matrices of the correction filter result as: $B_C=B_{C1}^T\otimes B_{C1}$, $A_C=A_{C1}^T\otimes A_{C1}$, where the symbol $\otimes$ denotes outer product of vectors. The correction filter has the following transfer function, where $\mathbf{z_1}=[1 \;\; z_1 \;\; z_1^2 \;\; z_1^3]$, $\mathbf{z_2}=[1 \;\; z_2 \;\; z_2^2 \;\; z_2^3]$:

$$H_C(z_1,z_2)=\left(\mathbf{z_1}\times\mathbf{B}_C\times\mathbf{z_2}^T\right)/\left(\mathbf{z_1}\times\mathbf{A}_C\times\mathbf{z_2}^T\right) \tag{30}$$

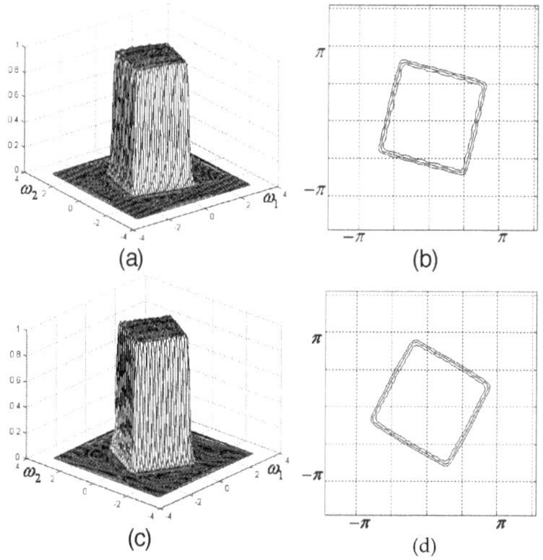

Figure 4. Diamond-type filters with orientation angle: (a), (b) $\varphi=\pi/12$; (c), (d) $\varphi=\pi/6$

The resulted 2D square-shaped correction filter characteristic is shown in Fig.3 (a) and is almost maximally-flat, as required. The corrected version of the diamond-type filter from Fig. 3 (b), (c) has the magnitude and the contour plot shown in Fig. 3 (d), (e). It can be easily noticed that the initial distortions have been eliminated. Another two diamond-type filters with orientation angles $\varphi = \pi/12$ and $\varphi = \pi/6$ are shown in Fig. 4 (a)-(d).

4. Fan-type recursive filters

In this section an analytical design method in the frequency domain for 2D fan-type filters is proposed, starting from an 1D analog prototype filter, with a transfer function decomposed as a product of elementary functions. Since we envisage designing efficient 2D filters, of minimum order, recursive filters are used as prototypes, and the 2D fan-type filters will result recursive as well.

In Fig.5 (a) a general fan-type filter is shown, with an aperture angle $\prec BOD = \theta$, oriented along an axis CC' and its longitudinal axis forming an angle $\prec AOC = \varphi$ with frequency axis $O\omega_2$. A particular case is the two-quadrant fan filter, shown in Fig.5 (b). Fig.5 (c) shows a DFB with 8-band frequency partition (Bamberger, 1992), an angularly-oriented image decomposition which splits the frequency plane into fan-shaped sub-bands (channels).

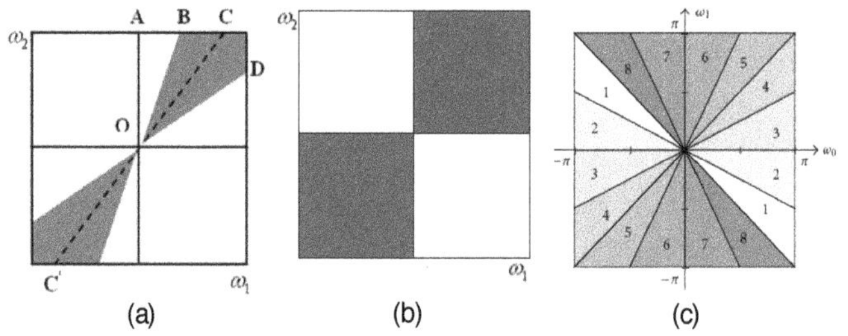

Figure 5. (a) Ideal fan filter with given aperture, oriented at an angle φ; (b) Ideal two-quadrant fan filter (c) 8-band partitions of the frequency plane

The 1D analog filter discussed in section 2 is used as prototype. The general fan-type filter can be derived from a LP prototype using the frequency mapping (Matei & Matei, 2012):

$$\omega \rightarrow f_\varphi(\omega_1, \omega_2) = a \cdot \left(\omega_1 \cdot \cos\varphi - \omega_2 \cdot \sin\varphi\right) / \left(\omega_1 \cdot \sin\varphi + \omega_2 \cdot \cos\varphi\right) \tag{31}$$

In (31), $a = 1/\text{tg}(\theta/2)$ is the aperture coefficient, where θ is the aperture angle of the fan-type filter. This frequency mapping in the complex variables $s_1 = j\omega_1$, $s_2 = j\omega_2$ is:

$$s \rightarrow f_\varphi(s_1, s_2) = j \cdot a \cdot \left(s_1 \cdot \cos\varphi - s_2 \cdot \sin\varphi\right) / \left(s_1 \cdot \sin\varphi + s_2 \cdot \cos\varphi\right) \tag{32}$$

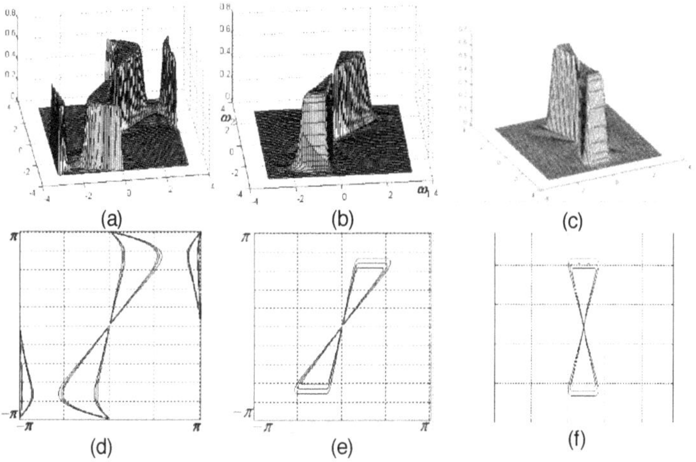

Figure 6. Frequency response magnitudes and contour plots for: (a) fan-type filter with aperture $\theta = 0.1\pi$ and orientation $\varphi = \pi / 7$; corrected filters with $\theta = 0.1\pi$, $\varphi = \pi / 7$ (b) and $\varphi = 0$(c)

Applying the same steps as in Section 3.2 in order to obtain a discrete form of the above frequency mapping, using relations (21), (22) and (32) we obtain the 1D to 2D mapping which includes pre-warping along both axes of the frequency plane:

$$s \to F_\varphi(s_1, s_2) = j \cdot a \cdot \frac{\left(s_1(1-0.05s_2^2)\cos\varphi - s_2(1-0.05s_1^2)\sin\varphi\right)}{\left(s_1(1-0.05s_2^2)\sin\varphi + s_2(1-0.05s_1^2)\cos\varphi\right)} \tag{33}$$

We now apply the bilinear transform (19) along the two axes and obtain the mapping $s \to F_\varphi(z_1, z_2)$ in matrix form, where $\mathbf{z_1}=[1 \;\; z_1 \;\; z_1^2]$ and $\mathbf{z_2}=[1 \;\; z_2 \;\; z_2^2]$:

$$s \to F_\varphi(z_1, z_2) = j \cdot a \cdot P_\varphi(z_1, z_2) / Q_\varphi(z_1, z_2) = j \cdot a \cdot \left(\mathbf{z_1} \times \mathbf{P} \times \mathbf{z_2}^T\right) / \left(\mathbf{z_1} \times \mathbf{Q} \times \mathbf{z_2}^T\right) \tag{34}$$

and the 3×3 matrices P_φ and Q_φ are given by:

$$\mathbf{P}_\varphi = \cos\varphi \cdot \begin{bmatrix} -1 & -3 & -1 \\ 0 & 0 & 0 \\ 1 & 3 & 1 \end{bmatrix} - \sin\varphi \cdot \begin{bmatrix} -1 & 0 & 1 \\ -3 & 0 & 3 \\ -1 & 0 & 1 \end{bmatrix}; \mathbf{Q}_\varphi = \sin\varphi \cdot \begin{bmatrix} -1 & -3 & -1 \\ 0 & 0 & 0 \\ 1 & 3 & 1 \end{bmatrix} + \cos\varphi \cdot \begin{bmatrix} -1 & 0 & 1 \\ -3 & 0 & 3 \\ -1 & 0 & 1 \end{bmatrix} \tag{35}$$

Substituting the mapping (34) into the biquad expression (2) with $b_1 = 0$, we get the 2D transfer function in matrix form $H_{W1}(z_1, z_2) = \left(\mathbf{z_1} \times \boldsymbol{B_2} \times \mathbf{z_2}^T\right) / \left(\mathbf{z_1} \times \boldsymbol{A_2} \times \mathbf{z_2}^T\right)$, similar to (26), where the vectors $\mathbf{z_1}$, $\mathbf{z_2}$ are given by (27). The 5×5 templates $\boldsymbol{B_2}$ and $\boldsymbol{A_2}$ are given by:

$$\mathbf{B}_2 = b_0 \cdot \mathbf{Q}_\varphi * \mathbf{Q}_\varphi - a^2 \cdot \mathbf{P}_\varphi * \mathbf{P}_\varphi \, ; \quad \mathbf{A}_2 = a_0 \cdot \mathbf{Q}_\varphi * \mathbf{Q}_\varphi - a^2 \cdot \mathbf{P}_\varphi * \mathbf{P}_\varphi + j \cdot a \cdot a_1 \cdot \mathbf{P}_\varphi * \mathbf{Q}_\varphi \tag{36}$$

The 2D transfer function for each biquad is complex. The characteristics of a fan-type filter designed with this method and based on the prototype filter of order 4 given by (6)-(7) is shown in Fig.6 (a), for the indicated parameters. As with the diamond-type filter analyzed in the previous section, the fan-type filter characteristic features marginal linearity distortions which can be corrected using a LP filter, similar with the correction filter used in Section 3.2 and having the frequency characteristic shown in Fig. 3 (a).

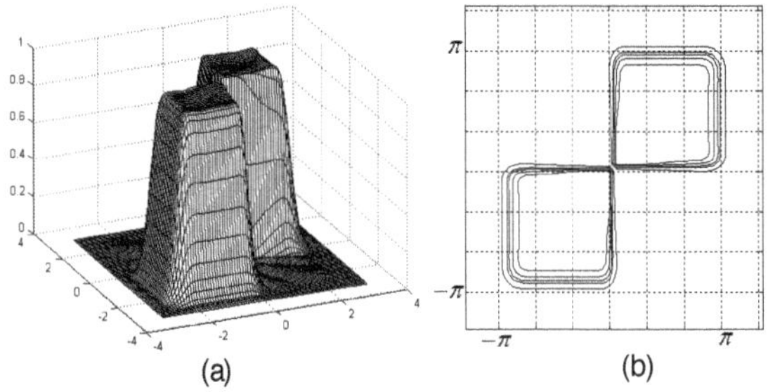

Figure 7. (a) Frequency response magnitude and (b) contour plot for the 2-quadrant fan filter

Two corrected fan-type filters with specified parameters have the magnitudes and contour plots shown in Fig.6 (b), (c). The initial distortions have been eliminated. With the same correction filter, we obtain the two-quadrant fan filter, shown in Fig. 7, by setting the aperture angle $\theta = \pi / 2$ and orientation angle $\varphi = \pi / 4$.

5. Very selective multidirectional IIR Filters

In this section a design method based on spectral transformations is proposed for another class of 2D IIR filters, namely multi-directional filters. The design starts from an analog prototype with specified parameters. Applying an appropriate frequency transformation to the 1D transfer function, the desired 2D filter is directly obtained in a factorized form, like the filters designed in the previous sections. For two-directional filters, an example is given of extracting lines with two different orientations from a test image. The spectral transformation used in the case of multi-directional filters is similar to the one presented in the previous section, derived for fan-type filters and given by (34), (35). In this section the design of two-directional and three-directional filters with specified orientation is detailed. The method can be easily generalized to arbitrary multi-directional filters.

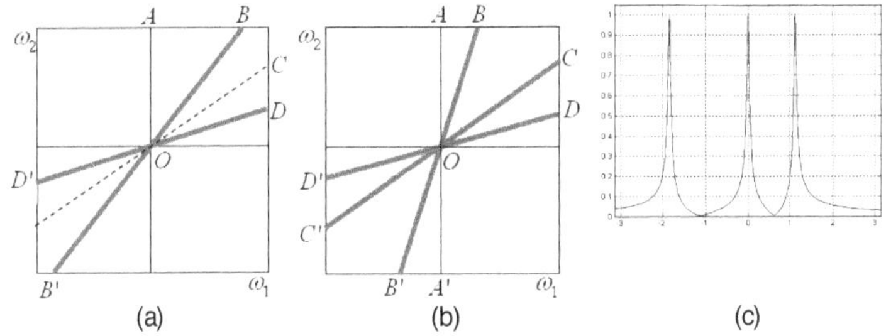

Figure 8. Ideal shapes of some directional filters in the frequency plane: (a) two-directional filter; (b) three-directional filter; (c) three-band selective filter with $\omega_{01} = 0.59\pi$, $\omega_{02} = -0.36\pi$

5.1. Two-directional fan-type filters

A two-directional 2D filter is orientation-selective along two directions in the frequency plane. It is based on a selective resonant IIR prototype as given in section 2. Applying the same frequency transformation $s \rightarrow F_{\varphi}(z_1, z_2)$ derived for fan-type filters and given by (33) to the prototype filter (9) we get the 2D two-directional transfer function $H_2(z_1, z_2)$ in matrix form $H_2(z_1, z_2) = (\mathbf{z}_1 \times \boldsymbol{B}_3 \times \mathbf{z}_2^T) / (\mathbf{z}_1 \times \boldsymbol{A}_3 \times \mathbf{z}_2^T)$, similar to (26), but the 5×5 matrices $\boldsymbol{B}_3$, $\boldsymbol{A}_3$ now have the form:

$$\mathbf{B}_3 = a \cdot \alpha \cdot \mathbf{P}_{\varphi} * \mathbf{Q}_{\varphi} \qquad \mathbf{A}_3 = a \cdot \alpha \cdot \mathbf{P}_{\varphi} * \mathbf{Q}_{\varphi} + j\left(a^2 \cdot \mathbf{P}_{\varphi} * \mathbf{P}_{\varphi} - \omega_0^2 \cdot \mathbf{Q}_{\varphi} * \mathbf{Q}_{\varphi}\right) \tag{37}$$

The denominator matrix A_3 has complex elements. In Fig.9 (a), the contour plot of the frequency response magnitude is shown for a two-directional filter with aperture $\theta = \pi / 6$ and orientation $\varphi = \pi / 5$. As with the previous types of filters, the marginal linearity distortions can be corrected using an additional LP square-shaped filter.

The templates $\boldsymbol{B}_{3C}$ and A_{3C} of the corrected filter result by convolution: $\boldsymbol{B}_{3C} = \boldsymbol{B}_3 * \boldsymbol{B}_C$ and $A_{3C} = A_3 * A_C$. In Fig.9 (b), (c) the frequency response magnitudes and contour plots are displayed for the corrected two-directional filter with specified aperture and orientation. The initial distortions have been eliminated.

The second two-directional filter in Fig.9 (d), (e) is a particular case, being oriented along the two frequency axes ($\theta = \pi / 2$, $\varphi = \pi / 4$), therefore can be used to detect simultaneously horizontal and vertical lines from an image, as shown in the next section.

5.2. Three-directional fan-type filters

In order to design a three-directional filter like the one depicted in Fig. 8 (b), we must start from an analog three-band selective filter, like the one with frequency response shown in Fig.

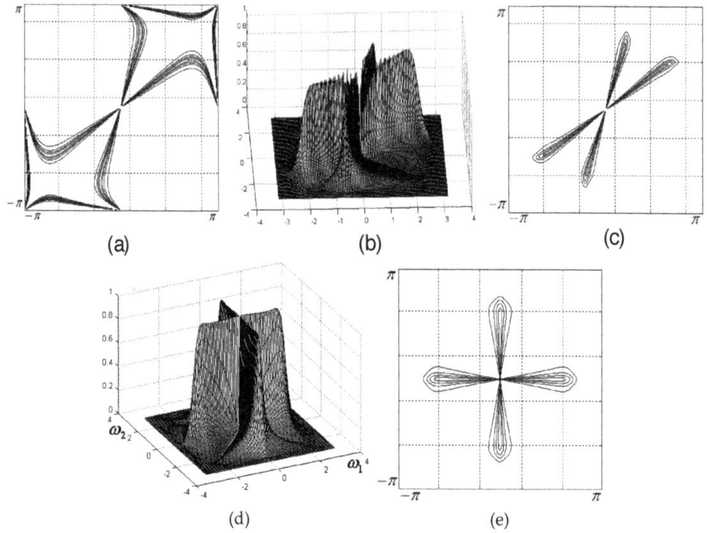

Figure 9. (a) Contour plot of a two-directional filter with $\theta = \pi / 6$, $\varphi = \pi / 5$; frequency response magnitudes and contour plots of two corrected two-directional filters with parameters: $\theta = \pi / 6$, $\varphi = \pi / 5$ (b), (c) and $\theta = \pi / 2$, $\varphi = \pi / 4$ (d), (e)

8 (c). For a three directional filter, the middle peak frequency can always be taken $\omega_0 = 0$, and the other two on each side at specified values. The prototype transfer function $H_P(s)$ in variable s will be in this case the sum of three elementary functions:

$$H_P(s) = \frac{B_P(s)}{A_P(s)} = \frac{\alpha}{s+\alpha} + \frac{\alpha}{s+\alpha+j\cdot\omega_{01}} + \frac{\alpha}{s+\alpha+j\cdot\omega_{02}} \tag{38}$$

The frequency response of a filter of this kind with parameter values $\alpha = 0.03$, $\omega_{01} = 0.59\pi$ and $\omega_{02} = -0.36\pi$ is shown in Fig. 8 (c). Substituting the mapping (34) into the expression (8) of the elementary function $H_{jS}(s)$, we get the 2D transfer function $H_1(z_1, z_2)$ in matrix form: $H_1(z_1, z_2) = (\mathbf{z}_1 \times \boldsymbol{B}_b \times \mathbf{z}_2^T) / (\mathbf{z}_1 \times \boldsymbol{A}_b \times \mathbf{z}_2^T)$, where $\mathbf{z}_1 = [1 \;\; z_1 \;\; z_1^2]$, $\mathbf{z}_2 = [1 \;\; z_2 \;\; z_2^2]$ and the 3×3 templates $\boldsymbol{B}_b, \boldsymbol{A}_b$ are given by:

$$\mathbf{B}_b = \alpha \cdot \mathbf{Q}_\varphi ; \mathbf{A}_b = \alpha \cdot \mathbf{Q}_\varphi + j \cdot \left(a \cdot \mathbf{P}_\varphi + \omega_0 \cdot \mathbf{Q}_\varphi \right) \tag{39}$$

Each of the three elementary terms in (38) corresponds to a pair of 3×3 templates $\boldsymbol{B}_b$ and $\boldsymbol{A}_b$ given by (39). If the three elementary filters are given by the pairs of templates $(\boldsymbol{B}_{b1}, \boldsymbol{A}_{b1})$, $(\boldsymbol{B}_{b2},$

A_{b2}) and (B_{b3}, A_{b3}), the templates of size 7×7 of the entire three-directional filter will result by summing up the convolutions of elementary templates:

$$\mathbf{B}_3 = \mathbf{B}_{b1} * \mathbf{A}_{b2} * \mathbf{A}_{b3} + \mathbf{A}_{b1} * \mathbf{B}_{b2} * \mathbf{A}_{b3} + \mathbf{A}_{b1} * \mathbf{A}_{b2} * \mathbf{B}_{b3} \; ; \quad \mathbf{A}_3 = \mathbf{A}_{b1} * \mathbf{A}_{b2} * \mathbf{A}_{b3} \tag{40}$$

The numerator $B_P(s)$ of $H_P(s)$ from (38) has the general form:

$$B_P(s) = \alpha \cdot \left(a_2 \cdot s^2 + a_1 \cdot s + a_0\right) \tag{41}$$

where $a_2=3$, $a_1=6\alpha + j \cdot 2(\omega_{01} + \omega_{02})$ and $a_0=3\alpha^2 - \omega_{01} \cdot \omega_{02} + j \cdot 2\alpha(\omega_{01} + \omega_{02})$.

We see that a_2 is real and a_0, a_1 are generally complex. The coefficients a_0, a_1 are real only when $\omega_{02} = -\omega_{01}$, i.e. for symmetric frequency values around the origin. Finally, for any specified set of values α, ω_{01}, ω_{02} the denominator factorizes as $B_j(s)=3\alpha \cdot \left(s + r_1\right)\left(s + r_2\right)$, where r_1 and r_2 are complex roots. Therefore the factorized prototype transfer function is:

$$H_j(s) = \frac{B_j(s)}{A_j(s)} = \frac{3\alpha \cdot \left(s + r_1\right)\left(s + r_2\right)}{\left(s+\alpha\right)\left(s+p_1\right)\left(s+p_2\right)} \tag{42}$$

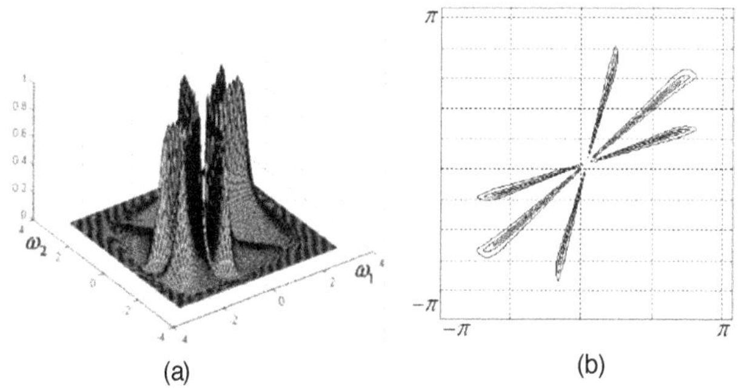

Figure 10. Frequency response magnitude (a) and contour plot (b) of a corrected three-directional filter with parameters: $\theta=0.23\pi$, $\varphi=0.27\pi$, $\omega_{01}=0.59\pi$, $\omega_{02}=-0.36\pi$

At the denominator, we denoted $p_1=\alpha + j \cdot \omega_{01}$, $p_2=\alpha + j \cdot \omega_{02}$. Applying to each factor the frequency transformation $s \rightarrow F_\varphi(z_1, z_2)$ given by (34), after some algebraic manipulations, we finally obtain the templates of the three-directional filter as discrete convolutions:

$$\mathbf{B}_3 = 3\alpha \cdot \mathbf{Q}_\varphi * \left(r_1\mathbf{Q}_\varphi + ja\mathbf{P}_\varphi\right) * \left(r_2\mathbf{Q}_\varphi + ja\mathbf{P}_\varphi\right) \tag{43}$$

$$\mathbf{A}_3 = \left(\alpha\mathbf{Q}_\varphi + ja\mathbf{P}_\varphi\right) * \left(p_1\mathbf{Q}_\varphi + ja\mathbf{P}_\varphi\right) * \left(p_2\mathbf{Q}_\varphi + ja\mathbf{P}_\varphi\right) \tag{44}$$

This implies the fact that the transfer function $H_3(z_1, z_2)$ of the 2D three-directional filter with templates B_3 and A_3 of size 7×7 results directly in a factorized form, which is an important advantage in implementation. As a general remark on the method, using an analog prototype instead of a digital one, as is currently done, simplifies the design in this case, as the frequency mapping results simpler and leads to a 2D filter of lower complexity. The designed filters result withcomplexcoefficients,howeversuchIIRfilterscanalsobeimplemented(Nikolovaetal.,2011).

6. Directional IIR filters designed in polar coordinates

We approach here a particular class of 2D filters, namely filters whose frequency response is symmetric about the origin and has at the same time an angular periodicity. The contour plots of their frequency response, resulted as sections with planes parallel with the frequency plane, can be defined as closed curves which can be described in terms of a variable radius which is a periodic function of the current angle formed with one of the axes.

It can be described in polar coordinates by $\rho = \rho(\varphi)$, where φ is the angle formed by the radius op with ω_1-axis, as shown in Fig.8(a) for a four-lobe filter. Therefore $\rho(\varphi)$ is a periodic function of the angle φ in the range $\varphi \in [0, 2\pi]$.

6.1. Spectral transformation for filters designed in polar coordinates

The main issue approached here is to find the transfer function of the desired 2D filter $H_{2D}(z_1, z_2)$ using appropriate frequency transformations of the form $\omega \rightarrow F(\omega_1, \omega_2)$. The elementary transfer functions (14) and (15) have the complex frequency responses:

$$H_1(j\omega) = \left(b_0 + b_1\cos\omega + jb_1\sin\omega\right) / \left(a_0 + \cos\omega + j\sin\omega\right) \tag{45}$$

$$H_2(j\omega) = \frac{b_1 + (b_2 + b_0)\cos\omega + j(b_2 - b_0)\sin\omega}{a_1 + (1 + a_0)\cos\omega + j(1 - a_0)\sin\omega} = \frac{P(\omega)}{Q(\omega)} \tag{46}$$

The proposed design method for these 2D filters is based on the frequency transformation:

$$F : \mathbb{R} \rightarrow \mathbb{C}^2, \omega^2 \rightarrow F(z_1, z_2) = B_f(z_1, z_2) / A_f(z_1, z_2) \tag{47}$$

which maps the real frequency axis ω onto the complex plane (z_1, z_2), defined by the real frequency mapping:

$$F_1 : \mathbb{R} \rightarrow \mathbb{R}^2, \omega^2 \rightarrow F(\omega_1, \omega_2) = \left(\omega_1^2 + \omega_2^2\right) / \rho(\omega_1, \omega_2) \tag{48}$$

In (48) $\rho(\omega_1, \omega_2)$ is initially determined in the angle variable φ as $\rho(\varphi)$ and can be referred to as a *radial compressing function*. In the frequency plane (ω_1, ω_2) we have:

$$\cos\varphi = \omega_1 / \sqrt{\omega_1^2 + \omega_2^2} \tag{49}$$

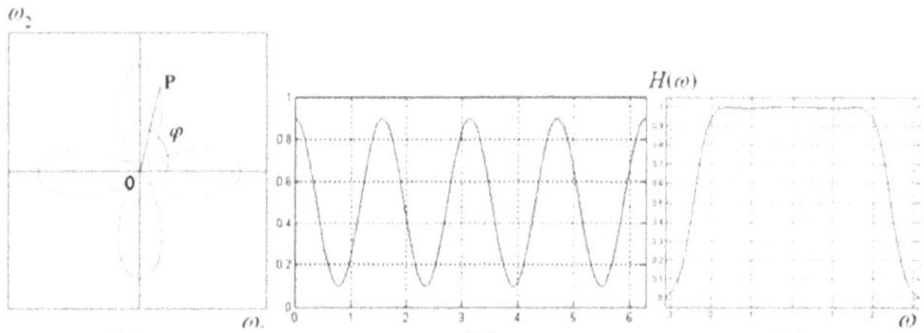

(a) (b) (c)

Figure 11. (a) contour plot of a four-lobe filter; (b) periodic function $\rho(\varphi)$; (c) LP prototype

If the radial function $\rho(\varphi)$ can be expressed in the variable $\cos\varphi$, using (49) we obtain by substitution the function $\rho(\omega_1, \omega_2)$. The function $\rho(\varphi)$ will result as a polynomial or a ratio of polynomials in $\cos\varphi$. For instance, the four-lobe filter with the contour plot given in Fig.11 (a) corresponds to a function:

$$\rho(\varphi) = a + b\cos 4\varphi = a + b - 8b\cos^2\varphi + 8b\cos^4\varphi \tag{50}$$

plotted in Fig.11 (b) on the range $\varphi \in [0, 2\pi]$. More generally, the 2D filter can be rotated in the frequency plane with a specified angle φ_0 about one of the frequency axes, e.g. $O - \omega_2$. For instance, in a four-lobe filter, two opposite lobes are oriented along a direction at an angle φ_0, and the other two at $\varphi_0 + \pi/2$, as shown in Fig.12 (b). It can be shown that the cosine of the current angle φ with initial phase φ_0 can be expressed as:

$$\cos^2(\varphi + \varphi_0) = \left(\cos^2\varphi_0 \cdot \omega_1^2 + \sin^2\varphi_0 \cdot \omega_2^2 + 0.5\sin 2\varphi_0 \cdot \omega_1\omega_2\right) / \left(\omega_1^2 + \omega_2^2\right) \tag{51}$$

For filters with an even number of lobes, the radial function $\rho(\varphi)$ is expressed in even powers of $\cos\varphi$ or $\cos(\varphi + \varphi_0)$. The frequency transformation (48) can be also expressed as:

$$\omega \to \sqrt{(\omega_1^2+\omega_1^2)/\rho(\omega_1,\omega_2)} = \sqrt{F_1(\omega_1,\omega_2)} \tag{52}$$

In order to obtain a rational expression for the frequency response of the 2D filter from an elementary 1D prototype of the form (45) or (46) by applying the frequency mapping (52), we need to derive rational expressions for the functions $\cos\sqrt{\omega}$ and $\sin\sqrt{\omega}$. Using the Chebyshev-Padé method in a symbolic computation software, the following second-order rational approximations were found:

$$\cos\sqrt{\omega} \cong \left(1.0559-0.086514\cdot\omega-0.1304\cdot\omega^2\right)\Big/\left(1+0.75\cdot\omega-0.110583\cdot\omega^2\right)=C_S(\omega)/A_S(\omega) \tag{53}$$

$$\sin\sqrt{\omega} \cong \left(0.167+1.46287\cdot\omega-0.259815\cdot\omega^2\right)\Big/\left(1+0.75\cdot\omega-0.110583\cdot\omega^2\right)=S_S(\omega)/A_S(\omega) \tag{54}$$

which are sufficiently accurate on the range $\omega\in[0,\pi]$. Since these functions are developed on the range $[0,\pi]$, their approximations result neither odd nor even. However, using the above approximations will lead to relatively complex 2D filters, described by templates of size at least 9×9. For the type of filters approached, namely selective two-directional (four-lobe) filters, the approximations for $\cos\sqrt{\omega}$ and $\sin\sqrt{\omega}$ need not necessarily hold throughout the range $[0,\pi]$, but only on a smaller range near the origin, corresponding to filter pass-band. Using now the Padé method we get the first-order approximations:

$$\sin\sqrt{\omega} \cong (s_0+s_1\omega)/(1+r\omega) \qquad \cos\sqrt{\omega} \cong (c_0+c_1\omega)/(1+r\omega) \tag{55}$$

with $s_0=0.0928$, $s_1=2.5218$, $c_0=1.0104$, $c_1=1.2193$, $r=1.979$, which hold only on a

narrower range around zero of the interval $\omega\in[0,\pi]$. Using (55) instead of (53), (54) will result in much more efficient 2D filters, which fully satisfy the imposed specifications.

We will use here a Chebyshev low-pass second-order filter of the general form (15). For this type of filter we have the coefficient symmetry $b_2=b_0$. According to (46) we can write:

$$H_2\left(j\sqrt{\omega}\right)=\frac{b_1+2b_0\cos\sqrt{\omega}}{a_1+(1+a_0)\cos\sqrt{\omega}+j(1-a_0)\sin\sqrt{\omega}}=\frac{P(\sqrt{\omega})}{Q(\sqrt{\omega})} \tag{56}$$

The numerator results real because the imaginary part is cancelled. Substituting the expressions (55) into this complex frequency response we get the rational approximation:

$$H_2\left(j\sqrt{\omega}\right)=\frac{b_1(1+r\omega)+2b_0(c_0+c_1\omega)}{a_1(1+r\omega)+(1+a_0)(c_0+c_1\omega)+j\cdot(1-a_0)(s_0+s_1\omega)} \tag{57}$$

which can be also written as:

$$H_2(j\omega)=\frac{b_1(1+r\omega^2)+2b_0(c_0+c_1\omega^2)}{a_1(1+r\omega^2)+(1+a_0)(c_0+c_1\omega^2)+j\cdot(1-a_0)(s_0+s_1\omega^2)} \tag{58}$$

The function (58) has even parity, since it is expressed as a rational function in ω^2.

6.2. Two-directional filter design

We approach now the design of a particular filter type designed in polar coordinates, namely two-directional (selective four-lobe) filters along the two plane axes or with a specified orientation angle. Let us consider the radial function given by:

$$H_r(\varphi)=1/\left(p\cdot\tilde{B}(\varphi)-p+1\right) \tag{59}$$

where $\tilde{B}(\varphi)$ is a periodic function; let $\tilde{B}(\varphi)=\cos(4\varphi)$. We use it to design a 2D filter with four narrow lobes in the plane (ω_1, ω_2). Using trigonometric identities, (59) becomes:

$$H_r(\varphi)=1/\left(1+8p\cdot(\cos\varphi)^2-8p\cdot(\cos\varphi)^4\right) \tag{60}$$

and is plotted for $\varphi\in[-\pi,\pi]$ in Fig.12(a). This is a periodic function with period $\Phi=\pi/4$ and has the shape of a multi-band ("comb") filter. In order to control the amplitude of this function, we introduce another parameter k, such that the radial function $\rho(\varphi)$ takes the form $\rho(\varphi)=k\cdot H_r(\varphi)$. We get using (49):

$$\omega^2\rightarrow F(\omega_1,\omega_2)=\left(\omega_1^4+(2+8p)\omega_1^2\omega_2^2+\omega_2^4\right)/\left(k(\omega_1^2+\omega_2^2)\right) \tag{61}$$

and the function $F_2(s_1, s_2)$ of the form:

$$F_2(s_1,s_2)=-\left(s_1^4+(2+8p)s_1^2s_2^2+s_2^4\right)/\left(k(s_1^2+s_2^2)\right) \tag{62}$$

Finally we derive a transfer function of the 2D filter $H(z_1, z_2)$ in the complex plane (z_1, z_2). This can be achieved if we find a discrete counterpart of the function $\rho(\omega_1, \omega_2)$, denoted $R(z_1, z_2)$. A possible method is to express the function $\rho(\omega_1, \omega_2)$ in the complex plane (s_1, s_2) and then find the appropriate mapping to (z_1, z_2) using the bilinear transform or the Euler approximation. Even if generally the bilinear transform is more often used, being a more accurate mapping, especially if a frequency pre-warping is applied to compensate for distortions, for the particular type of 2D filter approached here the Euler formula leads to more efficient filters and with better characteristics, at the same filter order. We will use the backward Euler method, which approximates the spatial derivative $\partial X / \partial x$ by $X[n]-X[n-1]$, replacing s by $s=1-z^{-1}$. On the two directions of the plane we have: $s_1=1-z_1^{-1}$, $s_2=1-z_2^{-1}$. The operators s_1^2, s_2^2 and $s_1 s_2$ correspond to second-order partial derivatives: $\partial^2 / \partial x^2 \leftrightarrow s_1^2=z_1+z_1^{-1}-2$, $\partial^2 / \partial y^2 \leftrightarrow s_2^2=z_2+z_2^{-1}-2$, $\partial^2 / \partial x \partial y \leftrightarrow s_1 s_2$. For the mixed operator $s_1 s_2$, using repeatedly the Euler formula, we get the expression (Matei, 2011 a): $2s_1 s_2=z_1+z_1^{-1}+z_2+z_2^{-1}-2-z_1 z_2^{-1}-z_1^{-1} z_2$. Substituting the above relations into (62) we obtain a frequency mapping similar to (47), with the templates:

$$\mathbf{B}_f=\begin{bmatrix} 0 & 0 & 1 & 0 & 0 \\ 0 & 2+8p & -8-16p & 2+8p & 0 \\ 1 & -8-16p & 20+32p & -8-16p & 1 \\ 0 & 2+8p & -8-16p & 2+8p & 0 \\ 0 & 0 & 1 & 0 & 0 \end{bmatrix} \mathbf{A}_f=k\cdot\begin{bmatrix} 0 & 1 & 0 \\ 1 & -4 & 1 \\ 0 & 1 & 0 \end{bmatrix}*\begin{bmatrix} 0 & 1 & 0 \\ 1 & -4 & 1 \\ 0 & 1 & 0 \end{bmatrix}=k\cdot\mathbf{A}_1*\mathbf{A}_1 \quad (63)$$

The template A_f results as a convolution of two 3×3 matrices. The last step in the design of this 2D filter is to apply the frequency transformation (61) to the frequency response (58) and we find the filter templates B and A as linear combinations of $\boldsymbol{B}_f$ and $\boldsymbol{A}_f$:

$$\mathbf{B}=(b_1+2b_0c_0)\cdot\mathbf{A}_f+(rb_1+2b_0c_1)\cdot\mathbf{B}_f \quad (64)$$

$$\mathbf{A}=\left(a_1+(1+a_0)c_0+j(1-a_0)s_0\right)\cdot\mathbf{A}_f+\left(a_1 r+(1+a_0)c_1+j(1-a_0)s_1\right)\cdot\mathbf{B}_f \quad (65)$$

where b_2, b_1, b_0, a_1, a_0 are the coefficients of prototype (15). Finally the 2D filter transfer function in z_1 and z_2 has the following expression, with $\mathbf{z}_1$ and $\mathbf{z}_2$ given by (27):

$$H_{2D}(z_1,z_2)=B(z_1,z_2)/A(z_1,z_2)=\left(\mathbf{z}_1\times\mathbf{B}\times\mathbf{z}_2^T\right)/\left(\mathbf{z}_1\times\mathbf{A}\times\mathbf{z}_2^T\right) \quad (66)$$

Let us design a two-directional filter following this procedure. As 1D prototype let us consider a type-2 low-pass Chebyshev digital filter with the parameter values: order$N=2$, stop-band attenuation $R_s=40$db and passband-edge frequency $\omega_p=0.5$ (1.0 is half the sampling frequency). The transfer function in z is:

$$H_p(z)=\left(0.012277\cdot z^2-0.012525\cdot z+0.012277\right)/\left(z^2-1.850147\cdot z+0.862316\right) \quad (67)$$

For a good directional selectivity we also choose $p=30$ and $k=10$. The 2D filter frequency response magnitude is displayed in Fig.12 (c) and shows a very good linearity along the two directions and practically no distortions in the stop band. The constant level contour in the plane (ω_1, ω_2) is given in Fig.12 (d). Calculating the singular values of filter templates for the above parameters we find the vectors S_A, S_B for the templates A, B:

$$\mathbf{S_A} = \begin{bmatrix} 2.09347 & 0.00225 & 0.0005 & 0 & 0 \end{bmatrix} \quad \mathbf{S_B} = \begin{bmatrix} 2.14297 & 0.01939 & 0.00387 & 0 & 0 \end{bmatrix} \tag{68}$$

Taking into account the fact that the first singular value of the templates A and B is much larger than the other four, the filter designed above can be approximated by a separable filter.

The singular value decomposition of a matrix M is written as $M = U \times S \times V$ where U and V are unitary matrices and S is a diagonal matrix containing the singular values. Thus we can write for the filter templates A and B:

$$\mathbf{A = U_A \times S_A \times V_A \quad B = U_B \times S_B \times V_B} \tag{69}$$

If U_{A_1} and V_{A_1} are the first columns of the matrices U_A, V_A, then A can be approximated by a matrix $A_1 = s_{A1} \cdot U_{A1} \otimes V_{A1}^T$, where s_{A1} is the largest singular value of A, U_{A_1} and V_{A_1} are the corresponding columns of U_A and V_A, $\otimes$ stands for outer product, T for transposition. Similarly for B we find $B_1 = s_{B1} \cdot U_{B1} \otimes V_{B1}^T$. For the specified filter parameters we obtain s_{B1}=2.14297 and for template B the column vectors U_{B1} and V_{B1} result identical: $U_{B1} = V_{B1} = [-0.00424 \;\; 0.3971 \;\; -0.82743 \;\; 0.3971 \;\; -0.00424]^T$.

For the template A we get s_{A1}=2.09347 and the vectors U_{A_1}, V_{A_1} have complex elements. The frequency response of the resulted filter is given in Fig.12 (e). As can be noticed, the effect of the above approximation is an "overshoot" at zero frequency. This should not affect the filter functionality in detecting lines parallel with the two axes. Moreover, since the marginal elements of the 5×1 vectors U_{A1}, V_{A1}, U_{B1}, V_{B1} have negligible values, by discarding them we obtain the 3×1 vectors: $U_{B2} = V_{B2} = [0.3971 \;\; -0.8274 \;\; 0.3971]^T$

$U_{A2} = [-0.315 \;\; 0.6334 \;\; -0.315]^T + j \cdot [0.2573 \;\; -0.5175 \;\; 0.2573]^T$

$V_{A2} = [0.4067 \;\; -0.818 \;\; 0.4067]^T - j \cdot [0.0024 \;\; -0.0048 \;\; 0.0024]^T$

We finally obtain a very selective two-directional 2D filter implemented with two minimum size (3×3) templates. The template B is real while A is complex. The frequency response magnitude of this filter is shown in Fig.12 (f) and is practically similar to the one in Fig.12 (e). Similarly we can design a two-directional (four-lobe) filter with a specified orientation angle. Using the previously described method and based on the Euler approximation, the expression

(10) of $\cos^2(\varphi+\varphi_0)$ corresponds to a frequency transformation in the complex variables z_1 and z_2, written in matrix form as:

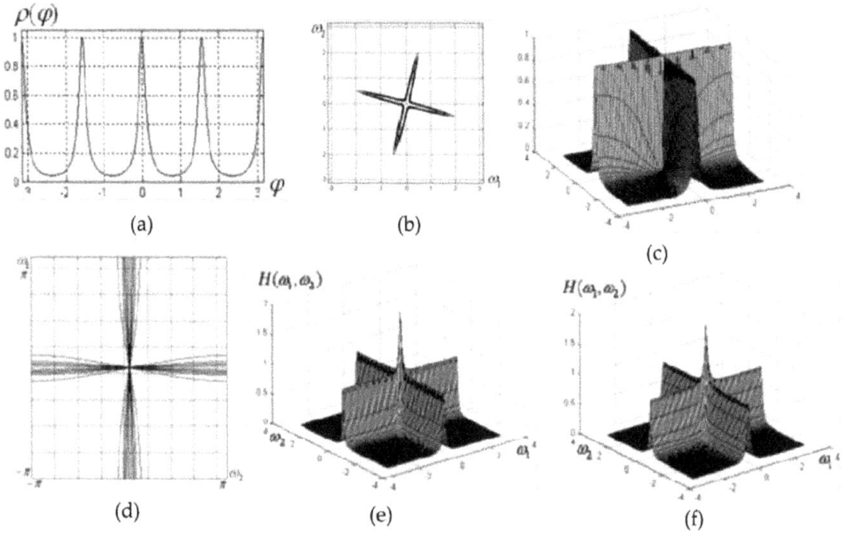

Figure 12. (a) Periodic radial function; (b) contour plot of a two-directional filter with orientation $\varphi_0=\pi/12$; (c) frequency response and (d) contour plot of the two-directional filter with 5×5 templates; frequency response of the filter with separable 5×5 (e) and 3×3 templates (f)

$$\cos^2(\varphi+\varphi_0)\rightarrow F(z_1,z_2)=B_{\varphi0}(z_1,z_2)/A_{\varphi0}(z_1,z_2)=\left(\mathbf{Z}_1\times\mathbf{B}_{\varphi0}\times\mathbf{Z}_2^T\right)/\left(\mathbf{Z}_1\times\mathbf{A}_{\varphi0}\times\mathbf{Z}_2^T\right) \tag{70}$$

$$\mathbf{B}_{\varphi0}=\cos^2\varphi_0\begin{bmatrix}0&1&0\\0&-2&0\\0&1&0\end{bmatrix}+\sin^2\varphi_0\begin{bmatrix}0&0&0\\1&-2&1\\0&0&0\end{bmatrix}+0.25\sin(2\varphi_0)\begin{bmatrix}0&1&-1\\1&-1&1\\-1&1&0\end{bmatrix} \tag{71}$$

and $A_{\varphi0}$ is identical to A_1 from (63). The 5×5 templates of the mapping (47) are given by:

$$\mathbf{B}_f=\mathbf{A}_{\varphi0}*\mathbf{A}_{\varphi0}\quad \mathbf{A}_f=\mathbf{A}_{\varphi0}*\mathbf{A}_{\varphi0}+8p\cdot\mathbf{B}_{\varphi0}*\mathbf{A}_{\varphi0}-8p\cdot\mathbf{B}_{\varphi0}*\mathbf{B}_{\varphi0} \tag{72}$$

The final filter templates result according to relations (64) and (65).

Regarding the proposed method, the frequency responses of this class of 2D filters can be viewed as derived through a radial distortion from a generic maximally-flat circular filter.

Indeed, referring to (48), the circular filter is the trivial case for which $\rho(\omega_1, \omega_2)=1$ and the mapping (48) reduces to $\omega^2 \to \omega_1^2 + \omega_1^2$, as expected. This method allows one to design 2D filters with a non-convex shape in the frequency plane. The proposed design method does not involve global numerical optimization techniques, but only a few numerical approximations. The method is more general and can be applied as well to design fan filters, diamond filters, multi-directional filters etc. (Matei, 2011 a).

7. Zero-phase FIR circular filters

Filters with circular symmetry are very useful in image processing. We propose an efficient design technique for 2D circularly-symmetric filters, based on the previous 1D filters, considered as prototypes. Given a 1D prototype $H_P(\omega)$, the corresponding 2D circular filter function $H_C(\omega_1, \omega_2)$ results using the frequency mapping $\omega \to \sqrt{\omega_1^2 + \omega_2^2}$:

$$H_C(\omega_1, \omega_2) = H_P\left(\sqrt{\omega_1^2 + \omega_2^2}\right) \tag{73}$$

The currently-used approximation of the circular function $\cos\sqrt{\omega_1^2 + \omega_2^2}$ is given by:

$$\cos\sqrt{\omega_1^2 + \omega_2^2} \cong C(\omega_1, \omega_2) = -0.5 + 0.5\left(\cos\omega_1 + \cos\omega_2\right) + 0.5\cos\omega_1 \cdot \cos\omega_2 \tag{74}$$

which corresponds to the 3×3 array:

$$\mathbf{C} = \begin{bmatrix} 0.125 & 0.25 & 0.125 \\ 0.25 & -0.5 & 0.25 \\ 0.125 & 0.25 & 0.125 \end{bmatrix} \tag{75}$$

Let us consider as prototype a LP analog elliptic filter of order $N=4$, pass-band peak-to-peak ripple $R_p=0.04$ dB, stop-band attenuation $R_S=40$ dB and passband-edge frequency $\Omega_p=\pi/2$. Its transfer function in variable s is:

$$H_P(s) = 0.1037 \cdot \left(s^4 + 19.864 \cdot s^2 + 84.041\right) / \left(s^4 + 3.2041 \cdot s^3 + 8.4315 \cdot s^2 + 13.126 \cdot s + 14.082\right) \tag{76}$$

Using MAPLE or another symbolic computation program and following the design steps described in section 2, we obtain a a polynomial approximation of the magnitude $|H_P(j\omega)|$ through Chebyshev expansion, which has the following factorized form, with $x=\cos\omega$:

$$|H_P(\omega)| \cong 48.6 \cdot (x+0.8491)(x+0.7717)(x-1.087)(x^2+1.9934x+0.994)$$
$$(x^2+1.0797x+0.318)(x^2-0.3849x+0.1766)(x^2-1.2882x+0.5314)(x^2-1.9338x+0.9726) \quad (77)$$

In order to obtain a filter with circular symmetry from the factorized 1D prototype function, we replace in (12) $\cos\omega$ with the circular cosine function (74). For instance, corresponding to (12), the filter template A results in general as the discrete convolution:

$$\mathbf{A} = k \cdot \mathbf{A}_{11} * \mathbf{A}_{12} * \ldots * \mathbf{A}_{1n} * \mathbf{A}_{21} * \mathbf{A}_{22} * \ldots * \mathbf{A}_{2m} \quad (78)$$

where A_{1i} $(i=1\ldots n)$ are 3×3 templates and A_{2j} $(j=1\ldots m)$ are 5×5 templates, given by: $A_{1i}=C+a_i \cdot A_{01}$ and $A_{2j}=C*C+a_{1j}\cdot C_0+a_{2j}\cdot A_{02}$, where A_{01} is a 3×3 zero template and A_{02} a 5×5 zero template with the central element equal to one; C_0 is a 5×5 template

obtained by bordering C with zeros. The above expressions correspond to the factors in (12).

The frequency response $H_C(\omega_1, \omega_2)$ of the 2D circular filter results in a factorized form by substituting $x=C(\omega_1, \omega_2)$ in (77). Even if the filter results of high order, with very large templates, next we show that using the Singular Value Decomposition (SVD), the resulted 2D filter can be approximated with a negligible error. For the filter template B we can write $\boldsymbol{B}=\boldsymbol{U}_B\times\boldsymbol{S}_B\times\boldsymbol{V}_B$. The vector of singular values $\boldsymbol{S}_B$ of size 1×27 has 14 non-zero elements:

S_{B1}=[0.50536 0.086111 0.032794 0.013627 0.00521 0.002937 0.001935 0.001061 0.000639 0.000451 0.000418 0.0000385 0.0000196 0.00000144]

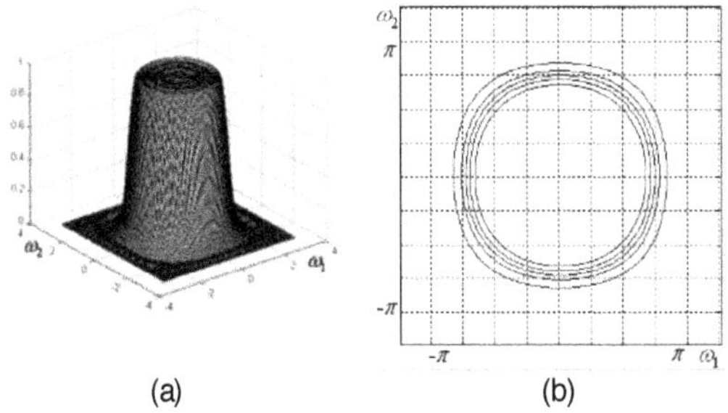

Figure 13. Frequency response magnitude (a) and contour plot (b) of a circular FIR filter

Let us denote the vector above as $\boldsymbol{S}_{B1}=[s_k]$, with $k=1...14$ in our case. The exact filter matrix B can be written as: $\boldsymbol{B}=\boldsymbol{U}_{B1}\times\boldsymbol{S}_{B1}\times\boldsymbol{V}_{B1}$, where $\boldsymbol{U}_{B1}$ and $\boldsymbol{V}_{B1}$ are made up of the first 14 columns of the unitary matrices $\boldsymbol{U}_B$ and $\boldsymbol{V}_B$. If we consider the first largest M values of the vector $\boldsymbol{S}_{B1}=[s_k]$, the matrix B can be approximated as:

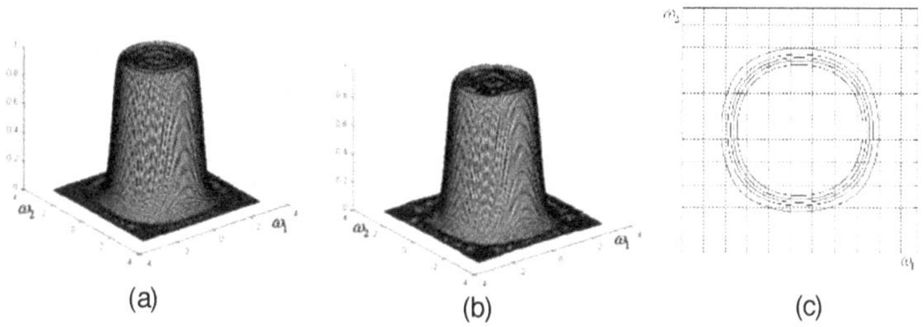

Figure 14. Frequency response magnitudes and contour plots for the circular filter resulted by taking into account the first largest: (a), (b) 8 singular values; (c), (d) 5 singular values

$$\mathbf{B} \cong \mathbf{B}_M = \sum_{k=1}^{M} s_k \cdot \mathbf{U}_{\mathrm{B}k} \otimes \mathbf{V}_{\mathrm{B}k}^T \tag{79}$$

Here B_M is the approximation of matrix B taking into account the first M singular values (in our case $M \leq 14$), while U_{Bk}, V_{Bk} are the k-th columns of the matrices U_{B1} and V_{B1}; $\otimes$ stands for outer product and the superscript T for transposition.

Fig.14 shows the frequency response magnitudes of the designed circular filter approximated by taking into account the first largest 8 singular values and 5 singular values. It can be noticed that even retaining only the first 5 singular values, the 2D filter preserves its circular shape without large distortions. In this case the filter template B is approximated by B_M from (79), for $M = 5$. Therefore, the template B can be written as a sum of only 5 *separable* matrices according to (79). This is an important aspect in the filter implementation.

8. Applications and simulation results

An example of image filtering with a two-directional filter is given. We use the filter shown in Fig.3(e), (f). This type of filter can be used in simultaneously detecting perpendicular lines from an image. The binary test image in Fig.15 (a) contains straight lines with different orientations and lengths, and a few curves. It is known that the spectrum of a straight line is oriented in the plane (ω_1, ω_2) at an angle of $\pi/2$ with respect to the line direction. Depending on filter selectivity, only the lines with the spectrum oriented more or less along the filter pass-bands will remain in the filtered image. In the output image in Fig.15 (b), the lines roughly oriented horizontally and vertically are preserved, while the others are filtered out or appear very blurred, due to directional low-pass filtering. The joints of detected lines appear as darker pixels and can be detected, if after filtering a proper threshold is applied.

Let us apply the designed fan-type filters, which can be regarded as components of a DFB, in filtering a typical retinal vascular image. Clinicians usually search in angiograms relevant features like number and position of vessels (arteries, capillaries). An angular-oriented filter bank may be used in analyzing angiography images by detecting vessels with a given orientation. Let us consider the retinal fluorescein angiogram from Fig.16(a), featuring some pathological elements which indicate a diabetic retinopathy. This image is filtered using 5 oriented wedge filters with narrow aperture ($\theta = \pi / 24$), designed using the method described in section 4. Fig.16 (b)-(f) show the directionally filtered angiography images.

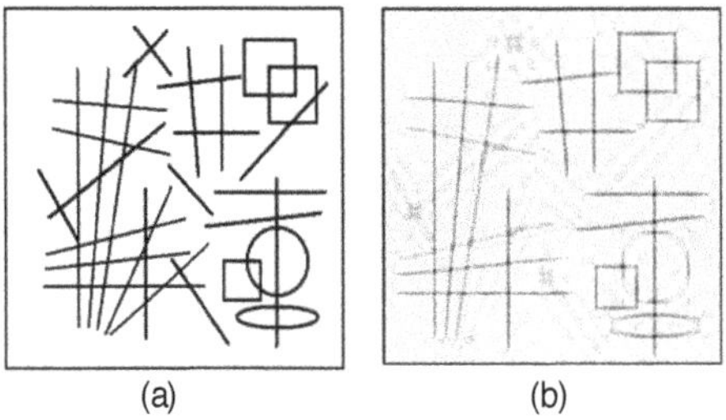

Figure 15. (a) Test image; (b) filtered image

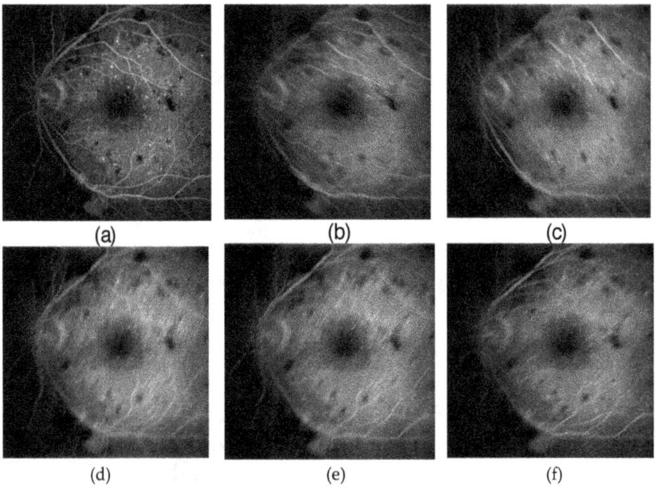

Figure 16. (a) Retinal fluorescein angiogram; (b)-(f) images resulted as output of five component filters of the fan-type filter bank

As can be easily noticed, the vessels for which the frequency spectrum overlaps more or less with the filter characteristic remain visible, while the others are blurred, an effect of the directional low-pass filtering (Matei & Matei, 2012). The directional resolution depends on the filter angular selectivity given by θ.

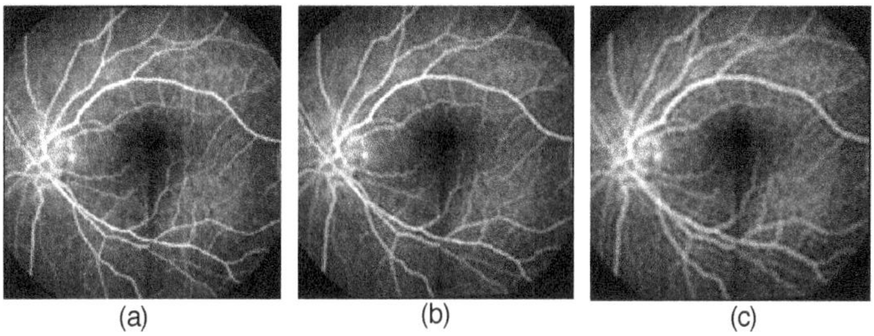

Figure 17. (a) Retinal fluorescein angiogram; (b), (c) results of image filtering with a FIR LP circular filter with cutoff frequency $\Omega_{C1}=0.16\pi$ and $\Omega_{C2}=0.08\pi$

The designed zero-phase circularly-symmetric FIR filters may be useful as well in pre-processing tasks on biomedical images, having a blurring effect on the image which depends on its selectivity given by the circular filter bandwidth. The effect is somewhat similar to the Gaussian smoothing, which is used as a pre-processing stage in computer vision tasks to enhance image structures at different scales. Applying the presented design procedure, a circularly-symmetric filter bank can be derived, with components having desired bandwidths. Let us consider another retinal fluorescein angiogram, displayed in Fig.17(a). In the simulation result shown in Fig.17 (b) and (c), the two circular filters introduce gradual blurring which is visible on the fine image details, like small vessels and capillaries. In the image in Fig.17 (c) all the finer details have been almost completely smoothed out.

9. Conclusion

The design methods presented in this chapter combine the analytical approach based on 1D prototype filters and frequency transformations with numerical optimization techniques. For the classes of 2D filters designed here, we have used mainly analog filters as prototypes, which turn out to make simpler the expressions of the derived frequency mappings, and therefore the complexity of the designed 2D filters is lower in the analyzed cases. The prototypes used here were both maximally-flat or very selective, either low-pass or band-pass. For each type of 2D filter, a particular spectral transformation is derived. An important advantage is that these spectral transformations include some parameters which depend on the 2D filter specifications, like bandwidth, orientation, aperture etc. Once found the specific frequency mapping, the 2D filter results from its factorized prototype function by a simple substitution in each factor. The designed filters are versatile in the sense that the prototype parameters (bandwidth, selectivity) can be adjusted and the 2D filter will inherit these properties.

An advantage of the analytical approach over the completely numerical optimization techniques is the possibility to control the 2D filter parameters by adjusting the prototype. Another novelty is the proposed analytical design method in polar coordinates, which can yield selective two-directional and even multi-directional filters, and also fan and diamond filters. In polar coordinates more general filters with a specified rotation angle can be synthesized.

The design methods approached here are rather simple, efficient and flexible, since by starting from different specifications, the matrices of a new 2D filter result directly by applying the determined frequency mapping, and there is no need to resume every time the whole design procedure.

Stability of the designed filters is also an important problem and will be studied in detail in future work on this topic. In principle the spectral transformations used preserve the stability of the 1D prototype. The derived 2D filter could become unstable only if the numerical approximations introduce large errors. In this case the precision of approximation has to be increased by considering higher order terms, which would increase in turn the filter complexity; however, this is the price paid for obtaining efficient and stable 2D filters. Further research will focus on an efficient implementation of the designed filters and also on their applications in real-life image processing.

Author details

Radu Matei[1] and Daniela Matei[2]

1 "Gh.Asachi" Technical University of Iasi, Romania

2 "Gr.T.Popa" University of Medicine and Pharmacy of Iasi, Romania

References

[1] Ansari, R. (1987). Efficient IIR and FIR fan filters, *IEEE Transactions on Circuits and Systems*, Aug. 1987, , 34, 941-945.

[2] Austvoll, I. (2000). Directional filters and a new structure for estimation of optical flow. *Proc. of Int. Conf. on Image Processing ICIP* 2000, Vancouver, Canada, , 2, 574-577.

[3] Bamberger, R, & Smith, M. (1992). A filter bank for the directional decomposition of images: theory and design, *IEEE Trans. Signal Processing*, Apr. 1992, , 40, 882-893.

[4] Berry, E. (2007). A Practical Approach to Medical Image Processing. Taylor & Francis, 2007

[5] Chakrabarti, S, & Mitra, S. K. (1977). Design of two-dimensional digital filters via spectral transformations, *Proceedings of the IEEE*, June 1977, , 65, 905-914.

[6] Danielsson, P. E. (1980). Rotation-invariant linear operators with directional response. Proc. of *5th International Conf. on Pattern Recognition*, Miami, USA, Dec. 1980

[7] Dougherty, G. editor). ((2011). Medical Image Processing: Techniques and Applications. Springer, 2011

[8] Frangi, A. F, et al. (1998). Multiscale vessel enhancement filtering, *Intl. Conf. on Medical Image Computing Computer-Assisted Intervention*, Berlin, 1998, , 1496, 130-137.

[9] Freeman, W. T, & Adelson, E. H. (1991). The design and use of steerable filters. *IEEE Trans. on Pattern Analysis and Machine Intelligence*, Sept. 1991, , 13, 891-906.

[10] Harn, L, & Shenoi, B. (1986). Design of stable two-dimensional IIR filters using digital spectral transformations. *IEEE Transactions on Circuits and Systems*, May 1986, , 33, 483-490.

[11] Hirano, K, & Aggarwal, J. K. (1978). Design of two-dimensional recursive digital filters. *IEEE Trans. on Circuits and Systems*, Dec. 1978, , 25, 1066-1076.

[12] Ito, N. (2010). Efficient design of two-dimensional diamond-shaped filters. *Proceedings of Int. Symposium ISPACS 2010*, Chengdu, China, 6-8 Dec. 2010, , 1-4.

[13] Jury, E. I, Kolavennu, V. R, & Anderson, B. D. (1977). Stabilization of certain two-dimensional recursive digital filters. *Proceedings of the IEEE*, , 65(6), 887-892.

[14] Kayran, A, & King, R. (1983). Design of recursive and nonrecursive fan filters with complex transformations, *IEEE Trans. on Circuits and Systems*, CAS-30(12), 1983, , 849-857.

[15] Khan, M. A. U, et al. (2004). Coronary angiogram image enhancement using decimation-free directional filter banks, *IEEE ICASSP*, Montreal, May 17-21, 2004, , 5, 441-444.

[16] Lim, J. S. (1990). Two-Dimensional Signal and Image Processing. Prentice-Hall 1990

[17] Lim, Y. C, & Low, S. H. (1997). The synthesis of sharp diamond-shaped filters using the frequency response masking approach. Proc. of IEEE Int. Conf. on Acoustics, Speech & Signal Processing ICASSP-97, Munich, Germany, Apr. 21-24, 1997, 2181-2184.

[18] Low, S. H, & Lim, Y. C. (1998). A new approach to design sharp diamond-shaped filters. *Signal Processing*, May 1998, , 67, 35-48.

[19] Lu, W. S, & Antoniou, A. (1992). Two-Dimensional Digital Filters, CRC Press, 1992

[20] Mastorakis, N. E. (2000). New necessary stability conditions for 2D systems, *IEEE Trans. on Circuits and Systems*, Part I, July 2000, 47, 1103-1105.

[21] Matei, R. (2010). A new design method for IIR diamond-shaped filters, *Proc. of the 18th European Signal Processing Conference EUSIPCO 2010*, Aalborg, Denmark, , 65-69.

[22] Matei, R. (2011a). New Design Methods for Two-Dimensional Filters Based on 1D Prototypes and Spectral Transformations, In: "Digital Filters", Fausto Pedro García Márquez (Ed.), IN-TECH Open Access Publisher, Vienna, 2011, 91-121.

[23] Matei, R. (2011b). A class of 2D recursive filters with two-directional selectivity, *Proc. of the WSEAS International Conference on Applied, Numerical and Computational Mathematics (ICANCM'11)*, Barcelona, Spain, Sept. 15-17, 2011, 212-217.

[24] Matei, R, & Matei, D. (2012). Vascular image processing using recursive directional filters, *World Congress on Medical Physics and Biomedical Engineering*, Beijing, China, May 26-31, 2012, IFMBE Proceedings , 39, 947-950.

[25] Mollova, G. S. (1997). Analytical least squares design of 2-D fan type FIR filter, *International Conference on Digital Signal Processing DSP'97*, July 1997, 2, 625-628.

[26] Nie, X, & Unbehauen, R. (1989). D IIR filter design using the extended McClellan transformation, *Proc. of International Conference ICASSP'89*, 23-26 May 1989, , 3, 1572-1574.

[27] Nikolova, Z, et al. (2011). Complex coefficient IIR digital filters, In: "Digital Filters", InTech, April 2011, , 209-239.

[28] Connor, O, & Huang, B. T. T.S. ((1978). Stability of general two-dimensional recursive digital filters, *IEEE Trans. Acoustics, Speech & Signal Processing*, , 26, 550-560.

[29] Paplinski, A. P. (1998). Directional filtering in edge detection. *IEEE Transactions on Image Processing*, Apr. 1998, , 7, 611-615.

[30] Psarakis, E. Z, et al. (1990). Design of two-dimensional zero phase FIR fan filters via the McClellan transform. *IEEE Trans. Circuits & Systems*, Jan. 1990, , 37, 10-16.

[31] Qunshan, G, & Swamy, M. N. S. (1994). On the design of a broad class of 2D recursive digital filters with fan, diamond and elliptically-symmetric responses, *IEEE Trans. on Circuits and Systems II*, Sep.1994, , 41, 603-614.

[32] Rydell, J, et al. (2008). Bilateral filtering of fMRI data. *IEEE Journal of Selected Topics in Signal Processing*, Dec 2008, , 2, 891-896.

[33] Semmlow, J. L. (2004). Biosignal and Biomedical Image Processing. Marcel Dekker, 2004

[34] Simoncelli, E. P, & Farid, H. (1996). Steerable wedge filters for local orientation analysis. *IEEE Transactions on Image Processing*, Sep 1996, , 5, 1377-1382.

[35] Tosic, D. V, et al. (1997). Symbolic approach to 2D biorthogonal diamond-shaped filter design, 21st Int. Conf. on Microelectronics, 1997, Nis, Yugoslavia, , 2, 709-712.

[36] Truc, P, et al. (2007). A new approach to vessel enhancement in angiography images, *Int. Conf. on Complex Medical Engineering CME 2007*, Beijing, 23-27 May 2007, , 878-884.

[37] Wong, W. C. K, et al. (2004). Trilateral filtering for biomedical images. IEEE Int. Symposium on Biomedical Imaging, 15-18 Apr. 2004, , 1, 820-823.

[38] Wu, D, et al. (2006). On the adaptive detection of blood vessels in retinal images, *IEEE Trans. Biomedical Engineering*, Feb 2006, , 53, 341-343.

[39] Zhu, W, & Zhenya, P. H. ((1990). A design method for complementary recursive fan filters, *IEEE ISCAS 1990*, 1-3 May 1990, , 3, 2153-2156.

[40] Zhu, W, & Nakamura, P. S. ((1996). An efficient approach for the synthesis of 2-D recursive fan filters using 1-D prototypes, *IEEE Transactions on Signal Processing*, Apr. 1996, 44, 979-983.

[41] Zhu, W, et al. (1999). A least-square design approach for 2D FIR filters with arbitrary frequency response. *IEEE Transactions on Circuits and Systems II*, Aug. 1999, 46, 1027-1034.

[42] Zhu, W, et al. (2006). Realization of 2D FIR filters using generalized polyphase structure combined with singular-value decomposition, *Proc. of ISCAS 2006*, Kos, Greece

Permissions

The contributors of this book come from diverse backgrounds, making this book a truly international effort. This book will bring forth new frontiers with its revolutionizing research information and detailed analysis of the nascent developments around the world.

We would like to thank Dr. Fausto Pedro García Márquez and Dr. Noor Zaman, for lending their expertise to make the book truly unique. They have played a crucial role in the development of this book. Without their invaluable contribution this book wouldn't have been possible. They have made vital efforts to compile up to date information on the varied aspects of this subject to make this book a valuable addition to the collection of many professionals and students.

This book was conceptualized with the vision of imparting up-to-date information and advanced data in this field. To ensure the same, a matchless editorial board was set up. Every individual on the board went through rigorous rounds of assessment to prove their worth. After which they invested a large part of their time researching and compiling the most relevant data for our readers. Conferences and sessions were held from time to time between the editorial board and the contributing authors to present the data in the most comprehensible form. The editorial team has worked tirelessly to provide valuable and valid information to help people across the globe.

Every chapter published in this book has been scrutinized by our experts. Their significance has been extensively debated. The topics covered herein carry significant findings which will fuel the growth of the discipline. They may even be implemented as practical applications or may be referred to as a beginning point for another development. Chapters in this book were first published by InTech; hereby published with permission under the Creative Commons Attribution License or equivalent.

The editorial board has been involved in producing this book since its inception. They have spent rigorous hours researching and exploring the diverse topics which have resulted in the successful publishing of this book. They have passed on their knowledge of decades through this book. To expedite this challenging task, the publisher supported the team at every step. A small team of assistant editors was also appointed to further simplify the editing procedure and attain best results for the readers.

Our editorial team has been hand-picked from every corner of the world. Their multi-ethnicity adds dynamic inputs to the discussions which result in innovative outcomes. These outcomes are then further discussed with the researchers and contributors who give their valuable feedback and opinion regarding the same. The feedback is then collaborated with the researches and they are edited in a comprehensive manner to aid the understanding of the subject.

Apart from the editorial board, the designing team has also invested a significant amount of their time in understanding the subject and creating the most relevant covers. They scrutinized every image to scout for the most suitable representation of the subject and create an appropriate cover for the book.

The publishing team has been involved in this book since its early stages. They were actively engaged in every process, be it collecting the data, connecting with the contributors or procuring relevant information. The team has been an ardent support to the editorial, designing and production team. Their endless efforts to recruit the best for this project, has resulted in the accomplishment of this book. They are a veteran in the field of academics and their pool of knowledge is as vast as their experience in printing. Their expertise and guidance has proved useful at every step. Their uncompromising quality standards have made this book an exceptional effort. Their encouragement from time to time has been an inspiration for everyone.

The publisher and the editorial board hope that this book will prove to be a valuable piece of knowledge for researchers, students, practitioners and scholars across the globe.

List of Contributors

Noor Zaman and Ahmed Muneer
College of Computer Sciences & Information Technology,King Faisal University, Saudi Arabia

Fausto Pedro García Márquez
ETSI Industriales, Universidad Castilla-La Mancha, Ciudad Real, Spain

Fausto Pedro García Márquez, Raúl Ruiz de la Hermosa González-Carrato and Jesús María Pinar Perez
University of Castilla-La Mancha, Spain

Noor Zaman
CCSIT, King Faisal University, Saudi Arabia

Jan Peter Hessling
Measurement Technology, SP Technical Research Institute of Sweden, Borås, Sweden

Alexey Mokeev
Northern (Arctic) Federal University, Russia

Shunsuke Koshita, Masahide Abe and Masayuki Kawamata
Department of Electronic Engineering, Graduate School of Engineering, Tohoku University, Sendai, Japan

Barmak Honarvar Shakibaei Asli and Raveendran Paramesran
Elecrtical Engineering Department, University of Malaya, KulaLumpur, Malaysia

Fumio Itami
Faculty of Engineering, Saitama Institute of Technology, Japan

Shunsuke Yamaki, Masahide Abe and Masayuki Kawamata
Department of Electronic Engineering, Graduate School of Engineering, Tohoku University, Sendai, Japan

Håkan Johansson and Oscar Gustafsson
Division of Electronics Systems, Department of Electrical Engineering, Linköping University, Sweden

Seyyed Ali Hashemi and Behrouz Nowrouzian
Department of Electrical and Computer Engineering, University of Alberta, Edmonton, Alberta, Canada

Radu Matei
"Gh.Asachi" Technical University of Iasi, Romania

Daniela Matei
"Gr.T.Popa" University of Medicine and Pharmacy of Iasi, Romania